Soil Remediation and Rehabilitation

ENVIRONMENTAL POLLUTION

VOLUME 23

For further volumes:
http://www.springer.com/series/5929

Helmut Meuser

Soil Remediation and Rehabilitation

Treatment of Contaminated and Disturbed Land

Springer

Helmut Meuser
Faculty A&L
University of Applied Sciences
Osnabrück, Germany

ISSN 1566-0745
ISBN 978-94-007-5750-9 ISBN 978-94-007-5751-6 (eBook)
DOI 10.1007/978-94-007-5751-6
Springer Dordrecht Heidelberg New York London

Library of Congress Control Number: 2012954787

Printed on acid-free paper

Springer is part of Springer Science+Business Media (www.springer.com)

Helmut Meuser

Soil Remediation and Rehabilitation

Treatment of Contaminated and Disturbed Land

Springer

Helmut Meuser
Faculty A&L
University of Applied Sciences
Osnabrück, Germany

ISSN 1566-0745
ISBN 978-94-007-5750-9 ISBN 978-94-007-5751-6 (eBook)
DOI 10.1007/978-94-007-5751-6
Springer Dordrecht Heidelberg New York London

Library of Congress Control Number: 2012954787

Printed on acid-free paper

Springer is part of Springer Science+Business Media (www.springer.com)

To my brother Wolfgang

Foreword

Over thousands of years, potentially harmful chemical compounds have been added to the upper soil layers, leading to soil and groundwater pollution. One phenomenon that sped up soil pollution was the Industrial Revolution which began in England and subsequently spread to several developed countries in Europe, the USA and Japan, from the turn of the eighteenth and nineteenth centuries. Additionally, large-scale mining and, hence, large-scale soil pollution came into existence in many parts of the world in the nineteenth century. Even bigger was the impact of the technological developments that took place mainly during the second half of the twentieth century. These developments were characterised by a more than proportional increase in emissions of contaminants into the environment. As a consequence, emissions of contaminants to soil increased, for example, through the large-scale use of fertilisers, expansion of industrial production, the use of fossil fuels and, as an overall impact factor, a huge increase in population growth. It was not only the bulk rate of production of contaminants that significantly expanded. It was also the enormous increase in variety of types of chemical compounds that were produced for public or industrial use, or produced as a by-product, and eventually entered the environment and the soil. Soil can often be considered as the ultimate sink for contaminants that enter the environment.

At the end of the twentieth century, the number of identified contaminated sites grew in most developed countries to six or seven digits. The Commission of the European Communities estimated the number of contaminated sites in the European Union at 3.5 million sites, affecting 231 million people and representing a market value of 57 billion Euros, in 2006. According to the European Environmental Agency (EEA), potentially contaminating activities are estimated to have occurred at nearly 3 million sites in 2007. Today, it is expected that this number has grown significantly. This European perspective is similar in other developed countries including the United States, Canada and Australia and many countries is Asia.

Soil pollution is one of the eight threats mentioned in the EU Thematic Soil Strategy. A contaminated soil map would roughly coincide with an anthropogenic map, since humans are generally recognised as the main polluters. Most of the contaminated sites are found in or close to cities. In the present day, most countries have become aware of the huge practical, social and financial impact of contaminated sites.

Since the earliest discovery of soil and groundwater pollution, there has been a race between bringing contaminated sites back into beneficial use without compromising human health and the environment and the design of cost-efficient remediation solutions. In parallel with the increase of the number of identified contaminated sites, the awareness of the need for cost-efficient solutions grew significantly. As a consequence, remediation technologies evolved in time from simple excavations for contaminated soil or abstractions for groundwater to sophisticated combinations of ex situ or in situ approaches with in situ technologies, within a time span of three decades.

In the late 1970s, remediation was often the same thing as complete removal of the contaminants and, hence, of the risks involved. Harsh remediation measures, such as dig-and-dump (remediation of the upper soil) and pump-and-treat (remediation of the groundwater), were the most popular mechanisms to achieve this goal. Alternatively, insulation of the contaminants, and hence of the risks involved, was used as a less strict but cheaper solution. Since the early 1990s, the general focus of remediation has evolved into the elimination of unacceptable risks, which does not necessarily mean complete removal of the contaminants. Today, the remediation objective is often set at a concentration where the risks for human health and the environment are acceptable.

Generally speaking, remediation focuses on the development of the strategies for controlling the risks from soil and groundwater pollution. It includes avoiding, mitigating or eliminating risks. From a cost-efficient perspective, the keyword in remediation is risk reduction. There are many ways to achieve risk reduction. Basically, remediation relates to removal or controlling of the source or to blocking the pathway from source to receptor. The challenge is to find the optimum balance between the most effective and most cost-efficient way of doing this by weighing the short-term advantages against the costs of aftercare. For example, source control by the application of barriers (blocking the source) could be cheaper but sufficient to reduce risk for human health or the environment. In some cases, compliance with policies requires more stringent measures than are necessary from a risk perspective.

The most simple and generally least expensive solution for contaminated site problems relates to changing the land use, or adapting the layout of the site within the same land use, in terms of blocking the major exposure pathways. The disadvantage of changing land use or the layout of the site with the same land use, however, is that concessions often have to be made concerning the ideal way the site is used. Therefore, remediation is often necessary.

Since remediation technologies underwent a dramatic development during the last decades, and certainly the last few years, it is of the utmost importance to have an overview of the state of art of existing remediation technologies. The present book presents such an overview from a technical and a practical perspective, including the most recent developments. I sincerely hope and expect that this book will have a substantial contribution to the design of sustainable remediation solutions, in which the protection of human health and the environment go hand in hand with cost-efficient solutions.

July 2012 Frank A. Swartjes

Acknowledgement

I would like to thank the University of Applied Sciences Osnabrück, Germany, for the financial support, Mr. Alan Hogg, MA, University of Applied Sciences, for proofreading the manuscript and my wife Christiane for drawing diagrams and figures as well as editing the complete text.

Contents

Abbreviations

Ag	silver
Al	aluminium
AMD	acid mine drainage
As	arsenic
Asl	above sea level
Au	gold
Ba	barium
BCF	bioconcentration factor
BTEX	benzene, toluene, ethyl benzene, xylene
C	carbon
Ca	calcium
Cd	cadmium
Ce	cer
CEC	cation exchange capacity
cfu	colony forming unit
CHC	chlorinated hydrocarbons
CN	cyanide
Co	cobalt
Cr	chromium
Cs	caesium
Cu	copper
D&G	drain-and-gate
DNAPL	dense non-aqueous phase liquid
DOC	dissolved organic carbon
DPE	dual-phase extraction
DTPA	diethylenetriaminepentaacetic acid
DW	dry weight
Dy	dysprosium
EC	electrical conductivity
EDTA	ethylenediaminetetraacetic acid
ENA	enhanced natural attenuation

EOX	extractable organic halogens
Er	erbium
Eu	europium
F&G	funnel-and-gate
Fe	iron
Gd	gadolinium
GDR	German Democratic Republic
H	hydrogen
HCB	hexachlorobenzene
Hg	mercury
Ho	holmium
hPa	hectopascal
HRC®	Hydrogen Release Compound
ISCO	*in situ* chemical oxidation
K	potassium
Kj	kilojoule
K_{OC}	distribution coefficient
Kwh	kilowatt
La	lanthanum
LNAPL	light non-aqueous phase liquid
Mg	magnesium
Mg	megagram
Mn	manganese
MNA	monitored natural attenuation
MPE	multi-phase extraction
Mt	megatons
N	nitrogen
Na	sodium
NAPL	non-aqueous phase liquid
Nd	neodymium
Ni	nickel
O	oxygen
ORC®	Oxygen Release Compound
P	phosphorus
P&T	pump-and-treat
PAH	polycyclic aromatic hydrocarbons
Pb	lead
PCB	polychlorinated biphenyls
PCDD/F	polychlorinated dibenzodioxins/polychlorinated dibenzofurans
PCE	pentachloroethylene
PCP	pentachlorophenol
Pd	palladium
PEHD	high-density polyethylene
PEP	phosphorus elimination plant
Pj	petajoule

PP	polypropylene
PPP	polluter pay principle
Pr	praseodymium
PRB	permeable reactive barrier
Pt	platinum
Pu	plutonium
PVC	polyvinyl chloride
REE	rare earth elements
S	sulphur
S/S	stabilisation/solidification
Sb	antimony
Sc	scandium
SEE	steam enhanced extraction
Sm	samarium
Sn	tin
Sr	strontium
SVE	soil vapour extraction
Tb	terbium
Tc	technetium
TCDD	tetrachlorodibenzodioxin
TCE	tetrachloroethylene
Ti	titanium
Tl	thallium
Tm	thulium
TNT	trinitrotoluene
TOC	total organic carbon
TPH	total petroleum hydrocarbons
U	uranium
V	vanadium
V	voltage
VCHC	volatile chlorinated hydrocarbons
WQ	storage volume
Y	yttrium
Yb	ytterbium
z0…z2	reuse of excavated material (German Waste Management Regulations)
Zn	zinc

Chapter 1
Introduction

Many factors can cause soil contamination in urban and rural environments. Large-scale soil contamination can be the result of naturally accelerated bedrock concentrations or dust deposition. Linear contamination can be found alongside traffic routes and infrastructure pipes or in floodplains. In rural areas fertilizing, sewage sludge application and the use of pesticides are typical reasons for contaminated sites. Particularly in urban areas, however, the main factor responsible for contaminated sites might be the presence of many derelict sites which were formerly used for industrial purposes, in addition to deposits such as mining heaps, dumps and filled natural depressions or quarries (Meuser 2010).

These contaminated sites are usually systematically recorded and integrated in a diagnostic assessment procedure which consists of three phases, a preliminary site assessment (I), a comprehensive site assessment (II) and the investigation and implementation of remedial measures (III). While the first phase will only provide a qualitative indication of the site based on a site inspection as well as the interpretation of data, e.g. geological and hydrogeological maps, aerial photographs, literature, reports of environmental agencies, interviews with contemporary witnesses, etc., phase II includes the sampling and physico-chemical analysis of different media such as soil, groundwater and soil vapour in order to identify exactly the contaminated locations. Normally, the third phase must be divided into two sections, whereby the first one expands the site investigation to define the contamination in more detail, e.g. by application of hydrological simulation models for the identification of the plume migration, by indoor-air investigations for the assessment of possible inhaled toxic air, by plant analyses for exact knowledge about the food chain pathway, etc. In the case of contamination affecting the different pathways the second section of phase III includes the planning and subsequent implementation of corrective actions (Asante-Duah 1995).

Phase III must answer the question whether decontamination or pollutant containment are preferred to solve the contaminated site problem. Remediation covers decontamination (destruction or detoxification) and containment-based

H. Meuser, *Soil Remediation and Rehabilitation: Treatment of Contaminated and Disturbed Land*, Environmental Pollution 23, DOI 10.1007/978-94-007-5751-6_1,

approaches, which leave the contaminants in place but prevent continuous migration. For both approaches a number of technical applications can be provided. In a long-term process which is still ongoing many techniques have been developed and advanced since the late 1970s (Swartjes 2011).

The aim of this book is to characterise the approaches in relation to different media such as soil, soil vapour, groundwater and surface waters. Apart from the techniques introduced, advantages and disadvantages, particularly restrictions to the application, are discussed.

Soils can also be adversely affected by man without any input of pollutants. The negative features are frequently of a physical nature such as sealing and insufficient water management in urban areas or removal of natural layers in quarries, open-cast coal mines or harvested peatland. Moreover, in mining areas chemical soil damage, e.g. acidification and metal contamination, contributes to environmental hazards. Sites affected in this way require rehabilitation measures (Genske 2003), which will also be introduced in this book. Consequently, the book will give comprehensive information about the current remediation and rehabilitation opportunities regarding predominantly soils of urban, industrial and mining areas.

In Chap. 2 rehabilitation measures for the particular problems in urban environments (e.g. de-sealing, rainwater management, possibilities with respect to urban mining) are discussed. The specific rehabilitation approaches of mining and raw material extraction areas are examined in Chap. 3. The interpretation of the rehabilitation strategies refers to large-scale affected terrain (e.g. open-cast lignite coal mining areas, peatlands), quarries, open pit mines and mining heaps.

General aspects in relation to the treatment of contaminated sites including soil management, site clearance operations and working safety are discussed in Chap. 4. On the basis of these aspects the opportunities for soil containment (Chap. 5) and soil decontamination (Chap. 6) are the content of the following chapters. All facets of soil containment extending to complete encapsulation are introduced and the common technical applications of decontamination operations, in particular the introduction of the most important technologies soil washing, bioremediation and phytoremediation, are presented. Approaches dealing with contaminated groundwater, soil vapour and surface water bodies are contained in Chap. 7. In particular, the multiple applications regarding the groundwater are mentioned in detail. Additionally, in Chap. 8 some approaches for soil and groundwater without complex technical input are presented. Finally, the remediation planning process is taken into consideration in Chap. 9. This section discusses, for instance, the requirement of soil and environment protection but also the continuously decisive time and cost factors. In Chap. 10, the last chapter, a brief attempt is made to present an outlook on the expected future development of soil rehabilitation and soil remediation approaches.

References

Asante-Duah, D. K. (1995). *Management of contaminated site problems*. Boca Raton: Lewis Publishers.
Genske, D. D. (2003). *Urban land – Degradation, investigation, remediation*. Berlin: Springer.
Meuser, H. (2010). *Contaminated urban soils*. Dordrecht: Springer.
Swartjes, F. A. (2011). *Dealing with contaminated sites*. Dordrecht: Springer.

Chapter 2
Rehabilitation of Soils in Urban Environments

Abstract In urban environments, the rehabilitation measures which do not directly concern soil contamination are strongly limited, but, nevertheless, there are some instruments which can improve the soil characteristics. These instruments are introduced in this chapter. They are mainly associated with urban brownfields, whose redevelopment is particularly important for the public authorities. The possibilities to integrate contaminated and uncontaminated brownfields into a proper town planning process in the context of defined quality criteria are described. Environmentally friendly methods such as de-sealing and roof planting are portrayed, irrespective of the current land use as brownfield, residential area or active industrial area. A high value is set on the rainwater management in urbanised and subsequent large-scale, sealed terrains. Moreover, this chapter includes the new discussion about so-called urban mining, which originates from the increasing shortage of resources. Whether urban deposits such as landfills are interesting for use as secondary raw material and what technology can be applied to realise this innovative approach is also discussed in the chapter.

Keywords Brownfield redevelopment • De-sealing • Landfill excavation • Rainwater management • Urban mining

2.1 Brownfield Redevelopment

2.1.1 Instrument of the Town Planning

Brownfields are defined as areas that have been mostly affected by a former industrial use and accordingly by soil contamination. Apart from industrial sites, other land-use types are included in the definition such as former military sites, derelict office blocks, disused railway areas, landfills and technogenic deposits. Currently, they are derelict and unused, but located in at least partly developed urban areas.

H. Meuser, *Soil Remediation and Rehabilitation: Treatment of Contaminated and Disturbed Land*, Environmental Pollution 23, DOI 10.1007/978-94-007-5751-6_2,

For this reason, sites with operating factories, houses and military sites in use, fallow agricultural land and burnt-out woodland do not fall under the definition. Brownfields require different measures in order to bring them back in an adequate, beneficial use (Nathanail 2011).

Apart from the re-use of remaining space in residential and commercial areas and the urban concentration, i.e. reduction of garden areas and establishment of high rise buildings, brownfield redevelopment belongs to the modern city development instruments. It is defined as remediation and reclamation of abandoned land usually carried out by regional development agencies which want to merchandise and sell the sites concerned. While remediation means the decontamination as well as containment procedures to eliminate or interrupt the contaminants pathways to different receptors such as humans, plants and groundwater (see Sect. 4.1.1), reclamation is focused on the establishment of physical conditions in which construction activities are able to occur. Financial support must be included in relation to brownfield redevelopment, particularly the evaluation of grant and funding sources. The land-use types are mostly rededicated and the sale takes place by piece of land.

Viewed ecologically, the redevelopment is targeted at the reduction of land consumption on the urban periphery, the decrease of sealed surfaces and the minimizing of landscape destruction. In urban agglomeration the shortsighted land consumption on the periphery is expected to be compensated by the introduction of land recycling.

The background for brownfield redevelopment observed in a number of European countries is of an economic nature. Some branches of industry disappeared in the post industrial era after 1990, especially in the former socialist countries and many industries moved from Europe to less developed countries with lower labor costs. Additionally, the re-use of brownfields became more important because of the limited space in the densely populated countries. The redevelopment was increasingly desirable, since economies broke down, the tax income of the municipalities decreased and the unemployment rate reached a high level associated with social conflicts.

The process of redevelopment requires a high level of financial investment that is usually covered by a funding scheme. Advantageously, the recycling of abandoned land is combined with the establishment of new clean and innovative industry such as trading and computer companies and thus with the reduction of unemployment. However, it should be noted that in most cases the substitution is not complete, since the original heavy industry employed much more workers than the new industry does.

Stakeholders are often interested in brownfields, because they are in a desired location close to the city centre or in structurally well developed suburbs. Whether private investors are willing to take part in the redevelopment, depends on the profit which can potentially be achieved. There are three economic opportunities based on an A-B-C model (CARBERNET 2006). The private sector will finance the remediation and reclamation in an investment project, if a later profit is assured (type A). This action, however, only occurs exceptionally. In most of the sites the perceived costs may exceed the expected benefits to the owners and investors. Hence, a financial intervention of the public sector is necessary to realize the redevelopment, e.g. in the form of a public-private partnership. The public sector is normally responsible for the financing of the infrastructure. A profit for the private sector can still be made and it is projected that the site value will rise (type B). If the costs are much higher or if the site value of the land cannot recoup the costs in an economically beneficial way, the redevelopment is financed by the public side only and the reclamation takes place for the public good (type C).

In most cases neither the public authority nor private investors are capable of financing the entire program on their own, so that a public-private partnership is frequently initiated. Besides, many owners are less interested in actively pursuing development of the land, since contamination can always be expected. The public-private partnership arrangement appears to be a good model for financing, but it is only successful if a coordinating public institution referring to distinct national levels (national, county, municipal) is present that ensures the different milestones of redevelopment. For instance, in the United Kingdom the London Docklands Development Corporation (LDDC) and in Germany the "Landesentwicklungsgesellschaft" of North-Rhine Westphalia (LEG) deals with the redevelopment of brownfields in association with public-private partnership projects. Of course, these institutions can only put the idea of brownfield redevelopment into practice based on a national or local funding system (Genske 2003).

Irrespective of the financing, promising brownfield redevelopment might only occur in relation to the political and administrative will. In this context, the current development in United Kingdom provides a good example. Based on the national brownfield strategy the target is that 60% of all new houses are constructed on previously developed land, which means predominantly brownfields. Indeed, the target has been met, since more than 60% of new buildings were constructed in formerly used areas, particularly post-industrial land. Accordingly, more and more brownfields were included in town planning in a number of British cities and this tendency is increasing. By the way, this development facilitated by the implementation of the Landfill Directive in 2004 that has exacerbated the disposal of contaminated soil, led to a boost in soil remediation technology in the United Kingdom (Nathanail 2011).

In cities influenced by economic structural change a high portion of former industrially used areas remain unused and cause increasing establishment of brownfields. These areas have normally a high percentage of sealed surfaces such as abandoned buildings, asphalt and concrete areas. In order to protect agricultural land in accordance with maintenance of the soil functions, in some cities the authority has begun to de-seal the sites and to re-use them for commercial, industrial, residential and recreational purposes (see Sect. 2.2.1). In Leipzig, Germany, for example, the consumption of cropland and pasture between 1996 and 2006 amounted to 1,470 ha in favor of settlement and traffic areas, though the population decreased. For this reason, the public authority decided to implement a compensation tool that preferred the redevelopment and re-use of abandoned land to the occupation of agricultural and forestry sites. A land register was introduced which collects data about unused former industrial areas available for re-use or at least environmentally friendly measures. Former industrial sites that, according to the planning, are not needed in the near future experience ecologically reasonable action, particularly de-sealing and plantation. Statistically, in one district with a

Table 2.1 Former and current land use of brownfields (11.6 ha) in Leipzig-Plagwitz (109 ha) to be redeveloped (Data from Stadt Leipzig 2008)

No.	Former land use	New land use
01	Construction company area	Residential area with extensive parks
02	Food warehouse	High tech industrial area
03	Spinning company area	Commercial area, exhibition space
04	Industrial area, bring-and-buy sale	Old-age home with extensive parks
05	Railway area	Footpaths
06	Zinc slag heap	Park
07	Metal processing area, zinc slag heap	Residential area, recreational area
08	Road maintenance area	Car repair shops
09	Trade company area	Residential area
10	Industrial site	Residential area
11	Residential area	Park
12	Trade area, basement garage	Trade area
13	Administrative building	Trade area
14	Industrial site	Park
15	Industrial site	Park
16	Power house	Residential area
17	Power house	Residential area
18	Railway bridge	Broadcasting building
19	Storage area	Broadcasting building
20	Production hall for tilling equipment	Commercial area
21	Residential and commercial area	Commercial area

size of 109 ha predominantly characterized by the metal processing industry in the past already 11.6 ha (48 branches of this industry) have been redeveloped and an additional 9.8 ha will be treated in this way soon. In Table 2.1 the areas redeveloped are listed and a distinction is made between the former land use and the current land-use type (Stadt Leipzig 2008).

Besides the financial support applied for, the participation of the public in the course of the projects concerned might be another important point for consideration in democratic societies. Detrimental circumstances such as dust development, vibration and noise associated with land management and remediation processes are often responsible for a reduced willingness of the people living near the area. Consequently, the redevelopment project should be implemented in a co-operative way. In the course of time the inhabitants should be involved in each important milestone of the project.

2.1.2 *Dealing with Contaminated Areas*

There is a suspicion that most of the brownfields are contaminated. For this reason, the redevelopment is often combined with the search for adequate remediation approaches based on detailed risk assessment that usually takes different remediation

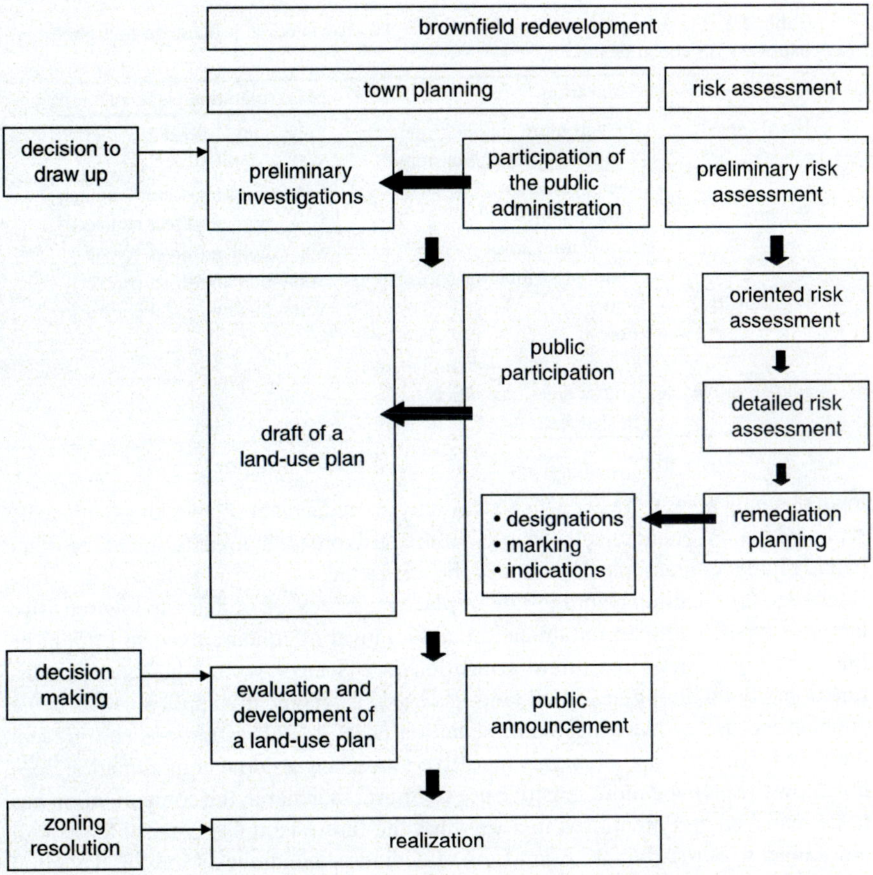

Fig. 2.1 Integration of contaminated brownfield risk assessment in the town planning process

options into consideration. The simple method of excavation and subsequent dumping at landfill sites should play a minor role, since this only re-locates the problem and postpones a solution. Hence, a consequent evaluation of the different remediation strategies must be the preferred way (see Sect. 9.1). One problem may exist with reference to the simultaneously occurring town planning and remediation planning procedures, in particular referring to the communication between different public authorities dealing with the planning procedures (e.g. town planning agency, environmental agency). In the end, different professionals must achieve an optimum solution, taking account of different perspectives and ways of understanding things. A flow chart exhibiting the course of action related to the town planning process and the parallel process of dealing with the contaminated soil is shown in Fig. 2.1.

Before the rehabilitation is planned the quality standards which have to be achieved must be defined. The risk acceptable for the given purpose must be taken within the constraints of budget, time and techniques applied. The standards depend on the sensitivity of the land use in future and the higher the sensitivity the more intensive the remediation which has to be applied. In general, with reference to the

Table 2.2 Sensitivity of different land-use types associated with the main migration pathways for contaminants

Classes	Utilisation	Main migration pathway
Highly sensitive	Playgrounds, school grounds	Soil – child (direct contact)
	Gardens, cropland, pasture	Soil – useful plant – human
	Water reserves	Soil – drinking-water – human
Sensitive	Sports fields	Soil – human (direct contact)
	Residential areas	Soil – human (direct contact)
	Outdoor swimming-pools	Soil – human (direct contact)
Low sensitive	Parks	Soil – human (direct contact)
	Forest	–
	Industrial and commercial sites	–
Non-sensitive	Traffic areas, car parks	–
	Sealed areas	–

town planning areas showing a high sensitivity are reduced to smaller areas of usually <3 ha which are economically viable with regard to decontamination. On the other hand, highly contaminated land will be restricted to low or non-sensitive use. Related to the clean-up planning safe in-place measures are oriented to low sensitive land-use types, while decontamination and removal of contaminated soil might be linked to highly sensitive future utilization (see Sect. 9.4). In conclusion, multi-functionality involving nearly all land-use types is technically applicable but never applied because of the expected remediation costs. Legal restrictions can prevent multi-functionality or even highly sensitive use, if the residual contamination does not allow the planned more sensitive use in future. In general, the contamination has to be managed and monitored in a way that the future land use is feasible without any danger to human health. Apart from remediation techniques, Monitored Natural Attenuation (MNA) can also be a part of the risk-based site management, particularly related to the groundwater pathway (see Sect. 8.2). The different sensitivities in association with the planned use are listed in Table 2.2. The distinct solutions for the clean-up and rehabilitation of contaminated brownfields are described in Chap. 9 in more detail.

The measures applied include a search for the people or institutions responsible for the contamination. In general, persons responsible for the hazard should finance the necessary measures based on the so-called polluter pay principle (PPP). If the polluter is not present any more (either dead or no longer in the country) or does not have the ability to finance the remediation (in debt or in prison), the present owner or user is supposed to pay. However, the reality is usually a different one, since the government of the country, the Federal State or the municipality might pay for the remediation and finance this from taxes. In contrast to virgin land, the redevelopment of brownfields appears to be more expensive, because a detailed site investigation and risk assessment is required and additional measures such as dismantling of ruined buildings and containment or decontamination of contaminated ground has to be taken into account, requiring a supplementary budget. On the other hand, some pre-requisites such as the availability of the infrastructure (public transport, traffic routes, electricity, wastewater pipes, etc.) which would otherwise have to be set up in a greenfield area are already provided. Hence, the real taxation of the market price for brownfields is not only derived from estate and the financing of the remediation, some costs which would be paid in the case of a greenfield usage can be saved.

2.1.3 Quality Criteria

Brownfield redevelopment for business and commercial purposes is based on compliance with important quality standards. Apart from dealing with the contaminated soil problem, important factors influencing the reclamation of waste land must be considered. Prior to the planning process knowledge of the wishes and needs of the stakeholders must be obtained.

Mechanical investigations have to be applied in association with the building ground assessment. Generally, the bearing capacity has to be investigated to enable the construction activities required. An eye should be kept on present foundations and shallow natural bedrock which may cause restrictions to the planned constructions such as subways and deep basements (see Sect. 4.3). Furthermore, damage derived from mining operations like subsidence has to be checked and treated (see Sect. 3.4). Moreover, the topography of the site might be an important factor, since slag heaps, for instance, or coal mining waste heaps decisively limit the redevelopment opportunities.

Sealed surfaces must be de-sealed to improve groundwater renewal rate, but the runoff water and seepage must not contaminate the groundwater quality. The percentage of sealed areas should be reduced as far as possible, for instance central alleys can be constructed instead of a differentiated road network. Apart from sealed areas previously used as traffic routes or car parks, the demolition of ruined buildings and foundations has to be carried out. However, attention must be paid to historic buildings or monuments which have to be protected, if present at the site.

Soil handling and the technical removal of foundations are operations which involve excavation, leveling and depositing of soil. These activities may cause another problem threatening human health. In some areas in European countries which were formerly industrially used but unfortunately also in a number of other countries worldwide that have not been influenced by the 2nd World War the removal of military remnants such as unexploded bombs and mines must be done carefully.

Present infrastructure pipes have to be inspected and new ones must be constructed where required. Mostly, it is supposed that the main infrastructure such as power and water supply as well as sewage systems has already been built, but possibly in a bad state. With reference to the planning procedures the availability of the public transport (bus, railway, tram, underground) and the accessibility of roads connecting the brownfield with the adjacent districts is another challenge for the town planner.

The attractiveness of the brownfield to be redeveloped must be considered together with the image of the location. Therefore, the site of redevelopment takes the entire area of influence into consideration. The redevelopment strategy must be compatible with the urban development approaches as well as the functional urban patterns. Brownfield redevelopment should follow a sustainable concept. Here there is an opportunity to implement modern environmental standards in the conversion from, for example, heavy industry which used to contaminate the earth to the establishment of new innovative enterprises. The building activities should be ecologically oriented. In the context of the city climate the rehabilitated brownfields used for business and commercial purposes in place of heavy industry avoid problematical emissions stemming from factory chimneys. Roof and frontage planting mean ecological advantages, enhancing the status of the redeveloped territory. The use of renewable energy by roof-integrated solar power installation and the rainwater management consisting of rainwater collection, storage in cisterns and re-utilization for garden watering and toilet flushing purposes (see Sect. 2.2) are appropriate technologies to use. The remediation requirement can also be met in a sustainable manner. For instance, brownfields where continuous groundwater remediation has to be applied can operate in a circulation system with extraction and infiltration wells. Apart from the optimized clean-up result (see Sect. 7.1.3), the recirculation system integrates a heat-cold-storage approach (the constant flow extracts heat or cold from groundwater with a constant temperature) and helps to reduce CO_2 emissions. In an example in Eindhoven, The Netherlands, carbon dioxide emissions were reduced by approximately 3,000 t $year^{-1}$ in this way (Schelwald-Van der Kley et al. 2011).

Furthermore, aspects of nature conservation, in particular the protection of valuable vegetation, and aspects of biotope quality and biodiversity should not be neglected. The planning should follow the idea of maintaining open spaces that involve both brownfield redevelopment and habitat networks. The measures allow the integration of landscape design (e.g. parks, green buffers) in the intended planning for the area previously used for industrial purposes and now abandoned. Consequently, the industrial character of the site does not form a contrast to a new landscape concept.

For assessment purposes of derelict land some biological criteria are generally of importance. A minimum size adapted to the plant and animal species which are to be protected has to be taken into consideration. This size depends upon the species themselves and accordingly differs between e.g. birds on the one hand and beetles on the other hand. The importance for the establishment of a habitat system, in particular in the presence of linear structures (habitat corridors), should be evaluated. Furthermore, the structural diversity might play a significant role, since with increasing numbers of ecologically valuable land-use types the diversity of species will increase simultaneously. In this context, the species spectrum and particularly the presence of rare species become important.

But not only spectrum and rarity of species is a decisive factor, but also the stage of maturity should be involved, because juvenile bioconoses (pioneer and tall forb communities) are normally given preference. This is caused by the species diversity which tends to be higher in juvenile ecosystems in contrast to older wooded and shaded areas. The adjacent areas should also be involved, since the migration potential of rare species to the derelict land is limited or completely non-existent, if the neighbouring sites are unvegetated and sealed. The ecological assessment, however, does not refer to the visible plant and animal species. With reference to abandoned and frequently contaminated brownfields the capability of soil microorganisms to degrade organic pollutants *in situ* might surely attract interest (see Sect. 6.3.1).

The structural change has been intensively observed in the Ruhr district, Germany. After 1870 the district changed into a densely populated area due to the discovery of hard coal and the subsequent industrial location of metal and steel processing factories. Until the 1960s the metropolitan region was characterised by smoking chimneys, slag deposits and coal gangue heaps. Afterwards, as a result of decreasing coal demand and increasing metal and steel production costs, a colliery and steel industry crisis began, leading to the closure of most of the heavy industry enterprises. With reference to the abandoned areas which had been used formerly for industrial purposes and had been heavily polluted, no further development took place, except for the dismantling and demolition of some industrial buildings and the areas remained untouched for a long time. Consequently, wilderness spread and, apart from buildings, the ground was covered with relics. Additionally, waste deposits and underground structures such as cable, piping and remaining foundations were left over. With increasing age the vegetation succession follows the plant community order:

- Pioneer community indicating bare soil and ruined buildings
- Tall forb community with first young-growth forest species (e.g. birch, willow, berry bushes)
- Young-forest vegetation, predominantly birches
- Woodland community with tree species such as maple, oak, birch, etc. and a few herbs and forbs.

Ultimately, the brownfields showed a particular character consisting of a mixture of wild growth, industrial artefacts, ruined buildings and widespread partially visible soil contamination (e.g. tar residues, unvegetated slag deposits), forming a special idyll that has never previously been discovered. In this way, a new, aesthetically unique landscape developed. Architects and artists showed increasing interest in this landscape, discovering new symbols of industrial or general human history. Accordingly, the brownfields in the Ruhr district attracted town planners, architects and particularly performers. In Fig. 2.2 a coal mining heap with a steel tetrahedron superstructure impressively exhibits an example of this development. In conclusion, as an alternative to the brownfield redevelopment, plans should be made for at least some areas to remain untouched, permitting natural succession together with the establishment of aesthetic monuments (USEPA 2010).

2.2 Environmentally Friendly Approaches

2.2.1 De-sealing

The main reasons for de-sealing approaches focus on relieving the burden on the sewerage in association with the protection from flood events, groundwater accumulation, improvement of the city climate, establishment of biotopes and

Fig. 2.2 Tetrahedral construction at the top of a coal mining waste heap in Bottrop, Germany, exhibiting aesthetic performance at a former industrial site

improvement of the living conditions for the inhabitants, in particular in urban environments.

Typical areas suitable for de-sealing are footpaths, pavements, pedestrian zones, car parks, inner courtyards and zones with limited traffic. Restrictions have to be taken into consideration with reference to contaminated land such as fills and deposits consisting of polluted substrates, areas influenced by atmospheric immissions and locations requiring high load-bearing capacity, e.g. areas used for truck and bus traffic. In relation to different land-use types, particularly sites displaying a high degree of paved area need to be included in de-sealing operations. For instance, road land (97%) and perimeter block districts (89%) have a high percentage of sealed surfaces, which can possibly be reduced. Industry plots with an average paved area of 92% are usually difficult to de-seal due to the truck traffic. Land-use types exhibiting average values between 39 and 46% like detached house districts, terraced house districts and exclusive residential districts are also sites which are interesting for de-sealing measures.

The potential of areas for de-sealing has been investigated in the city of Saarbrücken, Germany, a city with intensive heavy industry. Based on this study railway territory, roads with dense traffic, sports grounds and recreation facilities revealed a very low de-sealing potential. Furthermore, areas threatening groundwater contamination such as petrol stations were unsuitable for de-sealing as well. A low

Fig. 2.3 De-sealed road in Munich, Germany

to average opportunity was seen in relation to some industrial and residential areas, but the variation was considerable depending on the land-use type. In principle, the potential increased from terraced house districts to detached house districts and to exclusive residential areas. It was possible to use traffic areas showing low traffic intensity and car parks for de-sealing operations, but this depended on the frequency of use. The best opportunity to de-seal existed with respect to streams with consolidated embankments and service roads in areas used for agriculture or silviculture (Arweiler et al. 2000a).

It is possible to de-seal unused areas completely, including the removal of the asphalt or concrete layer and the sub-bases. Afterwards, soil has to be deposited, enabling future use, e.g. as a planted site. In Fig. 2.3 an example is shown where two lanes of an express motorway have been de-sealed and rehabilitated, since the traffic intensity decreased due to the closure of a cargo centre. The asphalt cover and the sub-base of the former lanes were removed and subsequently humic topsoil was deposited to recover the soil functions such as the filter and buffer capacity and to enable the establishment of vegetation.

As an alternative to a complete de-sealing procedure, the material used for sealing can be altered with the aim of an enhanced water infiltration and percolation. Hence, materials are chosen which may improve the water percolation to a great extent. In Table 2.3 the properties of alternative covering materials are listed. With reference to the load-bearing capacity the materials are only useable if the general vehicle use is strongly limited and truck and bus traffic can be excluded. Such a construction is feasible in the context of footpaths, inner courtyards, car parks and access roads for the fire brigade.

Some materials show vegetation growth, in particular pioneers or lawn vegetation, such as the water-bounded cover and the honeycomb-type paving stone. Small slabs indicate accelerated water percolation in the presence of wide joints filled with

Table 2.3 Properties of different sealing materials

	Water-bounded cover	Crushed stone cover	Bark much cover	Honeycomb-type paving stone	Porous perforated sealing material	Small slabs with wide joints
Construction (profile description)	5 cm sand/fine gravel	3 cm crushed stone	6 cm bark mulch	8–12 cm honey-comb-type stone	10 cm slab	10 cm slab, joints filled with gravel
	10–15 cm gravel/crushed stone	10–15 cm crushed stone/humic topsoil	10–15 cm gravel/crushed stone	5 cm crushed stone	5 cm sand	10 cm gravel/sand
	Anti-freeze course	10–15 cm gravel/crushed stone		5 cm fine gravel	10–20 cm crushed stone	10–20 cm crushed stone
		Anti-freeze course		15–20 cm crushed stone		
Vegetation	Pioneers	Lawn, pioneers, partially bare soil	None	Lawn	None	Joint vegetation
Load-bearing capacity	Low	Moderate to high	Low	High (low traffic intensity)	High (low traffic intensity)	High (low traffic intensity)
Maintenance	Periodic renewal	Periodic renewal	Frequent renewal (biodegrada-tion)	None	High pressure jet cleaning	Weed control
Application (examples)	Footpath, inner courtyard	Footpath, car park, track, fire brigade excess road	Footpath (park, cemetery)	Car park, track, fire brigade excess road	Footpath, inner courtyard, car park, track, fire brigade excess road	Footpath, car park, track, fire brigade excess road
Problems	Compaction, stone dispersion to adjacent sites	Compaction, stone dispersion to adjacent sites	Soiling to adjacent sites	Accessibility limited	–	Accessibility limited

gravel. For this reason, joints measuring 1–2 cm are favorable but wide joints can cause problems in association with the usability. This pavement solution in places in city centers such as shopping malls and theatre squares is possibly problematical because of the limited accessibility (high heels of ladies' shoes). A disadvantage is that the bark mulch cover tends towards biodegradation, resulting in a constant renewal of the surface. In addition, the usability is also limited in the proximity of buildings, which unfortunately people enter with dirty shoes. Furthermore, porous perforated sealing material reveals increasing blocking of the pore system, so that the percolation decreases significantly in the course of time. Thus, treatment using the high pressure water-jet technique is recommended to stop pore blocking.

2.2.2 Rainwater Management

A complete de-sealing is rarely carried out because of the costs for treatment and depositing of the removed layers. In urbanized areas, as an alternative to de-sealing, rainwater management is often taken into account to improve the conditions for rainwater infiltration and to reduce the burden on the sewerage. The purpose of rainwater management is to compensate for the reduced water infiltration into the ground and subsequent high runoff and discharge of rainwater to rivers. The latter is responsible for an increase of flood events and a general lowering of the groundwater table.

In urbanized areas a high percentage of the land surface is sealed and so rainwater percolation cannot take place. In the city of Saarbrücken, Germany, which has an annual precipitation of 849 mm on average, different land-use types were investigated in order to find out the yearly water percolation rates in consideration of the evapotranspiration, which may vary between the land-use types due to the presence and dominance of vegetation. Approximately 30% of the entire urban area did not show any infiltration at all, which was caused by the total sealing. Additionally, more than 10% indicated very low infiltration rates. These areas did not provide groundwater accumulation in the aquifer. Moreover, unsealed areas with a geological subsurface exhibiting limited rainwater percolation showed reduced infiltration rates. Soils on Holocene floodplains were not suitable for rainwater percolation because of the high groundwater table. The result was also applicable to loamy soils derived from limestone and marl as well as siltstone. Accordingly, in the city territory the areas suitable for rainwater infiltration were strongly limited to less than 350 mm (Helmes et al. 2000). It is the task of city authorities to search for alternative options to infiltrate rainwater and to accumulate the groundwater.

For this purpose, distinct techniques briefly defined as best management practices (BMP) can be offered but attention has to be paid to some assumptions in order to provide infiltration and downward percolation of the collected rainwater. First of all, the hydraulic conductivity should exceed $>10^{-5}$ m s^{-1}, which demands the texture classes gravel, sand, and silty sand. Subsurface stagnating water must not be present in order to ensure that the water percolates. Stagnic soils are negatively affected by impeded drainage and subsequent lateral water movement.

Due to the fact that in areas with slope inclination the soil in landscape depressions increasingly tends to moisten, special attention to the hydraulic conductivity should be paid. The slope gradient should be low to prevent erosion and runoff. Downslope positions are influenced by downhill drainage and runoff. An increased slope gradient is responsible for more and faster runoff which accords with a lower infiltration rate. Waterlogged soils are typical for the position at the foot of the slope.

In relation to the infiltration of dirty water stemming for instance from industrial areas beneficial conditions are visible in the case of a high hydraulic conductivity in the uppermost topsoil, because this horizon should be dry as soon as possible. The main reason is associated with the inflow of organic pollutants which are generally biodegradable in aerobic conditions (see Sect. 6.3.2) and which are only guaranteed by a fast downward percolation (Schleuss et al. 2000).

Depending on the texture, the distance between soil surface and groundwater level sufficient for rainfall infiltration has to be investigated beforehand. With reference to loamy to clayey soils the depth of the average groundwater table should be deeper than approximately 1.3 m to prevent capillarity. It is recommended to classify the required average groundwater table into three groups:

- Soils with a groundwater table >1.3 m are generally suitable for rainwater percolation
- Soils indicating 0.8–1.3 m have limited suitability
- Soils with a groundwater table <0.8 m are generally unsuitable (Arweiler et al. 2000b).

This stagnating water problem is demonstrated by an example in Woy Woy, Australia, an area with intensive residential development as well as a separate drainage and sewerage system. Consequently, drainage problems, in particular ponding of water over long periods and flood damage after storm occurrences, are present. Firstly, the devices constructed for rainwater drainage purposes do not achieve the level of effectiveness required, since compacted soils and high groundwater tables as well as an insufficient capacity of piped drainage systems are characteristics of the area of concern. Moreover, the dominant soils, sandy podsolic soils, have a hard sandy pan belowground consisting of cemented organic matter and iron associated with the prevention of water percolation. The design of the infiltration device requires deeper excavation, because the percolating water will be prevented from flowing downward because of the hard pan which exists relatively near to the surface (Hazelton and Beecham 2000).

In large-scale areas with permeable surfaces because of the sealing material and the soil below it infiltration of rainfall is feasible without the existence of interim storage. It might only be possible to apply the so-called area infiltration in rare cases to urban environments due to the lack of areas required. A runoff treatment does not satisfactorily occur, since the filtering humic topsoil covers only small portions of the catchment.

In areas exhibiting low subsurface hydraulic conductivity infiltration trenches and ditches are constructed. The trenches are filled with stones to capture and enable infiltration of stormwater runoff into the surrounding soil from the bottom and from

Fig. 2.4 Stormwater runoff storage system (swale) located between car park and market place

the sides of the trench. Below the trenches an additional storage is feasible leading to a slow infiltration rate. The trenches can be arranged in a network where many trenches are linked together.

An alternative adequate solution for stormwater infiltration is infiltration or detention basins, retention ponds and constructed wetlands, but these approaches require a lot of space, which is usually absent in urban areas. Infiltration basins and retention ponds are surface facilities which may provide for temporary storage of the stormwater runoff. Wetlands are constructed systems consisting of shallow marsh area combined with open water surfaces. The runoff is stored and simultaneously treated (see Sect. 7.2.1).

A simple technique for rainwater percolation is the construction of a swale consisting of humic topsoil underlain by the subsoil or parent material (Fig. 2.4). This approach is explicitly designed to capture and treat the stormwater runoff. The water may flow into the depressed area and percolate continuously and slowly downward. Subsequently, interim storage in zones with low hydraulic conductivity is included. If the subsoil does not achieve an acceptable water percolation, the system can be combined with a filter stripe belowground. This device is coated with a water permeable geotextile that prevents blocking and clogging of the filter stripe, and it is filled with gravel and a drainage pipe, if necessary (Fig. 2.5). The vegetative filter stripe design provides optimal biofiltering of the stormwater runoff. Swales and vegetative filter stripes are planted with hydrophilous vegetation, which can serve as an ecologically important biotope. A disadvantage that should be mentioned is that

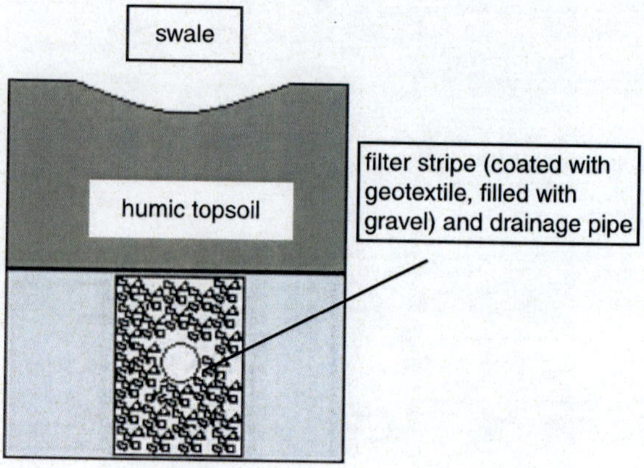

Fig. 2.5 Stormwater runoff storage techniques: swale with vegetative filter stripe

long-term siltation can occur, since fine particles tend to sediment in the course of time, finally blocking the pore system and subsequently interrupting the water percolation.

Both facilities wet swales and vegetative filter stripes show a high filter capacity in the presence of humic topsoil measuring a thickness of at least 20 cm due to the latter's high adsorption potential of the organic matter. It has to be expected that the rainwater originating from the roof gutter (roof runoff) is frequently contaminated. The problematical roof types are bitumen roofs (PAH), zinc and asbestos roofs (Zn, asbestos fibers), and roofs made of pantiles (Cu). Apart from the contaminants associated with the construction material used, deposited dust on large concrete and asphalt pavements which is washed away contributes to the contaminant inflow as well.

The area required for the water percolation device amounts to 10–20% of the total construction area. Accordingly, the future utilization of the land is limited. In particular, in residential areas consisting of detached buildings and one-family houses surrounded by small to medium-sized gardens opportunities for use are strongly restricted. In the case of stormwater occurrences the vegetative filter stripe solution is the best practice, because the water can percolate considerably quicker compared to the swale construction without filter stripe. Due to the more complex construction the filter stripe technique is the more cost-intensive version.

In densely constructed areas the solutions mentioned above are frequently not applicable, since the space is limited and necessary for more important issues. Thus, shaft constructions (cisterns) are preferred, as illustrated in Fig. 2.6. The roof runoff is collected in subsurface cisterns which are underlain by a temporary storage tank filled with gravel and coated with geotextiles which are water-permeable. The water flows continuously into the perforated storage facility below and then into the surrounding soil. Additionally, the collected water present in the cistern can be re-used for washing-machines and for toilet flushing purposes.

In general, this technique is applicable to subsoils with low hydraulic conductivity in the upper portion of the soil profile and to contaminated soils, since the percolating water does not get in touch with the contaminated layers. The roof runoff, however,

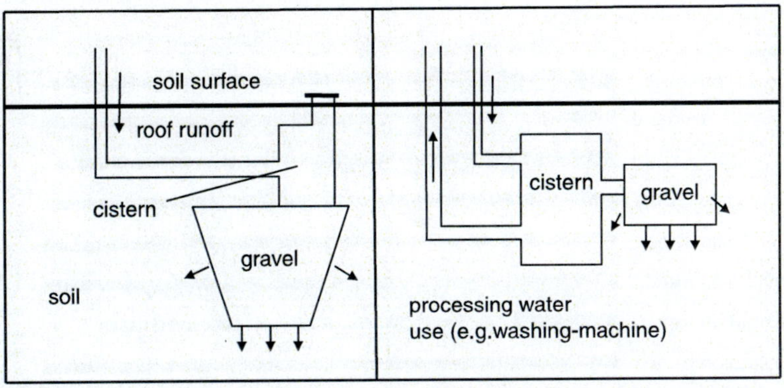

Fig. 2.6 Stormwater runoff storage techniques: cistern with (*right*) and without (*left*) process water use

is not filtered, leading to restrictions with regard to the re-use of the water in the private households in areas exhibiting high particulate matter concentration caused by air pollution. Moreover, the technical servicing is problematical, because the cistern is only accessible from the top, after opening the duct cover. The application might be relatively cost-intensive, but a long-term consideration of water re-use shows that the inhabitants save money on water consumption.

The effectiveness of any design for stormwater capture depends on questions about the filtering capacity of the frequently contaminated runoff before it enters the aquifer and on what percentage of the runoff water can be captured, which can vary drastically in magnitude. In the proximity of Seoul, Korea, different rainwater infiltration systems were investigated in detail. The runoff was generated from various land-use types including agricultural areas, residential sites and paved roads. The techniques swales, vegetative filter stripes, infiltration trenches, infiltration basins and wetlands were involved. The removal of contaminants, i.e. the concentration difference between inflow runoff and outflow effluent after passing the soil with its absorbing capacity, varied widely and depended on incoming pollutant concentration, rainfall pattern, facility maintenance frequency and length of time. In general, the retention of suspended solids and contaminants mainly originating from dust deposition appeared to play a significant role. The retention was a function of particle size distribution of the contaminated particles, the flow velocity of the percolating water and the residence time respectively. Slow water velocity and long residence time led to substantial sedimentation. The removal of totally suspended solids was usually high (Fig. 2.7) whereas the removal for nutrients stemming from the agriculturally used areas was moderate (Kim et al. 2008).

Before planning the stormwater runoff device, physical characteristics have to be taken into account. Knowledge about the physical parameters available, i.e. field capacity, air capacity and hydraulic conductivity, particularly of the subsoil, should be obtained. In general, a high level of variability of the hydraulic conductivity caused by macropores must be expected. The physical methods usually applied in

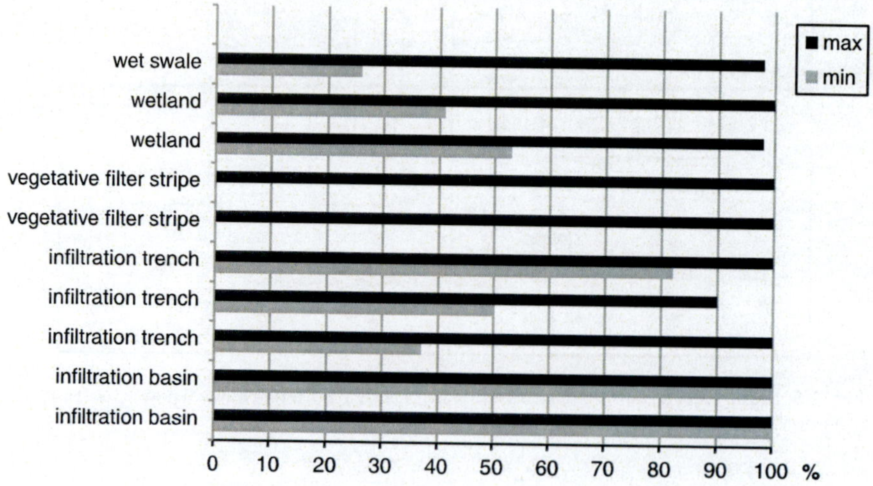

Fig. 2.7 Minimum and maximum removal efficiency (%) for different types of stormwater storage facilities in the Han River Basin in the proximity of Seoul, Korea (Data from Kim et al. 2008)

soil science such as borehole test, double-ring infiltrometry and lab analysis of the hydraulic conductivity are less accurate, since their sample volume is limited and accordingly a representative investigation of macropores does not occur. For instance, root and earthworm channels, which are responsible for high infiltration rates, cannot be determined exactly. Not only the biologically determined cracks, but also pedogenically developed voids, cause preferential flow that is difficult to estimate, because distinct patterns of preferential flow are generated. Such macropores may represent up to 15% of the soil matrix and they are partly connected to each other. It is assumed that they take part in the preferential flow by up to 50% (Winzig 2000).

The storage volume (WQ) required to capture the runoff of average annual rainfall depends on the volumetric runoff coefficient, the rainfall depth and the watershed area. In this context the watershed area is mainly focused on the roof area. The extent of the roof area responsible for the roof runoff must be calculated in order to recognize the dimensions of the construction planned. Furthermore, the climatic conditions must be included, in particular the average annual precipitation. Basically, WQ is calculated to capture 80% runoff of the average precipitation. Unfortunately, thunderstorms are difficult to estimate and consequently they can lead to uncertainties with regard to the rainwater treatment. WQ can be estimated by using the equation:

$$\mathrm{WQ}\left(\mathrm{m}^3\right) = \mathrm{P} \times \mathrm{Rv} \times \mathrm{A} \times 10 \tag{2.1}$$

whereby P = design rainfall depth (e.g. 30 mm h^{-1}), Rv = volumetric runoff coefficient and A = watershed area (ha) (Kim et al. 2008).

Most of the devices for stormwater runoff treatment need periodical maintenance. Particularly, soil erosion after heavy stormwater occurrences and the inflow of plant residues and rubbish might be the dominant problems. In Table 2.4 maintenance operations usually carried out are listed.

Table 2.4 Maintenance operations for stormwater facilities

Stormwater facility	Maintenance needed
Area infiltration (sealed surface)	Road maintenance service
	High pressure jet cleaning after pore blocking
Swale	Examination of the inflow
	Sediment removal
	Harvesting of plants
	Removal of unnecessary (exotic) plants
	Removal of garbage
Vegetative filter stripe	Examination of the inflow
	Sediment removal
	Harvesting of plants
	Removal of unnecessary (exotic) plants
	Removal of garbage
Infiltration ditch/ infiltration trench	Examination of inflow and channel water flow
	Sediment removal
	Harvesting of plants at the shore
	Removal of dead plants
	Prevention of erosion
	Removal of garbage
Infiltration basin/ Retention pond	Examination of inflow
	Sediment removal
	Harvesting of plants at the shore
	Prevention of erosion
	Removal of garbage
	Inspection of safety fence
Constructed wetland	Examination of inflow
	Sediment removal
Underground cistern	Examination of inflow (maintenance capabilities are hardly possible)

2.2.3 *Roof Planting*

On the one hand, rainwater treatment based on the best management practice as explained in Sect. 2.2.2 may help to prevent disasters such as floods, on the other hand the roof area itself may contribute to the environmental benefits in view of different aspects, when the roof is vegetated. Thus, roof and frontage planting may also be some of the environmentally friendly methods which are generally applicable. They increase evaporation and transpiration, leading to more pleasant climatic conditions, particularly in the warm seasons. The reduced radiation in wintertime and the pleasant indoor climate in summertime are linked to the energy conservation. Furthermore, dust is filtered and the burden on wastewater pipes is relieved. But not only do the abiotic properties benefit from the roof and frontage planting, the extensive roof plantation, in particular, affords the opportunity for the establishment of drought-resistant rare vegetation. A wide variety of succulents with high water

Fig. 2.8 Extensive roof planting construction

retention properties and slow growth is suitable. Specialized species which are reduced to a few biotopes in the natural landscape and are resistant to dryness, radiation, wind and frost are chosen. Only a minimum of aftercare is required.

A construction typical of an extensive roof plantation is shown in Fig. 2.8. Above the roof a thin protection layer consisting of polyethylene or bitumen is rolled out with the aim of making it impossible for roots to penetrate. Between the substrate and the protection layer a drainage layer must be planned to avoid waterlogging. This layer made of polyethylene or polystyrene includes storage hollows above and a drainage system below as well as openings for steam diffusion purposes.

Above, the substrate is filled, usually with a mixture of inorganic components such as rubble and pumice and organic components such as bark mulch and compost, indicating high air capacity, high available water capacity and high cation exchange capacity. It should be resistant to deflation and erosion. Regarding the inorganic materials the most favorable ones are perlite, bricks, natural pumice and uncontaminated blast furnace pumice, whereas the organic substrates used are, for instance, peat, compost, wood residues, bark mulch and rice hull. In a Hungarian study investigating different kinds of roof substrates a mixture of organic and inorganic materials in the ratio of 80:20 was found to be optimum. The organic percentage was responsible for a good structure, a continuous nutrient supply for plants, particularly in relation to nitrogen, and an enhanced biological activity of this more or less artificial environment. In contrast, the inorganic components are focused on the physical conditions, in particular the soil structure, since most of them are expected to be sterile. Because of their low specific gravity the materials are, in principle, suitable. Depending on the mixture compositions the pH value ranged from 6.2 to 6.9, the organic matter content from 13.3 to 39.5% and the total nitrogen content from 0.10 to 0.75% (Forro and Draskovits 2000).

The substrate is planted with the specialized vegetation as mentioned above with the aim of plant coverage of at least 60%. Altogether, the height of an extensive roof plantation measures 8–22 cm and the weight varies between 70 and 150 kg m^{-2}, achieving a result comparable with a gravel-filled roof that amounts to 90–150 kg m^{-2}. If intensive roof planting is carried out, e.g. on the roofs of office blocks and hospitals, a useful green space is created serving as recreation area for the people who work and live in these buildings (Earth Pledge Foundation 2005; Luckett 2009).

2.3 Urban Mining

2.3.1 Shortage of Resources

The term urban mining includes anthropogenically created deposits of useable raw materials and is not restricted solely to urban environments. Therefore, this term was originally associated with waste management and according to that landfill sites should be considered as anthropogenic deposits which may principally serve as economically important sites in future:

- Mining waste heaps, particularly coal and ore mining waste heaps
- Slag and ash heaps originating from blast furnace works, steel works, heavy metal works etc.
- Underground deposits in exploited mining shafts
- Deposits of solid household, commercial and industrial waste divided into regulated landfills and non-regulated waste deposits
- Deposits of construction debris and other inert materials
- Deposits of sludges (lagoons) originating from harbor dredging, sewage works, mining operations etc.
- Underground pipelines and installations
- Buildings and other constructions such as roads in use that will be demolished or removed at a later date.

In relation to the future use of raw materials buildings and installations which are still in use impose limitations to a certain extent. As long as the buildings are used, the raw materials must not be designated as deposits of raw materials, but after demolition useable materials can be obtained from many installations. In this context, the duration between construction and breakdown appears to be of importance and varies between relatively short periods, e.g. in conjunction with the extremely rapid city conversion in Chinese cities at the moment, and almost everlasting, e.g. archaeologically important places of interest such as the Acropolis in Athens.

The re-use of valuable materials can only be ensured, if selective breakdown of buildings is conducted (see Sect. 4.3.1). Apart from building materials that can be recycled such as concrete, steel and wood, secondary raw materials (e.g. copper tubes) should be included. On the one hand these are important substitutes, on the other hand they might cause demolition problems because of their toxicity, for example heavy metals and asbestos.

Table 2.5 Metal concentrations of near surface soil samples from the slag dumpsite near Liege, Belgium (Data from Ganne et al. 2006)

	Ag	As	Cd	Cu	Ni	Pb	Zn
	$mg\ kg^{-1}$					$g\ kg^{-1}$	
Mean	45	231	100	1,040	128	7.5	52.1
Standard deviation	3	159	74	891	43	6	30
Minimum	6	18	3	117	58	1	12
Maximum	146	903	403	6,182	289	31	162

Deposits containing inert materials mostly stemming from building demolition reveal a multitude of materials which can accumulate metals. Apart from bricks, mortar and gypsum, substrates are recognizably susceptible to enriched metal contents such as fly ash-based concrete, lead-based paints and tubes containing copper and lead.

Apart from the household waste deposits and landfills, deposits derived from metallurgic processes and inert material deposits such as construction debris contribute to the discovery of valuable constituents. Deposited mono-substrates contain extremely high metal concentrations depending on the component. For instance, steelworks slag heaps have high Cr and Ni concentrations, metalworks slag heaps exhibit high values for e.g. Cd, Cu, Pb and Zn but garbage incinerator fly ash deposits are also well-known for their enhanced metal concentrations. Moreover, metals usually not analyzed in conjunction with the assessment of contaminated land such as cobalt, antimony and vanadium showed increased concentrations as well (Meuser 2010).

The potential hazards of industrial waste deposits can be seen at a dumpsite of 53,000 m^3 close to Liege, Belgium. This site used in the context of former zinc extraction covers 8,000 m^2 and shows a thickness varying from 5 to 15 m. Because of the unavailability of waste treatment techniques the metal-bearing slag was dumped and is lying uncovered at the surface. The extremely high metal concentrations are summarized in Table 2.5. At the moment the leaching of metals is relatively low under the current conditions (pH value 6.5–8). However, based upon a long-term scenario (for instance 100 years) a release of 800 mg kg^{-1} Zn is expected because of natural acidification. Moreover, the spreading of contaminants by aeolian dispersion of the fine-grained slag dust as well as runoff causes a hazard for the environment and human beings (Ganne et al. 2006). In view of the economic significance of the potential dangers of metals in connection with environmental problems it would surely make sense to excavate and to re-use the slag deposit at this stage.

Furthermore, sludges filled into lagoons, e.g. sewage sludges and sludges resulting from the solution mining process applied in ore extraction, might be interesting sites. In relation to the sewage sludge the deposits can surely serve as a useable source for the macronutrient phosphorus which has been calculated to be sufficient for only 122 years (from 2008).

The rehabilitation opportunities in urban areas should involve the issues associated with urban mining. Many contaminated sites have to be remediated in combination with material excavation, transportation as well as treatment (see Sect. 4.2).

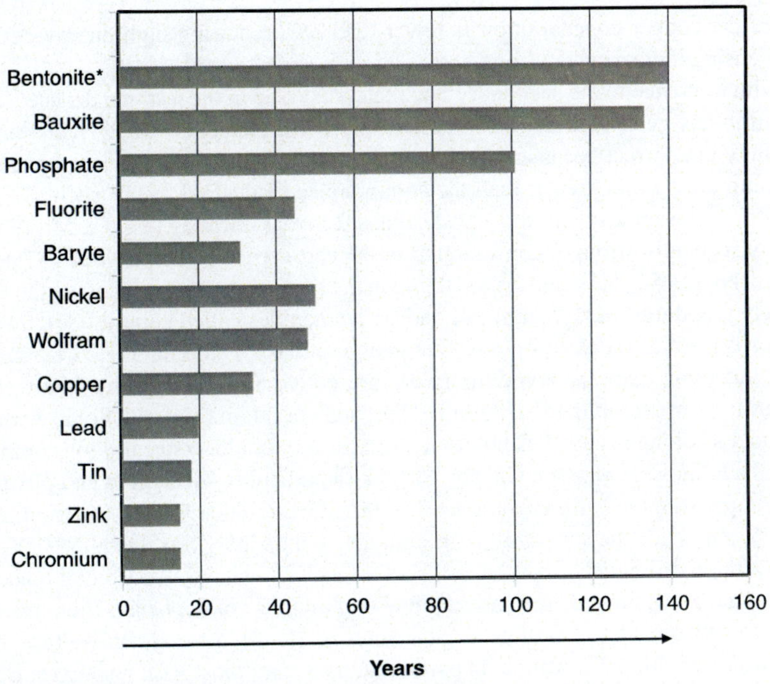

Fig. 2.9 Static life time (in years) of selected metals and minerals in relation to the reserves * used for instance for surface cover of contaminated land (see Sect. 5.1.1) (Data from Franke et al. 2010)

The excavation of usually contaminated land, e.g. unadjusted landfills, ore mining waste deposits and slag heaps, can also be considered in the context of urban mining in which the toxicity of the deposited materials is less interesting than their re-use. Accordingly, the toxic metals are only one side of the coin, while the other side refers to valuable metals that can be recycled irrespective of the toxicity and danger to human health and the environment.

The background to the contemporary issue urban mining might be the shortage of some resources. This has been well-known for a long time for fossil fuels such as crude oil and natural gas. In 2010 the energy reserves for oil and gas were estimated to be sufficient for 41 years (oil) and 59 years (gas) and in consideration of the potentially extractable resources that are currently not exploited the values are 64 (oil) and 134 (gas). The uranium reserves and resources used predominantly for atomic power are also relatively short, indicating enough for 365 years. In contrast, the hard coal reserves and resources are enough for 2,838 years and the lignite coal ones even for 4,276 years (Franke et al. 2010).

The main discussion about shortage of resources, however, might relate to the metals. Statistically, based on the consumption of 2008 some toxic metals are sufficient for only some decades, as shown in Fig. 2.9. The discovery of raw material must involve regions which had never been taken into consideration in former times. For instance, copper exploitation today takes place in geological deposits where the copper concentration is lower than 1%, requiring high energy demand and causing increasing environmental pollution (Franke et al. 2010).

With reference to the secondary raw materials issue in the last two decades much attention has been paid to the so-called rare earth elements (REE). This term is certainly not correct because, in most cases, the concentration of these elements in the earth crust is not greater than the commonplace industrial metals such as Cr, Ni and Zn. While some metals must be designated as rock-forming (e.g. Fe, Mn, Ti) and are consequently strongly concentrated in the earth crust, elements such as Ba, Co, Cr, Cu, Ni, Pb, Sn, Tl, V and Zn are found in the same magnitude as most of the REE, namely scandium (Sc), yttrium (Y) and 15 lanthanides called cerium (Ce), dysprosium (Dy), erbium (Er), europium (Eu), gadolinium (Gd), holmium (Ho), lanthanum (La), lutetium (Lu), neodymium (Nd), praseodymium (Pr), promethium (Pm), samarium (Sm), terbium (Tb), thulium (Tm) and ytterbium (Yb) and. Some elements that do not belong to the REE, however, even show a smaller concentration compared with the REE. Examples are Cd, Hg, Sb and, in particular, silver, gold and platinum. By comparison, the earth crust contains 0.0055% Cu, 0.008% Ni, 0.0016% Pb and 0.007% Zn, while the sum of all REE amounts to 0.0236% (Haxel et al. 2002).

The main difference might be the fact that REE only have a very small tendency to become concentrated in relatively few exploitable ore deposits. Thus, most of them come from a handful of geological sources. Hence, there are only a few countries which exploit REE. China, in particular, possesses most REE resources. Based on the 2009 data China's share is approximately 36%, followed by the Commonwealth Independent States (19%) and the USA (13%).

The fact that REE are so specific makes them significant from a technological, environmental and economic point of view. The technical application has multiplied enormously and consequently the demand has increased drastically. In 1960 only 2,270 t (total REE oxides) were exploited but in 2008 the value was already 124,999 t. The fields of application are summarized in Fig. 2.10 (Kingsnorth 2009; Haxel et al. 2002). Some examples are given here to clarify the context:

- Eu, for which there has been no substitute up to now, is used in colour cathode-ray tubes and liquid-crystal displays in computer monitors and television screens
- Ce oxides are suitable as polishing agents for glass; therefore, all polished glasses from primitive mirrors to precision lenses need to be treated with CeO_2
- Magnet technology is decisively improved by alloys containing Nd, Sm, Gd, Dy and Pr. They are responsible for a weight reduction in automobiles and are also used for computers, printers, DVD recorders, electrical clocks and electric domestic appliances
- Some rare elements such as Ce (catalytic converters, hybrid cars) and Tb (hybrid engines, fuel cell technology) are used in modern car manufacturing
- Energy-efficient fluorescent lamps use Ce, Eu, Gd, La, Tb and Y as a basis, permitting decreased energy consumption and consequently CO_2 reduction
- The optical industry requires La (specialized lenses) and Pr (sun glasses) and in laser technology Tm is of importance
- The element Y has optimized the quality of ceramic materials, for instance in the production of melting pots and smelters

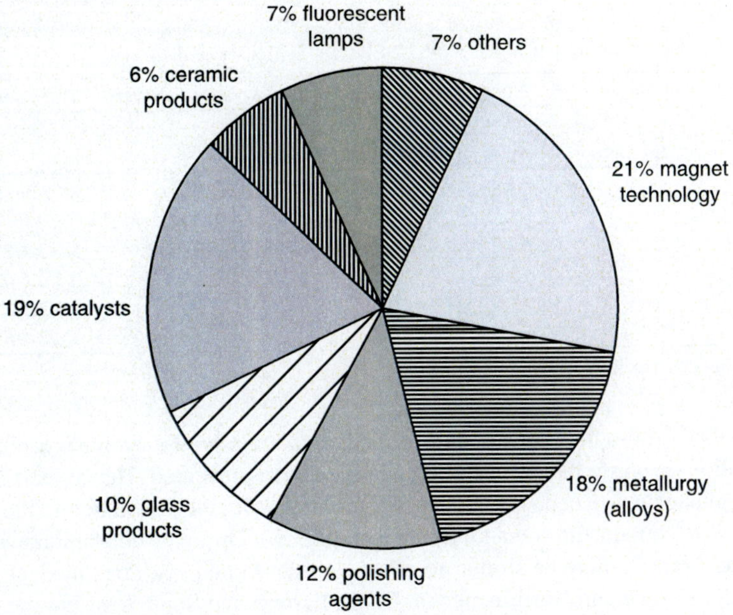

Fig. 2.10 Fields of application for rare earth elements (*REE*) (Data from Kingsnorth 2009 and Haxel et al. 2002)

- Some rare elements are used in medical treatment, e.g. Gd serves as contrast agent in the magnetic resonance tomography, Ho is used in laser surgery and Yb is a component of dental fillings
- Lanthanum-nickel-hydride batteries in communication technology, e.g. mobile phones and notebooks, are increasingly replacing the conventional Ni-Cd batteries, meaning enhanced energy density, better characteristics as well as fewer environmental problems related to disposal and recycling
- Communication technology makes use of Er in relation to the manufacturing of glass fiber cables.

2.3.2 Potentials of Landfills

The shortage of resources has caused an increasing preoccupation with man-made deposits which bear witness to the throw-away society. For example, in Germany until 1972, when the first Waste Management Act came into effect, every town and city district had its own non-regulated landfill in which every sort of waste was deposited in the vicinity of the urban areas. Accordingly, such waste deposits contain the whole spectrum of waste stemming from the period of the so-called "Wirtschaftswunder" (economic miracle). After 1986 some waste components such as glass, paper and iron metals were sorted and removed from the waste flow and

Table 2.6 Compositions of excavated landfills (in %) (Data from Göschl 2006; Prechthai et al. 2006; Bergstedt 2008; Rettenberger 2009; USEPA 1993; Meuser et al. 2011)

	Sharjah United Arab Emirates	Nonthaburi Thailand	Landskrona Sweden	Two sites Austria	Naples USA	Three sites India
Mineral fraction	55	31.8	60	55.2/82.7	60	85.2[a]/90.8[a]/86.2[a]
Plastics	17	52.2	30	35.4/14.3	2	5.7/3.4/5.0
Metals	1	3.0	2.5	3.5/1.2	2	0.2/0.4/0.2
Wood	6	9.0	1	5.8/1.8	–	5.3/3.0/2.4
Others	21	4.0	6.5	–	36	3.6/2.4/6.2

[a]Including construction debris

after 1994 the production was transformed so that waste was avoided. Since 2004 deposition of waste has been more or less entirely prevented. Hence, particularly non-regulated waste deposits and landfills coming from the period before the 1980s and 1990s contain different kinds of materials which are widely unknown. The German example may be similar to other well-developed countries in Central and Northern Europe and North America. But in Eastern and South East Europe nearly 100% of the household waste has been continuously landfilled up to now in many countries such as Bulgaria, Romania and Turkey and the situation in less-developed countries in Africa, South America and Asia might also not be very different (Williams 2005).

In the course of time the waste quality changed enormously. In the developed countries ash from coal combustion was predominant until the 1970s and decreased after gas fuels became important in the cities. Paper increased continuously due to packaging and advertising until modern recycling technology was introduced. After the 2nd World War plastics increased significantly, but in the last two decades recycling measures have reduced the percentage of plastics. The increase in the percentages of glass and metal occurred in a similar way. Organic materials such as food and garden refuse as well as wood were constantly significant and might be always present in waste deposits.

Nevertheless, the search for valuable raw materials present in waste deposits could be disappointing, since most of the waste material obviously consists of mineral soil. In Table 2.6 the main compositions of excavated landfills are listed. In most landfills presented the percentage of the mineral fraction has a magnitude of 55–60%, while metals indicate a small portion. Indeed, the idea of urban mining is mostly focused on the metal residues, which have to be calculated as low as <5%. The percentage of miscellaneous materials such as textiles, glass and rubber varied according to the specific site. Detailed investigation of an old dumpsite in Berlin, Germany, produced findings of 55.2% fine earth including minerals, 40.7% paper and textiles, 5.8% wood and 3.5% non-ferrous metals (Franke et al. 2010). These tendencies, however, do not seem to be characteristic of waste deposits in developed countries such as Austria, Sweden and USA, since, for example, analyses of three waste deposits in India revealed a comparable quantity with approximately 66% fine earth and a relatively low percentage of organic material, which animals like to consume at the sites (Meuser et al. 2011).

Regarding household waste deposits enriched with organic matter like kitchen and yard refuse the principal potential of the calorific value must not be forgotten. The calorific value could be an additional reason to re-open and excavate deposits and landfills, apart from the metals. In this context, one has to decide between the energy production based on the methane generation and use without any excavation procedures and energy production due to the incineration of the separated organic components, thus abandoning the option to use methane gas.

For instance, 5,014 closed landfills and, in addition, 88 refuse landfills with considerable proportions of high calorific content have been reported in Austria. Based on waste composition studies approximately 158 million Mg of deposited Municipal Solid Waste were calculated. The percentage of high calorific material was estimated as being 75 million Mg, divided into textiles (4 million Mg), composite materials (14 million Mg), wood (1 million Mg), plastics (23 million Mg) and paper and carton (33 million Mg). The estimation of the calorific value is difficult because of the decomposition process in landfills. For this reason, paper was supposed to have a half life of 12 years and wood 23 years, leading to a reduction in the calorific value. On the other hand, in old dumpsites a high value was discovered, since some materials such as plastics and wood showed a slow degradation process, in particular in reductive conditions which are normally present. The energy potential in Austria corresponds to 1,391 PJ (petajoule). This amount would replace approximately 10% of the annual primary energy consumption in Austria (Franke et al. 2010).

Waste deposits and landfills which have been already excavated and analyzed exhibited an energy potential between 7,000 and 22,000 kJ kg^{-1} (Hogland et al. 2004; Rettenberger 2009), but the energy balance looks much worse, because energy is consumed for excavation, transportation and recycling of the valuable materials.

In relation to the possible reuse of waste deposits the nutrient potential should also be involved. An extensive field survey was done at various waste disposal sites in the State of Haryana, India, to evaluate the general solid waste management procedure, present situation and nature and amount of waste being collected and deposited in and around the cities. For the study various dumping sites were assessed by onsite observations and finally three waste dumping sites, i.e. Rohtak, Jind and Karnal, were selected on the basis of variability of waste received. The waste deposits varied ranging from relatively fresh (3–4 years) deposits at Karnal to very old deposits (35–40 years) at Rohtak. The Jind site (8–10 years) consisted largely of construction debris and household waste and at present it is situated in the centre of the town.

It was not possible to recognize any significant difference in nutrient concentrations between the two size fractions >2 mm and <2 mm. There was a tendency, however, towards a slightly lower nutrient magnitude in <2 mm waste fraction samples. The following results refer to the fraction <2 mm, usually considered in soil analyses. The solid waste samples were analysed for total and plant available nutrients phosphorus (P), potassium (K), and sulphur (S) using standard methods.

Table 2.7 Total (%) and plant available (mg kg^{-1}) phosphorus, potassium and sulphur content in Rohtak, Jind and Karnal, India, for the size fraction <2 mm (Data from Meuser et al. 2011)

		P	K	S
Total concentration (%)				
Rohtak	Mean	0.68	0.74	1.36
	Range	0.25–1.09	0.52–1.09	0.92–2.02
Jind	Mean	0.31	0.61	3.91
	Range	0.25–0.38	0.54–0.72	0.37–4.09
Karnal	Mean	0.32	0.58	4.39
	Range	0.25–0.47	0.43–0.93	2.81–4.69
Plant available concentration (mg kg^{-1})				
Rohtak	Mean	218	2,086	606
	Range	136–383	1,005–3,538	294–960
Jind	Mean	209	1,969	443
	Range	98–328	1,403–2,758	290–618
Karnal	Mean	149	2,277	576
	Range	84–227	1,128–4,180	384–1,112

Total phosphorus content in samples of <2 mm fraction size was relatively higher in the Rohtak site than in Jind and Karnal (Table 2.7). While the total potassium concentration revealed comparable results between the three sites, clear differences for sulphur were discovered. It was interesting to note that the total S content was fairly high and greater than all other nutrients. In general, the results indicated that the wastes are rich sources of potentially available plant nutrients.

The plant available nutrients are predominantly of interest for use of the waste in agriculture. As shown in Table 2.7, the plant available concentrations of P, K and S mostly indicated higher values for Rohtak. If the ratio of the available value and the total amount based on average values is taken into account, available phosphorus has a value of 3.2–6.7% and sulphur of only 1.1–4.4%. The macronutrient potassium, however, even reached a value ranging as high as 28.8–39.1% (Fig. 2.11).

No definite trend of variation in nutrient content was observed with regard to different waste dumping sites. As expected, all the waste samples contained sufficient amounts of available P, K, and S as per the criteria outlined for agricultural soils in India. Taking a specific density of 1.5 g cm^{-3} and a depth of 20 cm into account, the values for available P should exceed 6.6 mg kg^{-1} and for available K 82 mg kg^{-1}. Accordingly, the measured nutrient concentrations of the waste samples reveal extremely higher values than the criteria defined for agricultural soils. Household waste deposits, however, result in densities of only about 0.5 g cm^{-3} (Meuser 2010), so that the nutrient criteria have to be multiplied by three, leading to higher values. In conclusion, the waste samples indicate a high nutrient supply in any case.

Therefore, it can be concluded that the dumping sites are nutrient storage places that should not be underestimated in future in the context of ideas about urban mining. Particularly, the potassium source for soil fertilization based on treated waste material could be of relevance, because this element showed high plant availability (Meuser et al. 2011).

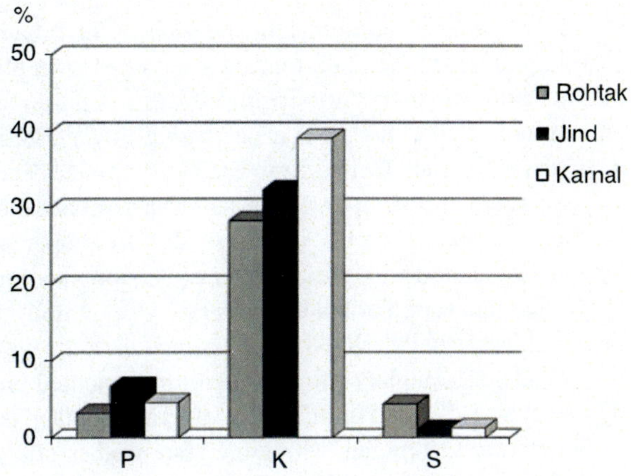

Fig. 2.11 Percentage of available *P*, *K* and *S* in relation to the total concentrations of P, K and S (%) in Rohtak, Jind and Karnal, India (Data from Meuser et al. 2011)

2.3.3 Technique of Landfill Excavation

The re-opening of unadjusted waste deposits and landfills is undoubtedly risky but at the same time offers new opportunities. The reasons for the re-use are as follows:

- Income is generated by the sale of the secondary raw materials
- Costs for assessment and remediation of the contaminated land and landfill are saved, particularly costs for the aftercare of registered landfills
- The volume to be filled increases significantly, since materials are removed and re-used in sections
- The waste excavation provides the opportunity to construct a barrier system at the sides and below before residual material is backfilled (see Sect. 4.1.2); hence, the urban mining process is associated with groundwater protection
- At least a portion of the designated deposit can be redeveloped in association with more important and better land use.

Some negative points should be taken into account. Firstly, the excavation and subsequent sorting is cost-intensive and time-consuming and will probably not be accepted by the inhabitants living in the neighborhood. Moreover, a lot of attention must be paid to the working safety for workers and engineers, because it must be assumed that toxic substances are compositions of the dumpsites concerned (see Sect. 4.4). Furthermore, a huge percentage of the former excavated material has to be backfilled. After the excavation and treatment larger waste pieces are usually crushed. Hence, the volume to be backfilled is fortunately reduced. In this context there will be some problems regarding the Waste Management Acts in some countries that do not allow this dubious practice and where it has never been carried out before. The main detrimental aspect, however, might be related to restricted knowledge about the composition of the waste and the secondary raw materials expected.

Nevertheless, in particular in the vicinity of urban agglomerations the deposit recycling provides the opportunity for a proper town planning process. For instance, in Sao Paulo, Brazil, six non-regulated waste deposits with a size of more than 10 ha are located in the city area, exacerbating any reasonable town planning strategy. Actually, in Sharjah, United Arab Emirates, a waste deposit containing 6.4 million m^3 was excavated in 2007 in order to establish a new residential district. The costs incurred for this were 49 million US $ (Göschl 2006). Comparable examples in Germany and Sweden associated with groundwater protection measures resulted in 4.2–9.1 million US $. The excavation of landfills has rarely been motivated by the idea of urban mining but it was usually related to the protection of the environment from contaminated landfill sites. Exemplary projects were implemented on the island of Sardinia, Italy (Cossu et al. 1995) and in Filborna, Sweden (Hogland et al. 2004).

Separation techniques which are also used in soil preparation approaches (see Sect. 6.1) range from simple and incomplete manual separation, based upon visual inspection of the waste stream and identification of recyclable materials, to more complex automatic systems utilizing techniques such as magnetic separation of ferrous metals, induction current separation of non-ferrous metals and density separation (Anastassakis 2007). The effectiveness and efficiency of such separation techniques varies widely. Manual sorting takes more time but it might be more exact. For instance, it is better to carry out manual separation to distinguish plastics, textiles and rubber or to distinguish wood, paper and paperboard. The use of relatively large quantities of water in many conventional separation approaches also creates an environmental water pollution problem.

A variety of separation technologies for excavated waste is available:

- Air classifiers, cones or cyclones use the spiral air flow action or acceleration within a chamber to separate or classify solid particles
- Density separators screen bulk materials or minerals based on the specific gravity, size and shape of the particles
- Electrostatic separators use preferential ionization or charging of particles to separate conductors from dielectrics
- Floatation systems separate hydrophobic particulates from hydrophilic particulates by passing fine air bubbles up. Afterwards the bubbles are collected
- Spiral and bowl classifiers use mechanical action to dewater and separate coarse bulk materials from finer materials
- Water classifiers such as hydrocyclones use settling to separate or classify materials based on particle size or shape.

Apart from rakes, the most common technique applied is vibrating screens and large rotary drums shaped with a grate-like surface with large openings in order to separate very coarse materials from bulk materials, e.g. coarse plastics from fine aluminum. Powerful magnetic fields to separate iron, steel, or other ferromagnetic materials from non-magnetic bulk materials are used. Rakes and drums, however, display operational problems in the presence of disturbing artifacts such as long plastic foils and pairs of tights. In connection with the separation technique a differentiation in size classes takes place. The preferred size classes are a fine fraction ranging from 0 to approximately 35 mm, a medium fraction from 35 to 80 mm and the coarse fraction relating to materials exceeding 80 mm.

To what extent the approach of opening the deposits and landfills and re-using the secondary raw materials will become important, depends predominantly on the economic influences. Major influences are the development of economic growth in the developing world, especially China and India, but also commodity-specific events such as tariff and usage changes and mine worker strikes, as well as the economic factors generally present and difficult to calculate such as wars and recessions. Ultimately, the prices will decide whether the re-use and recycling of e.g. metals including REE and the organic potential used for energy become objectives. Viewed on the long term, a rise in prices might be most likely. For this reason, it is currently impossible to predict exactly the future relevance of urban mining and consequently the future necessity to involve site-specific urban mining measures with reference to town planning processes. Nevertheless, it will probably play a major role worldwide.

References

Anastassakis, G. N. (2007). Solid waste disposal and recycling: An overview of unit operations and equipment in solid waste separation. In M. Kuzt (Ed.), *Environmentally conscious materials and chemicals processing*. Hoboken: Wiley.

Arweiler, F., Fetzer, K. D., Rupp, E., & Weyrich, J. (2000a). *Criteria of soil protection in respect to sealing and de-sealing investigations for the sample area of Saarbrücken (Saarland)*. Proceedings Vol. 2. Paper presented at 1st international conference on soils of urban, industrial, traffic and mining areas, Essen, Germany.

Arweiler, F., Drescher-Larres, K., Fetzer, K. D., Walter, T., & Weyrich, J. (2000b). *Assessing the suitability of soils for rainwater infiltration in the Saarland area*. Proceedings Vol. 2. Paper presented at 1st international conference on soils of urban, industrial, traffic and mining areas, Essen, Germany.

Bergstedt, E. (2008). *Landfill mining (LFM) – att ta fram begravda resurser.* Avfall Innlandet 2008, Sweden. (in Swedish)

CARBERNET. (2006). *Sustainable brownfield regeneration: CARBERNET network report*. Land quality management report, University of Nottingham.

Cossu, R., Motzo, G.M., & Laudatio, M. (1995). *Prelimnary study for a landfill mining project in Sardinia.* Proceedings. 5th International Landfill Symposium, Sardinia.

Earth Pledge Foundation. (2005). *Green roofs: Ecological design and construction*. Atglen: Schiffer Design Book.

Forro, E., & Draskovits, E. (2000). *Artificial soils for green roof system.* Proceedings Vol. 1. Paper presented at 1st international conference on soils of urban, industrial, traffic and mining areas, Essen, Germany.

Franke, M., Mocker, M., & Faulstich, M. (2010). *Resource potential of landfill mining* – A *national and regional evaluation*. Sulzbach-Rosenberg and Munich. www.iswa.org/fileadmin/galleries/…/Presentations/Mocker.pdf. Accessed 12 Feb 2011.

Ganne, P., Cappuyns, V., Vervoort, A., Buve, L., & Swennen, R. (2006). Leachability of heavy metals and arsenic from slags of metal extraction industry at Angleur (eastern Belgium). *Science of the Total Environment, 356*, 69–85.

Genske, D. D. (2003). *Urban land – Degradation, investigation, remediation*. Berlin: Springer.

Göschl, R. (2006). *Remediation of the old Sharjah landfill. Emirates Environmental Technology Co., Sharjah, United Arab Emirates*. http://www.baumaschine.de/Portal/download.php?w=TbgT&p1…p2. Accessed 26 Feb 2011.

Haxel, G. B., Hedrick, J. B., & Orris, G. J. (2002). *Rare earth elements – Critical resources for high technology*. USGS Fact Sheet 087.02.

Hazelton, P., & Beecham, S. (2000). *A field investigation of stormwater infiltration practices in New South Wales, Australia – Woy Woy test case.* Proceedings Vol. 2. Paper presented at 1st international conference on soils of urban, industrial, traffic and mining areas, Essen, Germany.

Helmes, T., Kubiniok, J., Fetzer, K. D., & Schmitt, C. (2000). *Analysis and evaluation of infiltration properties in the urban environment of Saarbrücken.* Proceedings Vol. 2. Paper presented at 1st international conference on soils of urban, industrial, traffic and mining areas, Essen, Germany.

Hogland, W., Marques, M., & Nimmermark, S. (2004). Landfill mining and waste characterization: A strategy of remediation of contaminated areas. *Journal of Material Cycles and Waste Management, 6*(2), 119–124.

Kim, Y., Yi, Q., Kim, L., & Min, K. (2008). *Introduction of test pilot best management practices for reducing non-point pollution in Korea*. Proceedings. Paper presented at the 2nd international conference on Asian-European environmental technology and knowledge transfer, Hefei, China.

Kingsnorth, D. J. (2009). *Industrial Minerals Company of Australia*. 5th international rare earth conference, Hong Kong.

Luckett, K. (2009). *Green roof construction and maintenance*. New York: McGraw-Hill Book.

Meuser, H. (2010). *Contaminated urban soils*. Dordrecht: Springer.

Meuser, H., Grewal, K. S., Anlauf, R., Malik, R. S., Narwal, R. K., & Saini, J. (2011). Physical composition, nutrients and contaminants of typical waste dumping sites. *American Journal of Environmental Sciences, 7*, 26–34.

Nathanail, C. P. (2011). Sustainable brownfield regeneration. In F. A. Swartjes (Ed.), *Dealing with contaminated sites*. Dordrecht: Springer.

Prechthai, T., Visvanathan, C., & Cheimchaisri, C. (2006). *RDF production potential of municipal solid waste.* 2nd joint international conference on sustainable energy and environment, Bangkok, Thailand.

Rettenberger, G. (2009). *Zukünftige Nutzung der Deponie als Ressourcenquelle*. 11th conference Münsteraner Abfallwirtschaftstage, Münster, Germany. (in German)

Schelwald-Van der Kley, L., De Fraye, J., Pentel, R., Buve, L., Jaquet, R., Slenders, H. L. A., Heasman, I., Wallace, S., & Ackermann, M. (2011). A stakeholder's perspective on contaminated land management. In F. A. Swartjes (Ed.), *Dealing with contaminated sites*. Dordrecht: Springer.

Schleuss, U., Buchenau, K., & Kuhnt, G. (2000). *Rainfall water management in a small town in northern Germany.* Proceedings Vol. 2. Paper presented at 1st international conference on soils of urban, industrial, traffic and mining areas, Essen, Germany.

Stadt Leipzig. (2008). *Leipzig – "Plagwitz" (Saxony)*. http://www.eukn.org/Germany/de. Accessed 6 Apr 2011.

USEPA (United States Environmental Protection Agency). (1993). *Evaluation of the Collier County, Florida landfill mining demonstration* (EPA/600/R-93/163). Washington, DC: Office of Research and Development.

USEPA – Environmental Protection Agency. (2010). *International brownfield case study: Emscher Park, Germany.* http://www.epa.gov/brownfield/partners/emscher.html. Accessed 28 Mar 2011. Accessed 15 Mar 2011.

Williams, P. T. (2005). *Waste treatment and disposal*. Chichester: Wiley.

Winzig, G. (2000). *The concept of stormwater infiltration.* Proceedings Vol. 2. Paper presented at 1st international conference on soils of urban, industrial, traffic and mining areas, Essen, Germany.

Chapter 3
Rehabilitation of Soils in Mining and Raw Material Extraction Areas

Abstract Coal, salt and ore mining areas and areas where other raw materials are extracted, such as quarries, pits of unconsolidated rock and harvested peatland, are anthropogenically strongly disturbed terrains which, apart from the ore mining sites, do not indicate contamination with toxic substances. Nevertheless, in relation to the pedogenically developed soil and the vegetation cover, they exhibit large-scale complete removal in the case of opencast mining, quarries, open pits and peatland. The formerly used areas must be rehabilitated, which can be arranged in different ways. Opencast lignite coal areas are either reclaimed for agriculture, forestry or recreational purposes or are restored in conjunction with the establishment of nature reserves. In this chapter, technical applications for ensuring reclamation and measures to initiate new vegetation are explained. The development of natural succession and treatment of controlled restoration are introduced in association with the natural reserves. The creation of mining lakes is also described. By way of analogy, rehabilitation measures in quarries and pits of unconsolidated rock, which mainly focus on natural succession of vegetation or implementation of woodland, are taken into consideration. Regarding ore mining areas which produce large-scale tailing ponds, special attention is paid to the heavy metal contamination and the opportunities to treat it. Furthermore, the restoration of differently harvested peatland types (bogs, fens) which is aimed at the regeneration of this landscape type is introduced in detail. The long-term processes of rewetting, re-naturation and regeneration are considered in different climates. Apart from the mining and raw material exploitation, which influences large areas, underground mining, which is associated with coal, salt and ore mining heaps and their environmental impacts, is discussed. The rehabilitation approaches for the mining heaps consisting of material with extreme physico-chemical properties form an important topic of this chapter. In this context, solutions for subsided mining terrains resulting from underground mining are considered as well.

Keywords Agricultural reclamation • Mining heap • Mining rehabilitation • Nature reserve • Open-cast mining • Peatland regeneration • Silvicultural rehabilitation • Tailing pond

H. Meuser, *Soil Remediation and Rehabilitation: Treatment of Contaminated and Disturbed Land*, Environmental Pollution 23, DOI 10.1007/978-94-007-5751-6_3,

3.1 Open-Cast Coal Mines

3.1.1 *Examples of Open-Cast Mining Areas*

Open-cast mining is mainly associated with the exploitation of lignite coal (brown coal) and sub-bituminous coal. The quantitatively more important lignite coal, which stems mostly from the Tertiary Period, is usually extracted in open-cast mines. Different coal seams located one below the other are extracted and used, while the intermediate overburden must be excavated and deposited on site. For this reason, a high percentage of waste material must be rehabilitated to a later date.

The handling of the excavated and backfilled soil is usually not linked to the dealing with contaminated material. The extracted overburden consists of natural soils which have only been disturbed but normally not polluted. As observed, for instance, in the Rhineland lignite coal open-cast mining area in Germany, the contaminant concentrations are mostly very low. The overburden containing loess exhibited values between 5.1 and 12.1 mg kg^{-1} for copper, between 1.3 and 7.0 mg kg^{-1} for chromium, between 3.8 and 12.2 mg kg^{-1} for zinc and <0.15 mg kg^{-1} for mercury respectively. Despite this, attention should be paid to heavy metals, in particular regarding deposits containing sandy material from the Tertiary Period that is strongly acidified, resulting in potentially enhanced metal mobility. Investigations in another German lignite coal mining area called Lusatia showed accumulated metal concentrations in leachates underneath deposited material which did not fulfil the quality standards for drinking-water. For example, the values were 0.12 mg L^{-1} for cadmium, 3.9 mg L^{-1} for chromium, 7.3 mg L^{-1} for copper, 0.4 mg L^{-1} for lead and 53.4 mg L^{-1} for zinc and the pH indicated strong acidity (pH 2.5) (Pflug 1998).

In eastern Germany (the former GDR) all efforts had been directed towards boosting industrial development until 1989, with a lack of attention being paid to the ecological conditions. Consequently, there was the largest lignite coal producer in the world, mining more than 300 million tons per year (central German lignite mining district and the district of Lusatia). The rehabilitation programme dealing with the negative impact of the large-scale coal extraction started after 1989 and for economic reasons the coal extraction dropped to some 100 million tons per year. In the context of the rehabilitation measures open pits have been flooded to form artificial lakes, which caused new ecological problems such as slope erosion at the shores and acidification of the lake water. Furthermore, remediation strategies had to be developed because of many open-cast mines which were used for waste disposal without any implementation of safety measures. Moreover, many former excavated areas had been used for the deposit of fly ashes stemming from the coal-fired power plants located close to the open-cast mines (UFZ 1999). In the Rhineland in the western part of Germany lignite coal production varied to a minor degree (90 million to 105 million tons per year) between 1970 and 2010. Until 2020 the mining process will continue to extract approximately 110 million tons per year (Pflug 1998).

A history similar to the development of the former GDR can be seen in the lignite coal areas of the Czech Republic. During the socialist era surface mining was concentrated on a few very large mines in the two coal basins North Czech Brown Coal

Basin and Basin of Sokolov. After 1989 a decrease in mining was ascertained under the new market conditions and the resulting political change. In a similar way, the re-cultivation efforts addressed to agriculture, forestry and recreation began. However, because of the deep mining activities reaching a depth of 200 m and the enormous thickness of the coal seams amounting to 50 m the number of refill opportunities was limited (Svoboda 1993).

The rehabilitation necessities in mining areas where lignite coal is produced might continue over long periods of time in future. As the example of the Polish lignite coal mining industry shows, which extracts more than 64 million tons per year, the rehabilitation of the destroyed terrain must be conducted on a wide scale. The establishment of forestry on waste heaps and the refilling of open pits take place at the same time as exploitation that will last at least until the end of the present century. It should be noted that the rehabilitation programme can be finished in the long term after the coal extraction has already stopped.

In some open-cast mining areas such as in the Rhineland mining areas the depth of the extraction reached up to 500 m. The overburden consists of distinct Pleistocene and Holocene material such as terrace gravel and sand, alluvial loam and clay, glacial loam, glacial drift, glacial sand, sanddrift, loess, etc. The ratio between extracted lignite coal and mining waste changed significantly from 0.3:1 (in the 1970s) to up to 6:1 (nowadays, in Germany). In some open-cast coalfields the average ratio even reaches 19:1, as noticed, for example, in the British surface mines (Pflug 1998; UFZ 1999).

The overburden of lignite coal mining sites can consist of clay marl, as found in the Tuzla coalfield in Bosnia and Herzegovina. This area covers approximately 20,000 ha with an increasing tendency (3,000 ha per year). Results from the utilisation are numerous craters with a depth of up to 200 m and a diameter of about 500 m. The rehabilitation approaches are different from those applied to deposited overburden derived from the Tertiary Period (Resulovic 2000).

3.1.2 *Effects of Large-Scale Mining Operations*

In the first instance, dewatering occurs with a series of wells (well batteries) which lower the water table below the deepest coal seam to be mined. After deep-well pumping the overburden is removed until the coal seams are reached. Crossing rivers must be treated. Larger streams must be transferred into new canals that are tightened and sealed to prevent water seepage into the previously dewatered underground. Smaller rivers and some wetlands are in danger of becoming dry during the long-lasting mining operation. In summary, the water balance of the entire landscape is completely changed, which has an impact on, for instance, forests and wetlands. The drying processes of overburden and layers containing seam mean that the material will come into contact with air, causing the oxidation of hydrogen sulfide FeS_2 (pyrite and marcasite) (see Sect. 3.1.3).

The open-cast mining and the subsequent rehabilitation affect the hydrogeological characteristics significantly. During the mining process the groundwater must be drastically lowered, reaching a depth of up to 500 m. Thus, huge groundwater cones

Fig. 3.1 Bucket wheel excavator in the Rhineland lignite coal open-cast mining area, Germany

of depression are created, influencing the adjacent large-scale areas. Biotopes adapted to wet soil conditions cannot survive and agriculturally used areas have to expect decreasing crop yield. Theoretically, the groundwater continuously pumped for the mining process could be transported to the endangered biotopes by pipelines based on a water circle system. This solution seems to be technically feasible, but mostly it does not make economic sense. Unfortunately, after the rehabilitation has started negative effects continue. For instance, the deposited material used for agriculture or woodland is heterogeneous, indicating enormous dislocation. Consequently, a continuous aquifer can hardly develop and hydraulic discontinuity prevails. Furthermore, in the post-mining period the amount of water required for the filling of aquifers and mining lakes is extremely high. In the eastern mining areas of Germany, for instance, 16 billion m^3 are needed to refill the aquifers (10.1 billion m^3) and the residual lakes (5.9 billion m^3). In contrast, only 5.4 billion m^3 of water are available by withdrawal from rivers in this mining catchment (Pflug 1998).

In open-cast mines big machines which cannot separate exactly the extracted mining waste of different geological epochs are in operation, especially bucket-wheel excavators (Fig. 3.1) and belt stackers. In particular, in mining areas indicating a lot of geological dislocations the required separation is not feasible. The layers stemming from different geological origins are cut by bucket wheel excavators and transported by conveyor belt bridges into the previously emptied part of the mine. The modern bucket-wheel excavators have enormous dimensions. They mine 240,000 m^3 day^{-1} and have a weight of 13,000 t, a length of 255 m and a height of 85 m (Pflug 1998). It is understandable that the material is deposited in a mixed manner, causing problems for the rehabilitation process (Fig. 3.2). The landscape appears strongly disturbed and is reminiscent of a moonscape (Fig. 3.3).

Fig. 3.2 Different overburden deposited by belt stackers in the Rhineland lignite coal open-cast mining area, Germany

Fig. 3.3 Open-cast lignite coal mining area north of Pernik, Bulgaria, reminiscent of a moonscape

The rehabilitation measures result from the different intentions of the affected persons and institutions, leading to a number of general interest conflicts. Agriculture prefers large-scale cropland and pasture with as few hedges as possible, the water authorities are interested in a small number of residual mining lakes located at a safe distance from the agriculture and forestry must decide between timber and recreational forest.

Because of the large-scale open-cast mining in which coal is also extracted in populated areas people are affected because of the necessity for resettlement. In order to generate large-scale open-cast mines villages and small cities were torn

Fig. 3.4 Removed cathedral of Most, Czech Republic

down. In general, only historical buildings of special significance were protected such as the cathedral of Most, Czech Republic, which was transported over a distance of several hundred metres (Fig. 3.4).

In principle, for all post-mining uses the soil must be intensively prepared by man. The term soil reconstruction is used for the necessary soil handling. Different land-use types may result from the rehabilitation of open-cast mines. In Germany the use of the rehabilitated land amounts to 34% agriculture, 49% forest, 8% surface waters and 9% built-up and traffic areas. The main tendency is for the cropland to decrease and for a relatively high percentage of mining lakes to increase simultaneously. For instance, in the Rhineland coal mining district only 44% was used for the previously existent cropland and the lake terrain indicated a tenfold increase (Pflug 1998). If one considers the re-use of open-cast mines worldwide compared to the situation before open-cast exploitation of coal occurred, the percentage of agriculture is decreasing, while both the forest and lake terrain show an increasing tendency. This development is due to the soil management which cannot rehabilitate the entire area to the same quality as in the past.

Compared to the previous utilisations considerable alterations can be found. For instance, in Yorkshire coalfields in the North of England, United Kingdom, where mine rehabilitation was preferentially linked to nature conservation, the woodland cover doubled and an increase of habitats including water habitats and their diversity occurred. The post-mining uses, however, changed in the last few decades for economic and environmental reasons. Whereas before 1990 in the British open-cast coalfields, located in the Scottish Lowlands, the North East, central East and central West England, in the Midlands and South Wales, 90% of the worked land was reclaimed for agriculture, nowadays there is an increasing tendency for the restoration to include environmental aspects. The land uses are focused on the creation of habitats in addition to the establishment of terrestrial leisure areas such as parks and golf courses and lakes used for sailing, windsurfing, water-skiing and fishing (UFZ 1999).

In the proximity of urbanised areas the open-cast mining also affords a redevelopment of the land use planning. In the open-cast mining district south of Leipzig, Germany, where the mining terrain adjoins the suburbs of the city, the post-mining utilisation included a simultaneous city expansion. In this district, approximately 240 km^2 of whose land surface have been exploited, the percentage of settlement areas increased from 7.7% to 16.6% whereas the percentage of cropland was reduced from 68.4% to 54.0% and, in particular, the environmentally valuable marsh areas disappeared almost entirely (reduction from 11.1% to 3.5%) (Pflug 1998).

The post-mining opportunities depend on a number of environmental characteristics. There is no doubt that the most important factor is associated with the properties of the deposited material. Nevertheless, climatic conditions such as precipitation, evapotranspiration and air temperature can also impact the utilisation options, since they determine the length of the growing season that influences, for instance, the agricultural use. Apart from the climate, the topography of the devastated land, in particular in the border areas where terminal heaps have been deposited, may influence the re-use of the disturbed terrain. Areas with steep slopes are predominantly of interest for woodland establishment.

3.1.3 Agricultural Rehabilitation

The geological origin of the deposited material becomes preferentially important. In the case of material stemming from the Tertiary Period, which is consequently sandy and contains pyrite, the re-use opportunities are strongly restricted and mainly focused on a silvicultural utilisation. Material stemming from the Pleistocene or Holocene, in particular alluvial sediments, sandloess and loess, however, are suitable for an agricultural use based on an optimum soil management. With reference to the texture deposits consisting of silt loams, loams and silty clay loams are preferred for agriculture.

Separated materials such as loess and sandloess are preferentially used for cropland. After depositing by belt stackers bulldozers try to grade the surface. In relation to planned agriculture at first a rough ground is deposited consisting of gravelly and sandy material that is easy to grade. Above, more loamy or silty material, e.g. loess and sandloess, is deposited in dry conditions reaching a thickness of up to 2 m. During the grading process attention must be paid to soil compaction. It is possible to minimise this, if the machines go over the ground as little as feasible.

In contrast to the establishment of cropland, pasture might be relatively easy to prepare. Pasture can be an option, if erosion must be expected and therefore a continuous vegetation cover must be guaranteed. After re-grading disk ploughs should be used to increase surface roughness and water infiltration. Accordingly, it is not possible to prepare a firm seedbed but in humid climates a mix of species consisting of slow growing grasses and faster growing perennials leads to optimised vegetation cover. The influence of the climatic conditions determines the choice of the plant species. For instance, under arid and semi-arid conditions in the western USA some *Agropyron* species, which have often been inappropriate in humid climates, are chosen (Thornburg 1982).

In relation to cropland the re-contouring of the landscape is of enormous relevance to enable sufficient water infiltration and limited runoff and subsequent soil erosion. The slope gradient should not exceed 1.5%. Additional runoff control measures like furrows and trenches can be included, whereas contoured terraces are only constructed as an exception. Bulldozers are problematical due to their high dead weight and the compressive force applied by the blade when pushing soil material. Earthmoving equipment with rubber tyres appears to be less suitable due to the compaction caused. For this reason, crawler excavators are preferred. With respect to the trucks rear-loading dump trucks minimise compaction to a minor degree (see Sect. 4.2.2) (UFZ 1999).

Alternatively to dry deposition silty materials are irrigated and afterwards pumped into lagoons surrounded by dams. A subsequent grading procedure does not have to take place. Macropores >50 μm are lacking in association with reduced hydraulic conductivity because the material in the lagoons which has a thickness of approximately 1 m shows different sink speeds of the texture classes leading to horizonation and accumulation of silty and clayey material in the lower part of the profile >30 cm (Table 3.1). Accordingly, rooting might be adversely affected (Pflug 1998).

The agricultural use requires a number of preparatory and support measures. In the first instance, obstacles, particularly large pieces like boulders, are removed, since they are disturbing factors in association with the soil ploughing procedures. The removal should be aimed at a maximum value of less than 10% rock fragments (Grandt 1988). The material is sorted and classified, for example with the help of screens. Afterwards, the area of concern is deposited and graded by bulldozers, as mentioned above. In this way, an artificial man-made landscape is created that should take the demands of the planned agriculture into consideration (Fig. 3.5).

By way of example, material used for agricultural and forest rehabilitation in Germany is introduced in Table 3.2. Overburden consisting of loess and alluvial loam exhibited a loamy-silty texture responsible for relatively high plant available water content (>16 mm dm^{-1}), while the air capacity tended to be less beneficial. As a research project in the coal mining area of South Wales, United Kingdom, showed, even 18 years after depositing and grading at shallow depths the porosity characteristics remained poor compared with adjacent undisturbed land (Table 3.3). Despite the installation of field drains, the reclaimed soils indicated continuously impeded drainage. Consequently, the net nitrogen mineralisation was lower in the reclaimed material as well. Simultaneously, the earthworm population of the insufficiently drained terrain was confined to shallow dwelling species. In conclusion, the adverse physical conditions also influenced chemical and biological parameters negatively. Careful and long-lasting aftercare observing a multitude of soil properties appears to be necessary in order to assure effective and productive agriculture (Scullion 2000).

Table 3.1 Physical properties of loess material washed into lagoons and deposited in dry conditions in the rehabilitated lignite coal mining terrain of the Rhineland, Germany (Data from Pflug 1998)

Treatment	Age (years)	Depth (cm)	Total pore volume (vol%)	Pores >50 μm (vol%)	Hydraulic conductivity (cm day^{-1})
Lagoon	10	0–30	42.5	7.3	15.6
		30–50	38.6	1.3	8.6
		50–70	38.6	0.6	4.7
		>70	39.9	0.9	6.3
	20/25	0–30	42.9	9.3	29.3
		30–50	39.0	3.1	7.2
		50–70	41.2	1.9	8.0
		>70	41.4	2.1	7.5
Deposit	10	0–30	43.4	9.3	15.7
		30–50	39.2	4.4	13.9
		50–70	39.6	4.5	10.2
		>70	40.3	5.9	16.3
	20/25	0–30	41.0	7.2	34.2
		30–50	36.3	3.0	7.1
		50–70	36.9	2.7	12.7
		>70	38.4	3.2	14.6

Fig. 3.5 Man-made landscape after rehabilitation of a lignite coal open-cast mining area in the Rhineland, Germany

As long as the used material such as loess has not been completely decalcified the pH value might be neutral and thus acceptable for future agricultural use. In general, pH values ranging from 4.5 to 8.5 should be sought for rehabilitated opencast mining areas which are used for cropland. The electrical conductivity should be less than 4 mS cm^{-1} and the percentage of exchangeable sodium should fall below 15% (Grandt 1988).

Table 3.2 Soil properties of material used for agricultural and forest rehabilitation of open-cast lignite coal mining areas in the Rhineland, Germany

Characteristic	Agricultural rehabilitation	Forest rehabilitation
Texture	Loamy silt	Slightly loamy sand
Humus content (%)	0.4–0.5	0.3–0.5
Bulk density (g cm^{-3})	1.55–1.75	1.4–1.8
Total pore volume (%)	35–52	32–48
Air capacity (%)	4–12	4–15
Available water capacity (mm dm^{-1})	16– >20	9–18
pH value	7.5–8.0	4.5–7.5
N total (%)	0.02–0.05	0.02
Available K (mg 100 g^{-1})	5–11	2–4
Available P (mg 100 g^{-1})	1.5–3	1–1.5
Available Mg (mg 100 g^{-1})	10–15	3–7

Table 3.3 Porosity in grassland soils reclaimed in different periods compared with undisturbed soils in British open-cast mining areas; texture: clay loam (Data from Scullion 2000)

Site	Macropores >30 μm (vol%)	Total pore volume (vol%)	Macropores >30 μm (vol%)	Total pore volume (vol%)
Depth	0–5 cm	0–5 cm	15–20 cm	15–20 cm
Reclaimed 2 years	6.0	45.5	0.3	38.7
Reclaimed 9 years	9.2	55.9	1.8	39.5
Reclaimed 18 years	8.4	54.1	0.7	41.4
Undisturbed area	18.5	64.9	13.4	58.7

The coal mining waste stemming from the Tertiary Period is different to the material described above. It reveals a skeleton-enriched sandy texture (sand to loamy sand) and contains rather few macronutrients amounting to <0.02% nitrogen and <1.5 mg 100 g^{-1} available phosphorus. According to the texture the available water capacity varies between 9 and 18 mm dm^{-1}, leading to water deficiency in the plants (Table 3.2).

The most disadvantageous feature, however, refers to the pH value. Because of the pyrite content of up to 2% and the subsequent pyrite oxidation in dry conditions sulphuric acids arise, causing a fast pH decrease to values of up to 2.5:

$$FeS_2 + 3.5O_2 + H_2O \rightarrow Fe^{2+} + 2SO_4^{2-} + 2H^+ \quad (3.1)$$

Apart from sulphuric acids (H_2SO_4, H_3SO_4), sulphate leaching must be taken into account. The SO_4^{2-} leaching may result in a salt pollution of the groundwater which corresponds to the residual mining lakes.

Afterwards, Fe^{2+} reacts biochemically (*Thiobacillus ferrooxidans*) with air-oxygen as follows:

$$Fe^{2+} + 0.25O_2 + H^+ \rightarrow Fe^{3+} + 0.5H_2O \tag{3.2}$$

Continuous reaction with water leads to the formation of ferric hydroxides ($Fe(OH)_3$) which remain water-soluble, cause a dark red watercolor and accelerate the proton delivery:

$$Fe^{3+} + 3H_2O \rightarrow Fe(OH)_3 + 3H^+ \tag{3.3}$$

All the chemical reactions can be summarised in a simple way as follows:

$$FeS_2 + 3.75O_2 + 3.5H_2O \rightarrow Fe(OH)_3 + 2SO_4^{2-} + 4H^+ \tag{3.4}$$

The macronutrients must be evaluated as problematic, since the overburden stems mainly from mineral layers without any organic matter. Though the material was deposited in a mixed manner and consequently would contain humus-rich layers derived from the upper parts of the extracted open-cast mines, the organic matter content normally amounts to less than 0.5%. Nevertheless, black material at the surface is often recognisable (Fig. 3.3) but this material is associated with the coal remnants simultaneously deposited.

A higher humus content is only present, if during the mining operation the topsoil is stored separately and afterwards put back onto the new surface but this beneficial approach has been applied extremely rarely. In the United Kingdom, where around 100,000 ha were affected by open-cast mining, topsoil and subsoil have been frequently stripped and stored separately, so that the replacement could occur in the appropriate order. In spite of this, during the long-lasting storage the topsoil degraded physically and the populations of earthworms and some other organisms reduced. In particular, more finely textured material revealed more detrimental properties (Scullion 2000).

In accordance with the lack of humus, the macronutrients closely correlated to the presence of organic matter such as nitrogen, phosphorus and sulphur appear to be deficient. In German mining areas nitrogen showed a low concentration (<0.05%) and the P content also ranged from 1 to 3 mg 100 g^{-1} plant available phosphorus (Table 3.2). With reference to the organic matter content the long-term agricultural use may enhance the value continuously but the detected rate was low at 0.02–0.08% (on average 0.03%) per year. For this reason and due to the no-till farming occurring in the first years the pedogenic development of the humic topsoil (A horizon with an initial C/N ratio of 6:1–8:1) needs at least 15 years (Pflug 1998). It may happen faster, if organic manure such as peat, sewage sludge, dung and straw is added.

In some mining areas like the Tuzla terrain in Bosnia and Herzegovina the deposited overburden consists of stony clay marl with a neutral pH (7.1–7.7), high $CaCO_3$ contents and an acceptable level of nutrients such as phosphorus and potassium. Although the humus content was initially low (0.6%), it increased to 2.5–4.5% within 20 years in the upper 30 cm. Accordingly, the macronutrients behaved in a similar way reaching low to moderate contents (e.g. P_2O_5 1.8–2.9 and K_2O 5.2–8.3 mg 100 g^{-1}) within the same time span. Moreover, some physical conditions tended to improve, because the skeleton-enriched clay marl was quickly disintegrated, forming an increasing percentage of small particles which caused better water retention. This process, however, was associated with accelerated hardening and compaction. Furthermore, erosion at slopes exceeding 5% and soil sliding in the context of formation of depressions and a wavy surface were discovered (Resulovic 2000).

The improvement of the physical soil conditions such as the air capacity in combination with the accumulation of organic matter and microbial activity can be achieved by using deep-rooting legumes like clover (*Melilotus* sp.), alfalfa (*Medicago sativa*) and birdsfoot trefoil (*Lotus corniculatus*). These plants should be predominantly planted within approximately the first 10 years and they should reach a percentage of up to 50% of the crop rotation. Alternatively, mixtures of grass vegetation (e.g. *Festuca* sp., *Dactylis glomerata, Bromus inermis, Cynodon* sp.) and sweet clover (*Melilotus officinalis*) are also adequate species that enrich the humus content and decisively improve the soil air household. The grasses may grow slower than the legumes but they are important for improving soil structure and reducing erosion. Plants such as *Phacelia tanacetifolia, Secale multicaule* and *Raphanus sativus* serve the same purpose (Grandt 1988).

In general, the crop rotation must be done in a soil-protective way including the implementation of catch crops such as Indian mustard (*Brassica juncea*), white clover (*Trifolium repens*) and rape, but fallow land should also interrupt the continuous crop growing. In the first decade some crops such as maize, potatoes and sugar-beet should not take part in the crop rotation process. In contrast, grain crops (e.g. wheat, rye, oat, barley) can be seeded on graded land formerly used for mining without any problems (Grandt 1988).

Mineral fertilizing takes place to a greater degree, in particular regarding nitrogen and phosphorus. For instance, with regard to the nitrogen fertilizing an increase of +50 kg ha^{-1} is recommended. Subsequently, the enhanced fertilizing results in an optimised root growth, which can accumulate the organic matter content additionally (Scullion 2000).

Apart from the cultivation of soil-protective plants and the application of mineral fertilizers the application of organic manure leads to humus enrichment as well. Appropriate fertilizers are sewage sludge, compost, dung and straw. Nevertheless, the user of sewage sludge and compost should have an eye for the potential contamination of the manures. Because of the sandy material to which sewage sludge, for instance, is added leaching of nitrate can be one of the ecological problems to be observed. Application of 10 t ha^{-1} up to 184 t ha^{-1} (dry weight) did not cause a significant increase of NO_3^- and NH_4^+ leaching. Analyses with ceramic suction cups at depths of 20 and 130 cm resulted in relatively low nitrate and ammonium concentrations after sewage sludge amelioration (19 t ha^{-1}) to plots vegetated with young pines and *Secale multicaule*. The values ranged between 0.1 and 0.6 mmol L^{-1} for NO_3^- and between 0.4 and 0.7 mmol L^{-1} for NH_4^+ respectively. The results were in the order of fertilizing with mineral manure (Schaaf et al. 2000). In the Most region, Czech Republic, the addition of high amounts of pulp waste (400 t ha^{-1}) and sewage sludge (200 t ha^{-1}) was not only responsible for an increase in the organic matter (4.8% on average compared to less than 1% based on the deposit of pure loess), but also microbial parameters such as microbial biomass carbon, potential and basal

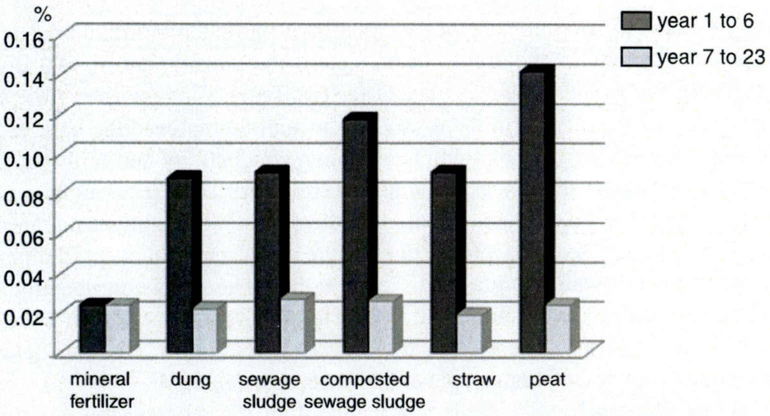

Fig. 3.6 Average yearly increase of organic matter content in rehabilitated loess topsoils after application of 3.8 t ha^{-1} a^{-1} organic manures (Data from Pflug 1998)

Table 3.4 Capability of different organic fertilizers for use in rehabilitated lignite coal mining areas (Data from Pflug 1998, modified)

Material	Stability against deflation	Mineralisation (without additional nitrogen application)	Input of nutrients	Input of seeds
Composted sewage sludge	++	++	O	––
Hay (alfalfa)	O	+	++	––
Hay (grasses)	+	+	++	++
Straw (wheat)	–	––	O	––
Bark mulch	++	––	+	–

++ very high
+ high
O moderate
– low
–– very low

respiration showed raised values. Unfortunately, the potential ammonification and nitrification also resulted in increased values which are possible to cause unwanted NO_3^- and NH_4^+ leaching (Ruzek et al. 2003).

Long-term studies in the German Rhineland mining area showed different impacts of the amendment with organic fertilizers on loess soils. Within the first 6 years the clearest increase in the organic matter content was associated with composted sewage sludge or peat whereas the most common agricultural organic fertilizers (dung, pure sewage sludge, straw) resulted in a moderate increase only. In the following 17 years the increase in the organic fertilizers did not reveal significant differences. By comparison, the effectiveness of mineral fertilizers has been considerably lower (Fig. 3.6) (Pflug 1998). In summary, the amendment of organic manure should beneficially impact the soil properties in a manifold manner (Table 3.4).

In semi-arid and arid climates the reclamation is particularly difficult because of the water deficiency. In the Ebro Basin, Spain, the amendment of two organic manures, pig slurry and straw, was tested for two types of overburden (fine spoil, coarse spoil). As the results, in the topsoil straw addition reduced the bulk density, while other physical properties such as hydraulic conductivity and infiltration rate showed no significant influence of the added substrates. The accumulation of the organic matter occurred to a greater extent in the deposit consisting of fine material coupled with higher biomass production of the cultivated plants (predominantly *Brassica scoparia*), which might result from higher water retention potential. The organic matter content was hardly influenced by the organic fertilizers, even after 2 years of investigation. The electrical conductivity (salinity), however, revealed higher values after amendment of pig slurry (Salazar et al. 2000).

Particular attention must be paid to the overcoming of soil compaction. Minimum tillage operations are recommended to avoid soil compaction and reductive conditions. There are many adverse impacts caused by compaction. Root growth is hindered by mechanical impedance and reduced aeration. The nitrogen availability is restricted by reduced nitrogen mineralisation and anaerobic decomposition of nitrate, the microbial activity is disturbed and even toxic substances can appear such as hydrogen sulfide. The tillage can be carried out without ploughing based on so-called no-till farming. The weight of the agricultural machines should be as low as technically possible by combining machines and using wide-base tyres. All tilling procedures should occur while dry to moderate soil moisture content prevails, as generally recommended with regard to loamy, silty and clayey soils (see Sect. 4.2).

It cannot be excluded that subsoil compaction becomes visible due to the subsidence of the deposited material. The soil must be rehabilitated again using deep cultivating ploughs that may reach a depth below the rooted topsoil. It is possible to minimise stagnating water, if drainage ditches are cut and the soil surface has a low slope gradient that must not exceed 1.5% (UFZ 1999).

Since the deposited material originates from subsoils, it does not contain many weed seeds. Accordingly, the weed control is reduced to extensive measures without use of herbicides. It is expected that the weed development will be a low one during the first decade after establishment of agricultural use.

In dry conditions the topsoil tends to be eroded by wind because the soil structure is hardly developed and the organic matter content rather low. If the agricultural sites are large-scale, windbreak belts consisting of shrubs and trees are urgently necessary. They reduce deflation considerably and provide the long-term soil fertility. The planted windbreak belts have a width of 3–5 m and are planted at a distance of 500 m each. The trees should be planted in 2–3 rows, commencing with one additional row of small trees or shrubs. They are beneficial with regard to reduced deflation and evaporation, accelerated soil temperature and are part of the biotope network. Furthermore, ecologically useful hedges can be implemented because many boulders and stones have been separated during the first rehabilitation measures (Fig. 3.5) (GOA 1995a; Pflug 1998).

In conclusion, based on the rules associated with adequate mineral and organic fertilizing, crop rotation including legumes, minimal tillage and involvement of windbreak belts, the crop yields appear to be favourable and it is assumed that they are as high as the previous yields before open-cast mining began (Grandt 1988).

The farmers' settlements are rebuilt with a design similar to that of the cropland and pasture. The farms must be reconstructed in a completely new way because all farms and villages were demolished in the course of the open-cast mining. On the one hand the loss of the familiar surrounding area is difficult to suffer for the inhabitants, on the other hand the rebuilt home provides new economic and social opportunities, if the resettlement is supported by public or private investors.

3.1.4 Silvicultural Rehabilitation

Strong acidification is difficult to compensate for in agriculture, so that silvicultural use is always preferred with regard to low pH values of the overburden (see Sect. 3.1.3). It has been found that the calculation to buffer the acidification must reach theoretically up to 2,500 dt ha^{-1} CaO or up to 3,000 m^3 ha^{-1} fly ash in order to enable agricultural use. The application of fly ashes that are calcareous and alkaline must be carefully planned and conducted because it is well-known that some fly ashes, particularly garbage incinerator fly ashes, are highly contaminated with heavy metals (Meuser 2010). Thus, the fly ashes used should be products from coal-fired power stations, if at all. The effectiveness of the application of liming agents is mostly overestimated, as investigations in the German coal mining area of Lusatia showed. Amelioration with 135 t ha^{-1} fly ash to sandy material from the Tertiary Period enhanced the pH value from 3.0 to 3.8 and the pH value of sandy material from Quaternary Period increased from 5.4 to 6.0 after application of 20 t ha^{-1} limestone at a depth of 0–30 cm (Schaaf et al. 2000).

Due to the enormous lime application with regard to agricultural use, forest management is mainly applied and surely appropriate because otherwise a spontaneous birch establishment occurs as a result of the fact that this tree species is adapted to the acid properties and will thus become predominant (Fig. 3.7). With reference to the planned silviculture gravelly and sandy material is deposited to a thickness of 3–5 m and afterwards graded by bulldozers. Excessive grading, however, also causes detrimental features regarding the land-use type forestry. Thus, a successful forestation is difficult to achieve, unless decompaction occurs. Decompaction includes ripping of the overburden and subsequent placement of topsoil in strips with delivery from the back of the trucks. In contrast to agricultural use, the material can consist of different substrates such as Pleistocene terrace gravel and terrace sand as well as Holocene loess, but also of sand derived from the Tertiary and Quaternary Period, because the requirements of the trees are essentially lower (Pflug 1998).

It is possible to establish forests on slopes, so that the heaps deposited during the coal exploitation at the edges of the open-cast mining area are also vegetated. The slope gradient for forest purposes is allowed to vary between 1:3 and 1:5 and is interrupted by the means of berms, if necessary, as also applied at hard coal mining waste heaps (see Sect. 3.2.1) and in the context of quarry rehabilitation (see

Fig. 3.7 Spontaneous birch growth at a mining waste heap stemming from an open-cast lignite coal mine close to Cottbus, Germany

Sect. 3.3.1). To condition the soil alfalfa (*Medicago sativa*), which is responsible for humus accumulation and improvement of the rooting depth, is used in the first 2 years (Pflug 1998).

Afterwards, the trees are densely planted (planting machine, hand planting) at a spacing of 2–2.5 m. In some cases the spacing was reduced to approximately 1 m. Alternatively, direct seeding by hand, tractor or helicopter is possible. During the early years it is recommended that the soil properties are improved with the help of organic materials and lime (UFZ 1999).

Depending on the climatic conditions and the soil characteristics the choice of tree and shrub should be made carefully. In temperate climates some tree species like black alder (*Alnus glutinosa*) and *Robinia pseudoacacia* are particularly favourable, because they grow quickly, produce a lot of litter, can use atmospheric nitrogen and are tolerant to acidic conditions. Apart from the birch (*Betula verrucosa*), which will generally grow on mine spoil sites, there are many tree and shrub species capable of growing in such conditions. While in former times poplar and pine were preferred, nowadays more demanding tree species like northern red oak (*Quercus rubra*) and Norway maple (*Acer platanoides*) are used. In the course of time further plants may invade from the surrounding undisturbed area. Nevertheless, some tree species such as the beech (*Fagus sylvatica*), which dominated the landscape prior to the mining period, have been drastically reduced. This species, for example, once covered 57% of the Rhineland mining terrain the percentage after rehabilitation amounts to 17% (Pflug 1998).

In a similarly way, in the course of time increasing knowledge about the capability of different tree species to impact in a soil-protective way (e.g. protection against erosion and deflation, adaption to the extreme soil properties in relation to low pH value and nutrient deficiency) caused constant changes of the chosen species, as documented in the German Lusetia coal mining area (Fig. 3.8). While in the past

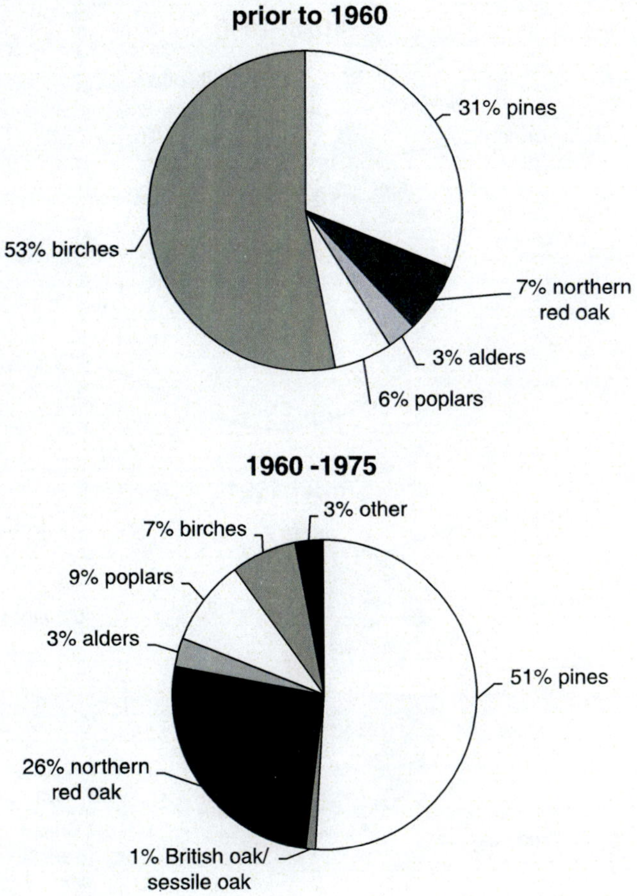

Fig. 3.8 Percentages of planted tree species at rehabilitated lignite coal mining sites in Lusatia, Germany in the course of time (Data from Pflug 1998)

birches and pines dominated prior to 1960 and pines and the northern red oak between 1961 and 1990, it is intended to plant a balanced mixture of species at present (Pflug 1998).

In Mediterranean conditions the species adequate for mining rehabilitation are different. For example, a study conducted in the brown coal mining area of Agacli-Istanbul, Turkey, showed that 17 years after forestation maritime pine (*Pinus pinaster*) and umbrella pine (*Pinus pinea*) prevailed. The time period was sufficient to create a forest floor revealing a fermentation layer apart from fresh litter and significant accumulation of organic matter as well as nitrogen. A faster formation of litter and fermentation horizon and a faster nitrogen accumulation was achieved under black locust (*Robinia pseudoacacia*) plantation, which is capable of nitrogen fixation from the atmosphere. Results from the plant observations are introduced in Table 3.5 (Makineci et al. 2011).

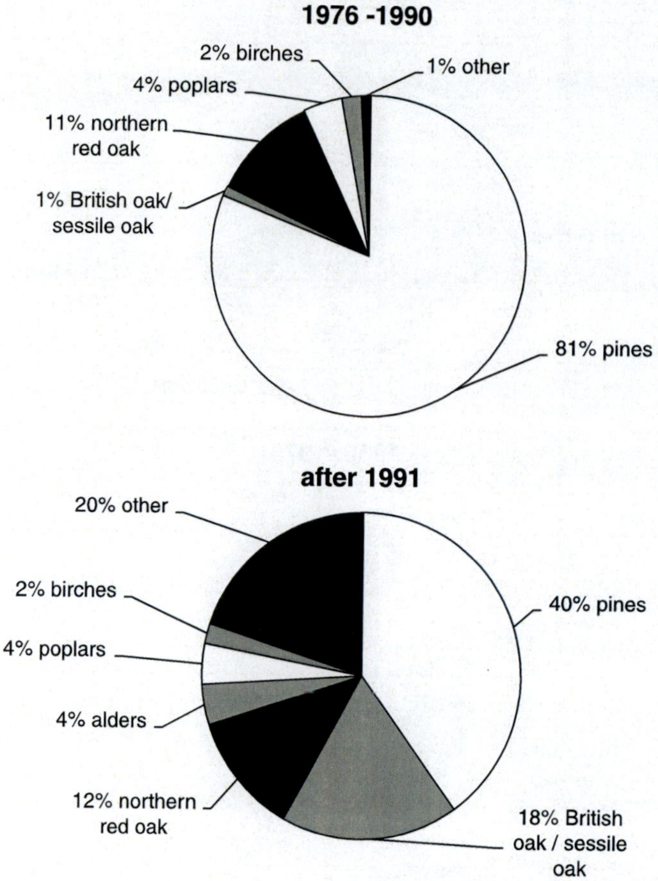

Fig. 3.8 (continued)

Table 3.5 Plant indices of brown coal mining sites 17 years after forestation in Agacli-Istanbul, Turkey (Data from Makineci et al. 2011)

	Soil cover (%)		Average height (m)		Total number
Species	Tree layer	Shrub layer	Tree layer	Shrub layer	of species
Pinus pinea	80	10	7	3.5	21
Pinus pinaster	60	40	9	2.5	19
Robinia pseudoacacia	70	50	9	2	23
Open area	5	5	–	–	19

3.1.5 Creation of Nature Reserves

In open-cast mining areas an alternative use is linked to the establishment of nature reserves. Whether the idea of ecologically valuable areas can be included depends on the morphological and pedological conditions. For instance, sandy soils on south-facing slopes, poor in nutrients and $CaCO_3$ and unaffected by groundwater, are typical sites accommodating the vegetation of sandy arid grassland. In contrast, gravel-free, mesotrophic and calcareous sites closely connected with the groundwater are present in species-rich woodland.

Temporary impoundment basins (species adapted to wet conditions, water habitats), highwalls (cliff nesting species), dry virgin soils, piles of stones and rootstocks (species adapted to dry conditions) may serve as secondary habitats. In any case, topographic manipulation and reconstruction of stream channels provide plant and animal diversity. The natural invasion in large-scale open-cast mining areas is restricted by the missing proximity of plant diaspores. The spontaneous succession can be an alternative to technical reclamation but it must be ensured that sufficient diaspores are definitely able to reach the mining sites (Makineci et al. 2011).

A lack of native seeds can be substituted with the reapplication of topsoil previously hauled and stored. To minimise deflation shelterbelts for very sensitive habitats should be involved. It should be noted that a minimum size is essential to implement the different vegetation structures. Moreover, in relation to the animals which have usually adapted to the site-specific vegetation the areas must, in most cases, be larger. They should amount to 0.5–10 ha for the edaphon, 10–20 ha for reptiles and amphibians smaller than 5 cm, 20–100 ha for greater reptiles and amphibians, 10–20 ha for beetles, 100–1,000 ha for mammals, 200–1,000 ha for birds and up to 10,000 ha for very great mammals and birds (Pflug 1998; UFZ 1999).

In arid conditions the establishment of secondary rangeland is possible, as investigations in the southwest of the USA showed. Water control and supply are the most important factors to be taken into account. The soil reconstruction should create small depressions which trap and slowly infiltrate water. Occasionally, the measures must be assisted by mulching and irrigation in the first years. Planting of trees and shrubs in addition to the natural invasion of species from the surrounding areas may provide a basis for habitat diversity and wildlife. There are different ideas concerning the use of rangeland in future. The recommended spacing for tree and shrub plantation in areas used for domestic livestock is approximately one plant per 10 m^2 but attention must be paid to overutilisation by livestock. In areas used as wildlife habitats one plant per m^2 in patches which cover 5–20% of the total rangeland is preferred. With reference to the chosen plant species an understanding of habitat composition and food chain of the targeted wildlife species is necessary (Boles 1983).

3.1.6 Creation of Mining Lakes

Due to the coal extraction and subsequent use in coal-fired power stations the backfilled material will not be sufficient to compensate for all the excavated holes and exhausted opencasts. The costs for re-handling overburden for backfilling of mined out sites are mostly too high, so that the creation of pit lakes appears to be economically viable, since in many cases the lakes created can be used for recreational purposes such as fishing, boating, swimming, etc. Consequently, in relation to landscaping some areas remain more or less untouched, indicating deep large-scale holes which are flooded after the pump wells continuously used during the coal extraction period have been switched off.

Two methods of filling the holes can be distinguished: rising groundwater or filling with surface water predominantly from rivers. With respect to the first one some advantages and disadvantages must be taken into consideration. Due to the low concentrations of dissolved organic matter (DOC), phosphorus and other nutrients, the water body will remain in a favourable trophic state, usually not reaching a heterotrophic and pathogenic level. On the other hand, the filling duration might be very long as a result of the disturbed aquifers in the surrounding area (see Sect. 3.1.2). Mechanically, the stability of the slopes is endangered because of the hydraulic gradient between the groundwater and the lake water. A particular adverse impact is linked to the pyrite oxidation in the overburden located adjacent to the developing lake. The pyrite oxidation is responsible for strong sulphuric acid creation, leading to acidification in the water body. Subsequently, the pH value in the water decreases to 2–3.5. The acidification is minimised when layers with high pyrite content are buried in the deeper portion of the deposited material without coming into direct contact with the lake bottom (Klapper 2003). Acidification causing clay mineral destruction and SO_4^{2-} formation may enhance significantly the electrical conductivity in the water. For this reason, in mining lakes exhibiting high salt concentrations at the bottom the stratification is stabilised, oxygen deficit is increased and hydrogen sulfide is accumulated in the deeper part of the lake (UFZ 1999).

The alternative filling of the holes with river water is quicker and might impact the slope stability to a small extent. Depending upon the water quality to be used it might show a distinction to poorly mineralised water but the water body will become more anoxic, in particular in the presence of oxygen-consuming substrates at the bottom. Thus, the danger to show a tendency towards eutrophication, whereby nutrient and contaminant cations will be immobilised in sulfidic form, is surely greater in comparison with the rising groundwater table option. One must be afraid of a number of detrimental effects related to lake eutrophication (e.g. mass development of algae, oxygen depletion, etc.) (see Sect. 7.3.1). Besides, pollutants dissolved in the surface water used for filling could be pumped into the lake water body created. Consequently, possible in-lake technologies for water treatment are required. In general, the approach to pump surface water from areas located outside the mining area through pipelines does not appear to be an adequate method, particularly because of the volume of water that must be transported.

Lake management is one of the most decisive challenges of lignite coal mining rehabilitation. In Poland, for instance, in the post-mining period more than 150 artificial lakes, some of them more than 100 years old and more than 150 ha each in area, covered the devastated area (UFZ 1999). The lakes exhibited high sulphate and iron concentration due to the pyrite oxidation. Investigations of 30 acidothrophic lakes revealed high sulphate concentrations ranging from 101 to 1,260 mg L^{-1}. The iron content varied between 0.3 and 182 mg L^{-1}. The waters, which were poor in organic matter, displayed relatively low nitrogen concentrations amounting to 0.1–11.6 mg L^{-1} NH_4-N and 0.1–1.8 mg L^{-1} NO_3-N. The extreme biotopes allowed only a small number of species to colonise the water (Jedrczak 1992). Analogous results of lake investigations in the German Rhineland coal mining area revealed accelerated sulphate concentrations in the water, varying between 70 and 1,620 mg L^{-1}. Moreover, the iron concentration ranged from 0.1 to 66.7 mg L^{-1}, whereas very low results were usually obtained for the nutrients such as ammonium (<0.7 mg L^{-1}), nitrate (<1.3 mg L^{-1}) and phosphorus (<1.2 mg L^{-1}) (Pflug 1998).

In the Eastern part of Germany more than 160 mining lakes were created. A statistical analysis of 39 examples showed that their size varied between 60 and 1,890 ha and their mean depth ranged from 10 to 40.7 m (UFZ 1999). The hollows remaining usually have steep slopes and the bed material is poor in humus. Consequently, the bioproductivity of the littoral zone might be low. A further statistical analysis of the lakes (n = 39) located in the southern part of the Rhineland coal mining area, where the coal extraction occurred close to the soil surface, showed that the lake area varied between 0.4 and 52.8 ha and the mean depth between 0.2 and 6.9 m (Pflug 1998).

In Alberta, Canada, the creation of artificial lakes was an adequate solution for the excavated holes as well. Four lakes with different surface areas (6.0–28.2 ha), maximum depths (7.6–70 m), mean depths (3.2–37 m) and shoreline lengths (1.34–1.98 km) which were left over were investigated in detail. The rehabilitation measures of the lakes involved re-sloping, grading operations on the shorelines, changes of the structures at the bottom, replacement of humic topsoil at the shorelines as well as seeding and planting of macrophytes. The shoreline and bottom configuration occurred irregularly to create habitats and increased the plant and animal diversity. The important littoral zone should cover more than 30% of the lake areas. Organic materials such as decomposed hay were placed in this zone to improve habitat conditions for the benthic organisms and the macrophytes. In proximity to the shore a series of small dams was constructed to create different pools which were filled with gravel behind the dams to provide spawning areas for fish. A few years later the measures were successful, because a sport fish population with a high growth rate had established itself, indicating high stability with the aquatic organisms (Luscar Ltd. et al. 1994). The results of the study show some important factors like size and depth of the lake are pre-determined during the excavation process but most of the parameters such as habitat design, shoreline structure, slope gradients, size of the littoral zone and sediment type must be created or anthropogenically manipulated in the context of creating the lake. It is expected that after approximately 2–5 years the desired aquatic communities are present.

Acidified lakes change their chemical characteristics in a natural way in the long term. The first period with very high acidity and a lack of macrophytes apart from some reed population is gradually altered until the pH value reaches 4–6 and floating leaf plants (e.g. *Potamogeton natans*), peat mosses and *Utricularia* species appear. In the course of time the pH will slowly increase, leading initially to precipitation of dissolved ferric hydroxide as reddish ochre flocs which appear as an ugly and slimy coating on the beach, excluding any type of recreational utilisation. At a later date, the pH value may achieve 6–7 and, apart from different floating leaf plants, more and more macrophytes will invade. It is assumed that within two decades the pH

Fig. 3.9 Operation to design the shoreline of an artificial mining lake by means of vibrating rollers near Cottbus, Germany

value will significantly increase after the pyrite oxidation process has been finished and subsequently the H_2SO_4 creation shows a decreasing tendency. Naturally, it takes a long time to reach adequate pH values of approximately 6–7, but this process can be accelerated by the means of added CaO and soda (Na_2CO_3) (Klapper 2003). The use of alkaline fly ashes was frequently applied, particularly in the acidified lakes in Eastern Germany but this must be considered problematical due to the possible heavy metal contamination.

To achieve an ecologically valuable lake environment diverse morphology should be established. A long shoreline in comparison to the lake area, deep and shallow parts in the lake, mixed configuration of steep and shallow banks, perhaps islands and bays would enable a high differentiation of the habitat structures and consequently a high plant and animal diversity. Otherwise, if the lake utilisation is predominantly oriented towards recreation (e.g. bathing, diving, fishing), the morphology should be planned differently. A great depth, steep shores (apart from the beach area), low shore development (vegetation) combined with small nutrient inflow and stable stratification appear to be favourable (see Sect. 7.3.1) (UFZ 1999).

Problems are associated with the unstable slopes contacting the water body, if the texture consists predominantly of fine to medium sand. In conjunction with the shore protection the material must be artificially compacted. In the upper part it is possible by means of vibrating rollers (Fig. 3.9), but belowground there is no effect as a result of this technique. Hence, explosives are used to densify loose, sandy soils. The shock wave and vibrations induced by the explosives cause similar results to those achieved by the vibratory compaction equipment. In particular, in dry or completely saturated conditions the effectiveness is high. In different depths boreholes are prepared and filled with dynamite, which is afterwards exploded (Fig. 3.10).

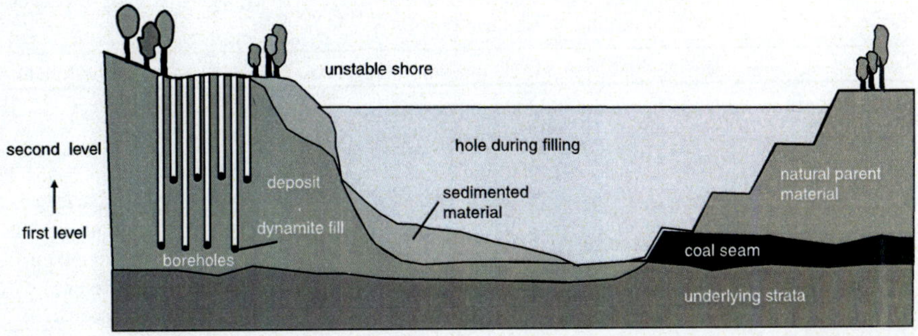

Fig. 3.10 Principle of shoreline compaction by the means of bulldozers (surface-near, not visible) and dynamite added to boreholes (belowground)

In this way, the overlying strata are lifted up in the short term. Subsequently, they slump down and compact the soil below. The compaction intensifies the slope stability to an acceptable degree. Furthermore, variations of the deposited material tend to produce a more uniform deposit after blasting (McCarthy 2007).

3.2 Mining Heaps

3.2.1 Coal Mining Heaps

The aboveground depositing of coal mining waste (coal gangue) occurs in the context of underground shaft mining operations, since the separation of exploited coal from likewise extracted mining waste cannot be carried out underground. Shaft mining is associated with the use of bituminous coal, anthracite coal and, particularly, hard coal.

Coal mining heaps represent an environmental problem in countries with intensive hard coal extraction. One of the biggest producers, for instance, is Poland, where until 1990 200 million tons of hard coal were extracted yearly and where even after the political transformation 84 million tons were still extracted, e.g. in 2008. For 1 t of produced coal 0.4 t of waste is generated. Accordingly, an enormous capacity for the depositing of coal mining waste was and is required. The rocky waste material originates from shaft deepening in a vertical direction, building of levels and rooms underground and driving drifts in coal seams, mostly in a horizontal direction. The processes are required to make the seams available for exploitation (Grzesik and Mikolajczak 2009).

In the German Ruhr area the annual coal mining waste generation amounted to 18 million tons in the middle of the 1990s. In the long course of hard coal mining about 300 heaps covering 25 km^2 were deposited. 76% of the mining waste is deposited aboveground, while only 1% is backfilled to the mining shafts (Schraps et al. 2000).

Table 3.6 Properties of coal mining waste heap types

	Barren cone	Table mountain	Landscape-adapted mountain
Date of origin	Up to approx. 1930	Up to approx. 1970/1980	Since 1970/1980
Height (average)	19 m	40 m	55 m
Size (average)	10 ha	55 ha	Frequently >100 ha, strict geometric contours absent
Plantation	None	Immediate afforestation (*Betula, Robinia*)	After soil deposit (20–50 cm) plantation of different shrub and tree species
Erosion potential	Very high	High	Very low
Disadvantages	Dust development, coal fires	Damage to the vegetation caused by subsidence and flower pot impact, coal fires	Long-lasting deposit, considerable land consumption

The waste rock derived from the Carboniferous and Permian Period consists petrographically of fireclay, clay shale, shale containing residual coal fragments, siltstone and sandstone. Normally, the particle size falls below 500 mm. Furthermore, waste of the coal processing industry, in particular the pre-treatment of raw coal, is simultaneously deposited in coal mining heaps, e.g. material from dense medium cleaning (>20 mm), from coal washing associated with the use of dense medium cyclones (0.5–10 mm) and from flotation (<0.5 mm). In addition, tailing ponds consisting of flotation remnants after drying by filter presses are constructed (Grzesik and Mikolajczak 2009).

Mining of hard coal produced different generations of coal mining waste heaps in the course of time. In Table 3.6 the heap generations are introduced. The heaps deposited until the 1930s caused problems to the people living in the neighborhood and to the environment. The lack of plants was responsible for erosion and coal dust development, contaminating the adjacent areas. Steep slopes tended to show rill and gully erosion (Fig. 3.11). Because of the low compaction and subsequent oxidation of pyrite as well as the high percentage of coal residues and timber derived from the seam protection, coal fires occurred to a great extent. The coal fires damage the roots due to the exothermal pyrite oxidation within the heap (Fig. 3.12).

The second generation of coal mining waste heaps also showed detrimental properties, in particular subsidence and the development of coal fires. These heaps, however, had been vegetated using shrub and tree root balls in addition to the spontaneous development of *Betula verrucosa* and *Robinia pseudoacacia*. After a short period of time the roots grew in the surrounding waste material, since the root ball behaved like a flower pot, resulting in a quick root die-off after touching the acid mining waste. Accordingly, after plantation the vegetation cover appeared disastrous and poor in species, leading to accelerated successive erosion.

As an alternative to the rootball approach, some heaps were covered with humic topsoil with the aim of optimising vegetation development. In the case of a 5–10 cm-thick soil deposit, which was usually preferred, the young plants became accustomed

Fig. 3.11 Gully erosion at a steep slope of a coal mining waste heap near Liege, Belgium

Fig. 3.12 Coal fire at the top of a flat-topped mining heap in Gelsenkirchen, Germany; water steam development and damaged trees with burnt and coaled roots are visible

to the detrimental soil conditions at an early stage. Nevertheless, the root development was strongly limited, resulting in damage to the vegetation caused by dry soil conditions and windblow due to thunderstorms. Deep soil cover of up to 1.8 m was alternatively tested after mixing of mining waste and topsoil in a ratio of 3:1 but this approach failed, since topsoil was mostly not appropriate in the calculated quantity.

Apart from some tree species such as *Acer platanoides, Acer pseudoplatanus, Acer campestre, Alnus glutinosa, Alnus rubra, Betula pendula, Populus* (hybrids), *Robinia pseudoacacia, Quercus rubra* and *Tilia cordata*, shrub species were planted as well, e.g. *Crataegus monogyna, Eleagnus angustifolia, Hippophae rhamnoides, Populus tremula, Rosa canina* and *Sorbus aucuparia* (Schulz 1996).

Many heaps were deposited over a very long period of time so that different generations are simultaneously visible. For example, the existing heap Debiensko in Poland covering a size of 140 ha is comprised of two conical heaps (1st generation) and one huge flat heap (2nd generation) in addition to four tailing ponds. The heap used until 2000 was also influenced by the philosophy of the current depositing technique.

Perhaps this heap is already associated with a possible 4th generation of heap management, since it is subject to a secondary exploitation aimed at the full recovery of the coal remnants originating from former periods and still contained in the waste (see Sect. 2.3.2). In general, the recovery of deposited stony mining waste appears to be of interest with regard to an after-treatment for construction purposes. So, parts of the heaps which predominantly contain relatively weathering-resistant sandstones are re-used for constructing bottom layers of highways or other roads. In the Polish example of the heap Debiensko, for instance, 3.5 million cubic metres were used for an adjacent motorway. According to the national plan of Poland until 2013 1,634 km highways, 3.032 km expressways and 3,993 km other roads are to be constructed with crushed stones from extractive waste deposits. In fact, this operation may also contribute to the possibilities to deal with mining heaps without any rehabilitation procedures (Grzesik and Mikolajczak 2009).

Nowadays, mining waste heaps are differently designed to blend into the adjacent natural landscape and to avoid disharmony with the surrounding areas (3rd generation). They reveal several heap tops corresponding to the mountainous landscape that frequently surrounds coal mining waste heaps. They do not show strict geometric contours, have a maximum height of 50–60 m and are planted as soon as possible after soil has been deposited. Disadvantageously, they require large areas and can reach a size of more than 100 ha. They are deposited from the centre outwards, thus requiring a longer period of time but this procedure is necessary to minimise dust and noise emission associated with loading, transportation and depositing of the mining waste.

The slope gradient amounts to 1:3. In the upper part of the heap this is even steeper (1:2.5) but in the part below it is less steep (1:>4). The different layers are constructed in terraced style to reduce erosion. The terraces are called berms, which are, for instance, 10 m wide, constructed at a distance of 8–12 m and provided with drainage ditches (Fig. 3.13). The construction is useful for reducing slope length and improving water infiltration. The structures are built up in layers of 0.5–4.0 m thickness. Heavy vehicles equipped with vibratory rollers which can compact the

Fig. 3.13 Mining waste heap configuration with berms for erosion protection purposes in Osnabrück, Germany

material considerably compact the soil during the operation. The aim of this intensive compaction is mainly to prevent oxygen from entering the heap and causing coal fires. The plantation occurs separately immediately after completion of each berm until the different plateau sectors of the heap are completed. For this reason, the outer layer of the heap of approximately 2 m should be deposited loosely, because rainwater infiltration and water percolation within the root zone are required (Schulz 1996).

Ultimately, the heap landscape is preferentially used as forest that helps to compensate for and to replace areas which were lost due to the creation of the mining heaps. In urbanised areas afforested heaps can serve as recreational places in and around cities. In this context walking trails, toboggan runs as well as viewing points and catering at the heap tops are interesting options.

In Pernik, Bulgaria, where sub-bituminous and bituminous coal is extracted by shaft mining, a suburban park ranging over the Sofia-Pernik agglomeration was established on the deposited waste heaps. The heaps consisting of clay schist, marl and coaly shale showed different soil properties associated with the geologically distinct origin of the material. The pH value ranged from 3.1 to over 7 and the degree of base saturation (> 90%) and phosphorus content (15–22 mg 100 g^{-1}) were favourable to a large extent. In such conditions it is possible to carry out afforestation successfully. The heaps were planted with 22 shrub and 19 tree species at the beginning of the 1960s. Coniferous species were used moderately. 26–27 and 38–39 years later biometric data of the planted vegetation were investigated (Table 3.7). Generally, the plant height increased more in the first period, whereas the stem diameter revealed differences among the species. In this context some species were initially slower in the growth of the diameter (*Betula pendula, Quercus robur, Tilia* sp.), the other plants behaved conversely. Species like *Robinia pseudoacacia* had a period of up to 11 years of slow growth due to adaptation to the soil conditions,

Table 3.7 Biometric data of tree species grown on coal mining waste heaps in Pernik, Bulgaria (Data from Sokolovska et al. 2000)

	1961	1961–1983		1983–1998		1998
		Mean annual increment		Mean annual increment		
Species	Height (m)	Height (m)	Diameter (cm)	Height (m)	Diameter (cm)	Height (m)
Betula pendula	1.2	0.53	0.72	0.05	0.86	13.5
Fraxinus americana	1.1	0.32	0.50	0.06	0.36	8.7
Gleditsia triacanthos	1.0	0.45	0.57	0.73	0.25	19.0
Quercus robur	0.7	0.42	0.99	0.09	1.32	11.0
Robinia pseudoacacia	2.1	0.45	0.58	0.10	0.13	13.5
Tilia sp.	–	0.42	0.53	0.17	0.78	12.4

while other species grew rapidly after a few years, indicating increase of radial growth and plant height. Plants cultivated on southern and western slopes grew faster. During the observed period forest litter began a progressive evolution, leading to the presence of different horizons (fresh litter, organic horizons), apart from slopes with a declination of more than 30°.

The forest development strongly depends on human influences. For instance, the effects in relation to the late growth of birches and oaks might be caused by a strong thinning of the stand. In principle, after approximately 20 years an intervention aimed at the regulation of the plant composition is required, in particular if ornamental shrubs are desired in the context of a recreational park character (Sokolovska et al. 2000).

The physico-chemical properties of the hard coal mining waste originating from the Carboniferous Period are generally extremely adverse to plants. Only a minor part of the texture exceeds 2 mm in diameter and the clay content is usually lower than 1%. Fine material is slowly created in association with the weathering processes. Absence of clay minerals in addition to a lack of humus results in low cation exchange capacity as well as water retention potential. Moreover, high infiltration rates and high hydraulic conductivity are responsible for unfavourable growth conditions for the vegetation. Moreover, the soil is susceptible to erosion, in particular on heaps with steep slope gradients amounting to 1:2 or 1:3, because of the less developed soil structure (Schraps et al. 2000).

The black colour of the heap material, which predominates if the surface is unvegetated and bare, influences the soil temperature, resulting in a considerable heating up in the summertime. Surface temperatures of up to 70 °C in sunny periods are reached. The south-facing slopes tend to exhibit an extremely dry soil initiated by evaporation (Schulz 1996).

Some nutrients are deficient, particularly phosphorus, which indicates average values of 0.2 mg 100 g^{-1} plant available phosphorus only. In contrast, the cationic nutrients such as potassium (average value: 7 mg 100 g^{-1}) and magnesium (average value: 6 mg 100 g^{-1}) are classified as sufficient. Because of the low organic matter content microbiological activity and nitrogen content are also deficient. In the course of time the nitrogen supply increases due to both bioaccumulation in conjunction with the development of the vegetation and microbial fixation of atmospheric nitrogen by some species such as *Robinia pseudoacacia.* In the first 3 years in the uppermost horizon (0–5 cm) extremely low ammonium (mean value <1 mg 100 g^{-1}) and nitrate (mean value approx. 5 mg 100 g^{-1}) rates are detectable. After 20–30 years, however, at least nitrate shows enhanced values (38 mg 100 g^{-1} on average, maximum 97 mg 100 g^{-1}). At this time nitrate has already accumulated in deeper layers between 5 and 40 cm (42 mg 100 g^{-1} in maximum) (Schraps et al. 2000).

The main problem in chemical conditions relates to the strong acidification caused by pyrite (FeS_2) oxidation, which results in a pH decrease to 2–3. Subsequently, the coal mine drainage exhibits elevated concentrations of sulphates and iron. Some coal mining heaps contain further pyrite compounds such as arsenopyrite (FeAsS) or chalcopyrite ($CuFeS_2$) which cause enhanced metal solubility. As shown in the Tula coal region south of Moscow, Russia, the pyrite oxidation resulted in a pH decrease to 1.8–2.5 and accelerated values for coal mining drainage. Subsequently, 0.6–1.4 mg L^{-1} copper, 1.0–1.4 mg L^{-1} cobalt, 1.3–2.0 mg L^{-1} nickel, 0.5–1.1 mg L^{-1} lead and 10.7–19.7 mg L^{-1} zinc respectively were found (Komnitsas et al. 2001). Normally, higher levels of metals in hard coal mining heaps are not detectable. For instance, in the German Ruhr area the average total metal concentrations of the mining waste were low (e.g. copper <50 mg kg^{-1}, lead <30 mg kg^{-1}, nickel <80 mg kg^{-1}, zinc <70 mg kg^{-1}) (Schraps et al. 2000).

Contaminated mining heaps require not only rehabilitation strategies. Decontamination strategies also appear to be desirable. The measures can be oriented towards the separation of pyrite before depositing, because *in situ* approaches might be less practicable:

- Pyrite can be removed by flotation with the help of soda in coal tailings, which are produced during the pre-treatment of raw coal. Soda increases the pH value to approximately 11 but subsequent neutralisation is required.
- Sulphuric compounds in the mining waste can be chemically removed with sodium carbonate, sodium hydroxide or potassium hydroxide; again, neutralisation must occur afterwards.
- Desulphurisation can also take place biologically; the bacterial cultures may oxidise pyrite as long as toxic substances such as heavy metals are not problematical; moreover, the bacterial activity is strongly reduced in the winter period and the produced effluent is assumed to be contaminated (Komnitsas et al. 2001).

In summary, nearly all soil characteristics mean bad growth conditions for the vegetation. Therefore, in the past much attention was paid to planting the heap edges fast to make the disturbing, black and unvegetated heap disappear in urbanised areas. Shrubs and trees should cover the heap to give it a better appearance. For this reason, hedges were planted in close proximity to the heap to shield its unsightly appearance.

Nowadays, immediate heap plantation is included in heap rehabilitation. For instance, based upon the Australian guideline for mining heap rehabilitation the soil must be pre-treated before the plantation occurs. Already during the heap formation in the upper part every layer is keyed into the layer underneath to avoid a physical barrier which the roots of the plants would not be able to grow through. In order to improve water infiltration and to reduce compaction of the uppermost layers, which results from the heavy trucks used for depositing of the heap, ripping on the contours to a depth ranging from 0.5 to 1 m is applied. Ripping, however, can cause large stones to be brought to the surface. In general, one opportunity to create the final cover relates to use and careful management of humic topsoil with its favourable biological and structural properties and its native seed bank, which improves the possibility for intensive plant growth. On steeper slopes at least a shallow topsoil deposit is required. Because of limited topsoil supply in some zones it is carried out in strips along the contours of the heap (GOA 1995a).

In the case of a lack of topsoil the ploughing-in of organic material such as compost to a shallow depth, possibly combined with liming and fertilizing, are optional procedures. Afterwards, a relatively coarse seed bed with a lot of furrows is created, which traps runoff, retains rainwater and minimises undesired surface crusts. Fertilizing is reduced to nitrogen and phosphorus, since the other elements are usually sufficiently present. In Germany 40–60 kg ha^{-1} N and 10 kg ha^{-1} P are recommended for native plants for the initial fertilizing of hard coal mining heaps (Jochimsen 2001). In Australia a higher level of 50–100 kg ha^{-1} N and 30–60 kg ha^{-1} P is normally used for coal mining dumps. The lower value is associated with native plants, which have lower nutrient requirements than exotic species (GOA 1995a). Additional fertilizing at a later date is sometimes required, depending on the status of plant growth.

Some species will invade from the surrounding areas, providing natural succession. The natural colonisation, however, takes a relatively long time because of the adverse site conditions such as low pH, low water retention potential, coarse material hindering root growth, etc. Furthermore, a fast vegetation development is hindered by the lack of pioneer plants, since they tend to be exterminated by man, in particular in the proximity of urbanised areas, where many mining sites are located. Nevertheless, some pioneer plants are capable of a fast spread, enabling a spontaneous colonisation of the mining heaps. From the ecological point of view, which favours maintenance of natural colonisation, the main problem seems to be the search for sufficient seeds of adequate plant species (Jochimsen 2001).

Nevertheless, it is certain that spontaneous vegetation will appear in the course of time. Initially, some herbs develop, the soil remains more or less bare and poor in species. After one to two decades the first shrubs are established, usually birches. Afterwards, a birch forest rich in grasses can be discovered that will change into secondary woodland containing some demanding tree species like oaks. Without any man-made manipulation the development of natural vegetation is restricted to a great extent. Thus, the succession should be controlled and monitored but a lot of theoretical possibilities are less practice-oriented. For instance, irrigation of coal mining waste heaps to accelerate the soil moisture or liming to buffer the acidification do not appear to be realistic options. The amendment of seeds combined with an initial moderate fertilizing would surely facilitate natural succession and accordingly it is worth considering. The fertilizing that is applied twice can be reduced to 40–60 kg ha^{-1} N, 10 kg ha^{-1} K and 10 kg ha^{-1} P at any one time (Jochimsen 2001).

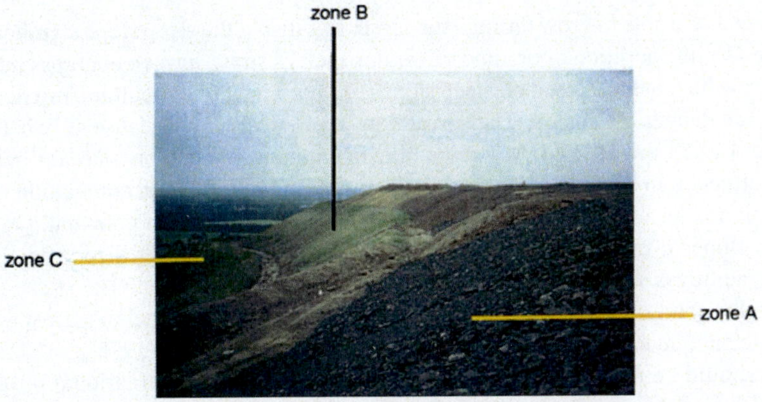

Fig. 3.14 Rehabilitation of an anthracite coal mining waste heap in Ibbenbüren, Germany. *Zone A*: untreated deposited mining waste. *Zone B*: pre-treated soil indicating legume and grass vegetation. *Zone C*: lower part with dense vegetation cover and shrub plantation

Despite the possibility of natural succession it is recommended that immediate cover with fast-growing grasses (percentage: 50–80%) and legumes (percentage: 20–50%) is carried out to reduce erosion (Fig. 3.14). In principle, native grasses provide optimised and long-term slope stability. The choice of the species depends on the climatic conditions. For example, in Australia erosion control can be beneficially achieved by using fast-growing stoloniferous species such as Rhodes grass (*Chloris gayana*) and deeper-rooted species such as buffel grass (*Cenchrus ciliaris*) (GOA 1995a). Under Central European conditions the legume species *Lupinus polyphyllus,* in particular, appears to be a suitable plant. In addition, some species such as the legume alfalfa (*Medicago sativa*) have more or less ubiquitous application. Whereas grasses are completely dependent on the nutrient supply of the soil, legumes can benefit from the nitrogen in the air due to the symbiotic relationship with the *Rhizobium* bacteria. Subsequently, a grass-legume mixture might be the best option to cover the soil quickly. In summary, the combination of ripping, fertilizing and seeding should be taken into account to avoid environmental problems and to reduce the costs.

The seed used is mostly coated. This seed pelleting means the application of a layer around the seed which is inert and contains important growth factors such as fertilizers, growth hormones, mycorrhizal fungi and occasionally pesticides. To establish the desired vegetation seed quality, particularly purity and absence of undesirable weed seeds, must be checked. The sowing rate depends on this quality in addition to the soil properties. Grasses, for instance, can be sown at 10–15 kg ha^{-1}. Seeds used for effective stabilisation of slopes, however, need a higher rate amounting to 70–100 kg ha^{-1} (GOA 1995a).

Steep slope gradients require particular treatment. Nevertheless, if strong erosion such as gully erosion is visible, remedial action must be taken. Jute meshes that are laid after soil preparation and before seeding are effective means to reduce erosion.

The seeds are able to grow through the mesh, stabilising the steep slope significantly. Moreover, the addition of organic substrates such as straw, hay, woodchips and bark mulch, which are also used on steep slopes in the context of agriculture, might minimise the danger of erosion. The spread rate is relatively low because only between 2 and 4 t ha^{-1} are applied. Moreover, the amendment of such organic fertilizers is also effective in relation to protection against climatic effects, accumulation of the organic matter content and improvement of the water retention potential. On very steep slopes a mixture (up to 300 kg ha^{-1}) of water, seeds, fertilizer, mulch and some additives which can stick the sprayed components is used. The sticky agent (e.g. glue solutions, bitumen) helps to reduce the erosion potential to a great extent. This technique is termed hydro-mulching (GOA 1995a; Schulz 1996).

It should be noted that with reference to utilisation for recreational purposes a fast soil cover is also required. This groundcover should exceed at least 60% before access is granted to people. Otherwise, trampling feet would surely destroy the vegetated surface (GOA 1995a).

After a period of 2–3 years the pre-treated soil can be planted with shrubs and trees. Attention should be paid to the competition between the grasses and legumes which had already been seeded to establish dense vegetation cover with reduced erosion and the shrubs and trees which will follow. In particular, sown exotic grasses compromise the growth of the wood plants. Thus, the grass seed rate and the fertilizer rate used for the grasses should be reduced. With regard to the choice of shrubs and trees local native species are preferred, since they have proven suitability for the climate as well as resistance to pests and diseases. Again, it might be favourable to consider nitrogen-fixing tree species. In Australia, for instance, they include different species belonging to the genera *Acacia* and *Casuarina,* which are capable of nitrogen fixation (GOA 1995a).

Normally, shrubs and trees are not sown. The best results are achieved by planting tubed stocks which are a few months to 3 years old. They should stem from nursery-raised stocks in local distribution areas. Around these seedlings organic material such as mulch originating from the previous grass and legume cover or pasture hay is placed to maintain soil moisture, suppress weed growth and reduce runoff. The plants are cultivated with a spacing of 1 m only (Fig. 3.15). After planting, it is recommended to fence the plants in as soon as possible in order to give protection from animals such as rabbits and deer. Furthermore, the creation of a small basin around the shrubs and trees collects the rainwater, which is important with respect to the water supply. Whether periodic irrigation is even necessary depends upon the climatic and weather conditions, which can vary from year to year. Usually, irrigation of mining heaps is not very common but within the first few years there can be a necessity for this, if the summers are very dry and hot (GOA 1995a; Schulz 1996).

The chosen tree and shrub species depend on the various regional climates. For instance, in Germany the tree species *Acer pseudoplatanus*, *Alnus glutinosa, Betula pendula, Carpinus betulus, Populus tremula, Populus nigra, Populus canescens, Quercus rubra,* and *Salix caprea* are recommended. Some shrub species such as *Sorbus aucuparia* (rowan), *Corylus avellana* (hazel-nut), *Prunus spinosa*

Fig. 3.15 Top of a mining waste heap showing narrow plantation of shrubs after the preparatory grass seed application in Osnabrück, Germany

(blackthorn), *Frangula alnus* (buckthorn), *Amelanchier ovalis* (juneberry), etc. are also frequently used. The attempt to plant Austrian pine (*Pinus nigra austriaca*), which has adapted to mountainous sites with steep slopes due to its origin with a 3–5 m spacing pattern, was doomed to failure, since the roots were not able to survive in the acid soils (Schulz 1996).

3.2.2 Salt Mining Heaps

Different salts are extracted, namely calcite ($CaCO_3$), dolomite ($CaMg(CO_3)_2$), gypsum ($CaSO_4 \times 2H_2O$), anhydrite ($CaSO_4$), sodium chloride (NaCl), kieserite ($MgSO_4 \times H_2O$), potassium chloride (KCl), kainite ($KCl \times MgSO_4 \times 3H_2O$) and carnallite ($KCl \times MgCl_2 \times 6H_2O$). Residues of the mining process are deposited aboveground in so-called salt mining waste heaps. With reference to the potassium-based mining processes the deposited material is also called potash mining waste. In Germany 67% of the solid waste of the potash industry is dumped, 25% recycled and 8% backfilled (BRGM 2001). Salt mining heaps are occasionally termed slag heaps, although no residues from metal working plants are parts of the deposit (Meuser 2010).

The quality of the mining waste is closely linked to the terms salinity and sodicity. The first one refers to the presence of soluble salts in the waste material, the latter one describes a high percentage of sodium ions relative to other cations such as calcium, magnesium and potassium. Sodium normally comprises a small portion of the cations because it is not adsorbed to the mining waste to a significant extent.

Fig. 3.16 Salt mining heap in the proximity of Hanover, Germany

The most important anions are chloride and sulphate, which are susceptible to rapid leaching. However, in the mining waste compounds of low solubility are also present, e.g. calcium sulphate (gypsum) and calcite remnants (GOA 1995b).

Regarding the heap configuration barren cones and flat-topped table mountains (Fig. 3.16) can be distinguished (see Sect. 3.2.1). Their sizes differ considerably. For instance, two salt mining waste heaps in Cardona Diapir (Spain) covered an area of 12.6 and 25.7 ha and reached a height of 90 and 150 m. Accordingly, 3 million tons and 7 million tons were dumped. The heaps were deposited from 1925 to 1972 and from 1972 to 1990 respectively. Salt mining heaps consist of remnants which are produced during the solubilisation and crystallisation of the raw material (brine) and of solid residues stemming from the mining shaft formation, which in the Spanish example extended up to 1,340 m below the soil surface (Lucha et al. 2008). The relationship between solid material and sludgy brine varies. In the German heap Bleicherode, for example, the relationship solid material to sludgy brine amounted to 4:1 (Heiden et al. 2001).

Just after the deposit various chemical compounds, predominantly sodium chloride, anhydrite and kieserite exist in the mining waste heap. The constituents of the heap alter considerably in the long term, since most of the compounds are soluble and consequently it is easy for them to percolate downwards. In particular, sodium chloride and kieserite migrate relatively rapidly. This results in a residual accumulation of anhydrite, which cannot be solubilised so well. The time-dependent chemical changes of a salt mining waste heap are presented in Table 3.8. The remaining anhydrite consists of a loose and light material which can be eroded by wind and accordingly can contaminate the surrounding areas (Kahl et al. 2000).

A lot of physico-chemical properties of the mining waste heaps are assessed as being extreme. In contrast to the coal mining heaps, the material colour is white to pink, darkening in the course of time and finally changing into grey due to dust deposition. The upper part of the soil exhibits a very loose structure associated with a low

Table 3.8 Constituents (%) of a salt mining heap in the course of time (example Bleicherode, Germany) (Data from Lücke 1997; Heiden et al. 2001)

	Fresh deposit	After 6 months	After 1 year	Old deposit
$CaSO_4$ (anhydrite)	14.3	37.2	80.9	79.7
$MgSO_4$ (kieserite)	10.5	0.1	0	0
NaCl (sodium chloride)	65.0	58.1	2.0	0
KCl (potassium chloride)	1.5	0	0	0
$MgCl_2$	2.5	0	0	0
Water content	5.0	1.3	10.0	14.5
Miscellaneous	1.2	3.2	7.1	5.8

specific gravity of <1.0 g cm^{-3}, which decreases with depth. The weathered loose upper part progresses downwards at a rate of 1–6 cm a^{-1} (Heiden et al. 2001). After achieving a thickness of approximately 30–50 cm the first vegetation appears.

After depositing a permanent leaching occurs, impacting the physical stability of the heap enormously. At the surface rill erosion is visible. The erosion increases in the presence of a high sodium proportion, since sodium is prone to dispersion. Theoretically, the erosion can be minimised with the help of a calcium amendment (lime), since the dispersion decreases with an increasing Ca/Na ratio. Also gypsum can be used to provide a source of calcium, which will reduce the relative proportion of sodium (GOA 1995b). The loose surface looks like the sea bottom where the soil is formed by waves washing over sandy sediments, the so-called ripple marks (Fig. 3.17).

The salt mining heaps are low-density bodies which tend towards deformation. The deformation occurs not only in the layers near the surface, internal erosion processes are also simultaneously present, which lead to a wide spectrum of sink-hole types and small-scale subsidence. At the surface collapses with funnel-shaped structures appear. Invisible deep-seated gravitational slope movement also takes place. For this reason, the heaps have to be fenced in to prevent people from entering the heap landscape.

The continuous leaching is responsible for the extremely high salt concentration detectable below the heaps in the course of time. For instance, below one German heap (Bleicherode) the salt concentration showed values of 260–390 mg L^{-1}. Maximum chloride concentrations of up to 16 g L^{-1} were found. The salty plume had a length of approximately 1,000 m downstream (Heiden et al. 2001). Though the rainwater percolates predominantly downward because of the high hydraulic conductivity, runoff happens additionally, requiring the construction of surrounding ditches (Fig. 3.18) to collect the salty water and to treat it in a waterworks. The intensive leaching means a quick decrease in mobile sodium chloride, which amounts to less than 0.1% already a few years after depositing. In general, the rate of salt release will slow down in time. The time scale depends on the site characteristics and might last years to centuries. The high leaching leads to a continuous long-term reduction of the mining waste but one must reckon with centuries until the salt mining waste completely disappears. For natural salt outcrops it is assumed

Fig. 3.17 Loose surface at the slope of a salt mining heap in proximity to Hanover, Germany

that the surface lowering amounts to several centimetres per year, reaching a maximum of 10 cm $year^{-1}$. The reduction rate of an old salt mining heap in Cardona Diapir (Spain) was estimated at 5 cm $year^{-1}$ (Lucha et al. 2008).

Under moist conditions a single grain structure prevails, because the particles tend to disperse. During dry periods, however, the structure alters because of crystallised iron oxides and aluminium oxides to a compacted and cemented one, which is unfavourable for water infiltration. The soil surface becomes dense and cloddy, since the aggregation is badly destroyed (Lucha et al. 2008). The constant change of the soil structure is responsible for the poor growth of the vegetation.

The texture of the material is sandy (silty sand, medium sand and coarse sand) to gravelly, even if the investigation of the texture seems to be difficult, since an analysis based on the sink speed normally applied is never feasible due to the salty material that is solubilised during the conducting of the analysis. Because of the predominantly sandy texture the hydraulic conductivity shows values higher than 10^{-4} m s^{-1} but the heterogeneity causes very different data within the mining heap. Areas with more sandy texture allow lower water percolation than coarse-grained material.

Fig. 3.18 Ditches to collect the salty runoff at the sides of the salt mining heap in the proximity of Hanover, Germany

If coarser material is associated with 'hot spots' where the salt concentration is extremely high, the risk of high salt leaching must be taken into account (Heiden et al. 2001; Lucha et al. 2008). According to the texture, the available water capacity is also strongly restricted, causing hazards to the plants. Moreover, attention must be paid to the extreme temperature of the heap topsoil which can reach up to 55°C in summertime on south-facing slopes. Frost is generated, if the temperature falls below −15°C, since the salt content decreases the freezing point significantly (Lücke 1997).

Chemically, the heaps reveal detrimental features, as investigations in two German salt mining heaps showed (Table 3.9). For instance, the cation exchange capacity varied between only 0.2 and 2.9 $mmol_c\ kg^{-1}$. Furthermore, the nitrogen content was lower than 0.1% and phosphorus also appeared to be deficient, indicating <2.0 mg 100 g^{-1} (plant available P). Surprisingly, even the potassium (<1.5 mg 100 g^{-1}) and magnesium (0.1 mg 100 g^{-1}) supply were not sufficient for plant growth (Lücke 1997). Furthermore, the high salt level influences negatively the availability

Table 3.9 Physical and chemical properties of the upper part (about 0–60 cm) of salt mining heap soils (Data from Lücke 1997)

Parameter	1st heap	2nd heap
Bulk density (g cm^{-3})	0.57–0.95	0.51–1.00
Total pore volume (vol%)	64–65	65–73
Available water capacity (vol%)	4.0	2.2–6.9
Phosphorus (mg 100 g^{-1})	0.7	0.5–1.5
Potassium (mg 100 g^{-1})	0.8	0.4–1.4
CEC ($mmol_c$ 100 g^{-1})	0.2	0.7–2.9
TOC (%)	na	0.2–2.1
Total N (%)	na	<0.01–0.1
pH	6.5–7.5	6.5–7.7

na not analysed

of essential plant nutrients. In conclusion, nearly all parameters showed a negative impact on plant establishment, unless the pH value is taken into account, because this characteristic was the only one which is more or less beneficial to plants (6.5–7.7) (Kahl et al. 2000). Higher pH values ranging between 8.5 and 10 are measurable in the case of non-saline sodic material (GOA 1995b).

Because of the adverseness to the vegetation only mosses and lichens are present, but after more than two decades in humid and temperate climates some tolerant species such as birches (*Betula pendula*), pines (*Pinus sylvestris*) and especially aspens (*Populus tremula*) immigrate gradually to the heap landscape. Thus, the immediate plantation of these tree species may improve the possibility of a rapid plant cover. Alternatively, grass and herb seeding (seed rate 30 g m^{-2}) can be carried out, resulting in a plant cover amounting constantly to <20% (Heiden et al. 2001). In warmer climates, for instance in semi-arid regions of Australia, the tree species *Casuarina glauca* (swamp oak), *Eucalyptus camaldulensis* (river red gum) and the *Tamarix* (tamarisk) species *Tamarix aphylla* and *Tamarix pentandra* are mostly chosen for salt mining heap plantation. At mining heaps in dry climates a surface-near accumulation of salts occurs, whereas the deeper horizons tend to release salts. For this reason, grasses with a shallow root development cannot survive. Deep rooting trees, however, are able to survive (GOA 1995b). Salt-tolerant species enable a first establishment of vegetation, which improves the soil conditions progressively. Consequently, in the course of time more less salt-tolerant species are able to invade the terrain.

Due to the nutrient deficiency the succession should be promoted by application of organic matter such as compost and composted sewage sludge. However, attention must be paid to the nutrient leaching, in particular nitrate. As found in a German experiment on a salt mining heap, the amendment of 100 t ha^{-1} compost or 100 t ha^{-1} composted sewage sludge caused for a short time an acceleration of nitrate up to 65 mg L^{-1} and 105 mg L^{-1} respectively. The application of 200 t ha^{-1} increased the nitrate concentration of the leachate up to values between 88 and 345 mg L^{-1} in the first year and up to 50 mg L^{-1} in the following half year. Nevertheless, after compost amendment it was possible for vegetation to establish itself, whereby a reduction of the water percolation by 20–30% and a subsequent reduction of the salt leaching occurred (Heiden et al. 2001).

Investigations have been carried out with the amendment of dried or composted sewage sludge that was mixed with the mining waste at a ratio of 70 vol% anhydrite and 30 vol% organic manure. The mixture improved the water storage, in particular if hydrogel was additionally applied, so that it was possible to overcome drought periods. The enhanced nutrient supply for the planted vegetation caused by the high nutrient concentration of the organic fertilizer was found to be the main advantage. Poplars, for instance, showed nutrient contents indicating ranges which are typical for the species in uncontaminated areas (e.g. nitrogen 2.81%, phosphorus 0.23%, potassium 2.83%, magnesium 0.21%). However, the plants showed smaller habitats, which was probably due to the elevated salt concentration. Some shrubs, which also accumulated nutrient concentrations comparable with natural soils, did not even manifest any growth restrictions (e.g. *Elaeagnus angustifolia*, *Symphoricarpos albus*) (Kahl et al. 2000).

Apart from organic fertilizers, geotextiles, in particular jute sheets, were tested on slopes to minimise rill erosion. Simultaneous use of jute sheets and grass seeding (e.g. species of the genera *Agrostis, Festuca, Poa*) enabled a 10–15 cm deep rooting and a vegetation cover of 60–70%. More laborious is soil amelioration with sewage sludge or fly ash from coal power stations which have been ploughed to a depth of 20 cm but this approach showed intensive vegetation cover as well (Heiden et al. 2001).

When choosing plants the plant salt tolerance is the most important factor to be considered. At a low soil salinity rating (EC < 1.9 dS m^{-1}) only sensitive or moderately sensitive plants are chosen, at a medium to high salinity (EC 1.9–7.7 dS m^{-1}) moderately tolerant to tolerant species are of interest and at very high salinity (EC 7.7–12.2 dS m^{-1}) only very tolerant vegetation is accepted. At extremely high salt concentrations (EC > 12.2 dS m^{-1}) too saline conditions, generally speaking, are present for the establishment of vegetation (GOA 1995b).

The rehabilitation can be related to the natural succession of the vegetation, but this process will take up too much time and is usually not preferred in order to make the white heap invisible as soon as possible. An adequate approach appears to be the deposit of distinct materials to enable the re-use of the heap area in an optimum way (Fig. 3.19a, b). Above the salt mining waste a layer of 1–2 m in thickness is deposited. This may interrupt the capillarity and consequently consists of coarse material such as construction debris. The rising salty water cannot reach the soil deposited above the capillary breaking layer. The soil above used in a thickness of at least 1 m also contains coarse materials mixed with subsoil. Consequently, the hydraulic conductivity should range between 10^{-5} and 10^{-6} m s^{-1}. In addition, shallow humic topsoil is deposited over this layer to provide for the good establishment of vegetation afterwards. It should be noted that subsidence resulting from leaching at a later date can cause disturbances of the layer performance (GOA 1995b; Kahl et al. 2000; Heiden et al. 2001).

The slope gradient should not exceed 1:1.3 (capillary breaking layer) and 1:2.5 (topsoil). Otherwise, berms must be formed, as introduced in Sect. 3.2.1. Because of the material transport the ground of the berms must be stabilised with construction debris, slag or crushed stones in order to increase the load-bearing capacity. The necessity of berm construction is also expected on very long and steep slopes exceeding 50 m and 38° (Heiden et al. 2001).

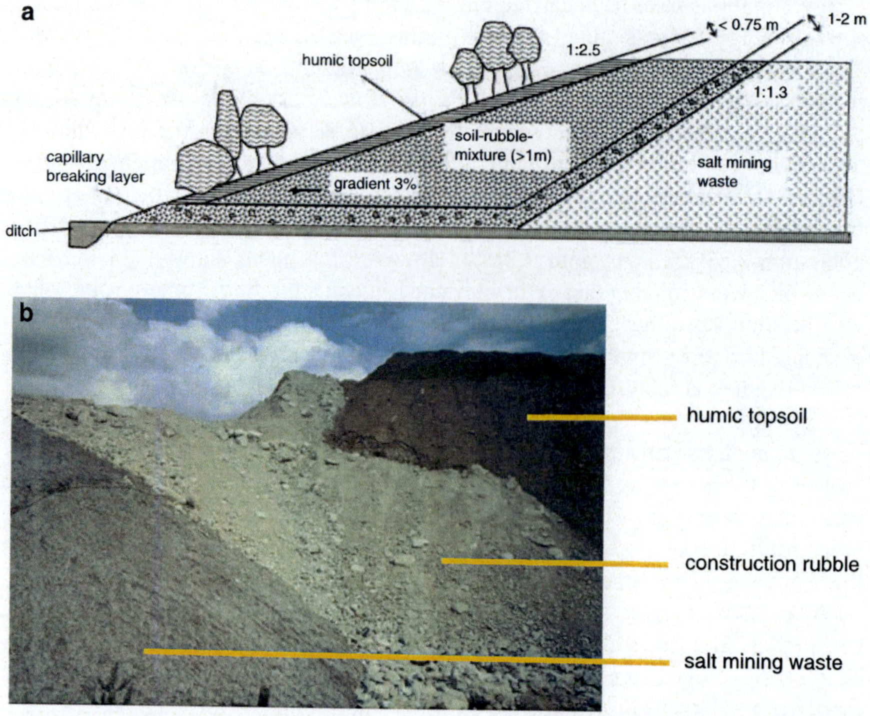

Fig. 3.19 (**a**) Schematic construction for the cover of salt mining heaps. (**b**) Photograph illustrating the cover process of a salt mining heap in the proximity of Hanover, Germany

Besides, the complete covering of the heap landscape requires a large foreland and use of sufficient material (e.g. construction debris). Close to urban agglomerations it is possible to realise the complete covering of a salt mining heap within 15–20 years, because it is probable that many construction activities will be carried out. In rural catchments, however, longer periods must be calculated. These can extend up to 30–50 years and more. For this reason, rapid slope stabilisation using grass seeding or temporary plantation of willows and poplars as well as amendment of organic fertilizers should be carried out (Heiden et al. 2001).

In any case, the heap design allows the establishment of, for example, forest and grassland. Species like *Alnus glutinosa, Betula pendula, Elaeagnus angustifolia, Hippophae rhamnoides, Pinus sylvestris, Populus* hybrids, *Prunus avium, Quercus petraea, Robinia pseudoacacia* and *Tilia cordata* have been successfully planted. Most of the plants should be coppiced to reduce the proper weight of the plants (Heiden et al. 2001). With regard to their high transpiration and interception potential the use of conifers could also be an option. As an alternative to the wood plants, grasses adapted to the potentially salty soil such as *Calamagrostis epigejos* are able to grow. By using trees and grasses a plantation cover of 90–100% can be achieved. Therefore, the erosion potential and the water percolation rate will also decrease

Fig. 3.20 Vineyard at the south-facing slope of a salt mining heap in the proximity of Hanover, Germany

significantly. For instance, at the German salt mining heap Bleicherode a cover only 0.75 m thick reduced the water percolation by 80% and excluded erosion at the surface (Heiden et al. 2001). An eye should be kept on the nitrogen and phosphorus content, if the deposited topsoil indicates a lack of nutrients. It is peculiar, but even vineyards have been grown at a salt mining heap near Hanover, Germany, as illustrated in Fig. 3.20.

3.2.3 *Metallic Ore Mining Heaps*

Most of the ore mining heaps consists of originally natural materials like topsoil, overburden and waste rock with low modifications such as crushing. The ore mining process is usually associated with ore processing or ore enrichment procedures using problematical inorganic and organic additives. The results from the processing phases are generated tailings, which are deposited separately (see Sect. 3.3.2). The enrichment of ores occurs by separating the minerals from the gangue, whereas the creation of concentrates requires further processing steps apart from the separation into ore and gangue. The subsequent further processing to extractive metallurgy should enable the metals to be accumulated with a certain degree of purity.

In the context of metallic ore mining a distinction is made between exploitation of ferrous and non-ferrous metals. The ferrous metallic ore extraction contains less hazardous components (mostly iron as single component). Ferrous metals used in the steel industry are additionally comprised of potentially toxic elements such as chromium, nickel and cobalt. Non-ferrous metallic ore extraction includes a vast number of partially toxic metals (e.g. arsenic, cadmium, chromium, copper, mercury, lead, zinc). Moreover, some precious metals such as gold, silver and platinum are exploited and one metal used for energy production (uranium) is of importance. In contrast, in this sense industrial minerals such as potash and other salts (see Sect. 3.3.2) as well as stones and unconsolidated rock including clay minerals (see Sect. 3.3.1) do not belong to this definition. In Europe metallic ore mining is currently concentrated in some Mediterranean countries, Ireland and Scandinavian countries. In Belgium, France, Germany, Italy and United Kingdom almost all ore mining sites have been shut down. In the past intensive metallic ore mining took place in these countries and was associated with a high number of ore mining waste deposits (BRGM 2001). The solid waste of former mining activities might be more enriched with the desired metals due to the lower level of technological progress at that time.

In the case of ore deposits which are difficult to reach from aboveground underground mining is favoured. For this purpose, a series of passages (shafts, inclines, drifts) must be constructed. These must be connected to one another and to the surface and perform different tasks such as removal of ore, drainage of water, ventilation, access for persons and machines, etc. The underground variation should produce less waste per unit of ore than the open pit approach. Normally, open pit mining causes ten times more waste compared with underground mining activities. Ultimately, the stripping ratio depends on the ore grade of the exploitation area. For instance, in European mines the copper grade varies from 0.4 to 5.0%. Whilst in 1900 the average ore grade amounted to approximately 4%, 75 years later the value was reduced to 0.5% (Cooke and Johnson 2002). In addition, underground mining is only feasible, if subsurface collapse or other structural deficits are ruled out, as far as this is humanly possible to predict.

Apart from temporary stockpiles of lean ore, which are principally used depending on the market situation, solid waste rock is deposited. In Europe it is assumed that more than 4.7 billion tones of solid mining waste are stored (Cooke and Johnson 2002). Neither the solid rock waste nor the tailing material is backfilled underground to a noteworthy extent because of the high costs. The potentially contaminated waste rock is piled in immediate proximity to the shafts in order to reduce transport costs.

Due to the expected contamination there should be a sufficient area around the dumps for trenches which can collect runoff and seepage. More information about the environmental impacts, in particular in relation to the groundwater pathway, and the possible establishment of metallophytes can be found in Sect. 3.3.2 which discuss the impacts of metallic ore mining pits, in which smaller piles of mining waste are usually deposited as well.

3.3 Quarries and Open Pit Mines

Definitions for Rehabilitation

Open pit mines including quarries affect vast areas on a global scale. For instance, until the beginning of the twenty-first century in the USA 3.7 million ha and in China 2 million ha of land were degraded mainly by open pit mining (Cooke and Johnson 2002).

Open pit mining is a temporary land use, since the ore deposit is finite and completely exhausted at any time. Therefore, prior to the planning of the exploitation rehabilitation considerations should be incorporated into the mining process. There are generally different approaches to rehabilitation. Reclamation means that the area is prepared for a return to beneficial use in a productive way, in particular in relation to agricultural, forestry and recreational use. The term restoration is defined as an ecologically guided approach which promotes the recovery of ecological integrity. The term rehabilitation is differently used but frequently focused on the reinstatement of the original ecosystems. Unfortunately, in practice the terms are used interchangeably (Cooke and Johnson 2002). In this book reclamation refers only to the establishment of agricultural, forestry and recreational uses, whereas restoration should be defined as approaches which are aimed at the creation of ecologically valuable biotopes. Rehabilitation should be used as a generic term including all opportunities for subsequent use in areas influenced by mining activities.

The general question of future use for the large explored areas, which is also related to open-cast mines (see Sect. 3.1), can be answered by outlining two processes:

- The area of concern becomes a natural landscape that develops and regulates in a natural way (natural succession) or in a semi-natural way (controlled restoration)
- The area is prepared for an agricultural and forest use and subsequently planted with crops or trees (reclamation).

Establishment of Vegetation

In association with the development of the vegetation generally three versions are feasible. They are described as natural succession without any human measures, anthropogenically influenced controlled restoration and man-made reclamation (Table 3.10). Prior to the rehabilitation procedure, the goals for re-vegetation must be clearly identified, since possible ecosystem service must be provided. While the human impact on the site development is completely missing in the natural succession approach, in the second variation involving contour alterations and constant biotope management the human being permanently manipulates the landscape but the vegetation establishes itself in a nearly natural way and consequently will consist of rare species and ecologically valuable biotopes. Quarries and pits of unconsolidated rock offer a number of biotopes associated with the mining waste but it must be considered that modern extraction methods leave less mining waste than in former times, whereby the biotope diversity appears to be decreasing with regard to today's open pits (Monaghan 2007). The third option dealing with relief design, deposit of soil material, plantation and seeding does not allow the development of natural vegetation, since the areas of concern are used for agriculture and forestry purposes. In quarry terrains forest management appears to be predominant, because the agricultural use is topographically limited to a few level areas only.

Table 3.10 Natural succession, controlled restoration and reclamation as potential options for the treatment of unused quarries and open pits

	Natural succession	Controlled restoration	Reclamation
Soil handling	Untreated manifold relief	Rehabilitated relief (biotope management)	Relief design, deposit of soil, particularly humic topsoil
Pioneer stage	Extremely dry to moderately moist	Extremely dry to moderately moist	Plantation, sowing
Transitional stage	Xeroplant grassland, (dry) grassland	Controlled stabilisation of biotopes	Agricultural and forest management
Climax	Different biotopes (tendency towards woodland)	Different biotopes requiring biotope management	

The controlled restoration enables manifold vegetation cover such as pioneer vegetation, dry grassland and planted vegetation, and it includes a number of measures which are of significant importance in relation to the favoured vegetation (AG 2006):

- Immediate cover with grasses and legumes using slurries which include seed, mulch, fertilizer, water and binding agents – these are pumped out by special hydroseeding machines in order to protect the different quarry and open pit areas from rock slide and erosion
- Amendment of plant residues resulting from cutting or mowing to accumulate the seed potential
- Seeding of indigenous and rare species as far as the acquisition of the seeds is possible at all – the invasion of native seeds from adjacent areas by wind and fauna only takes place, if these areas provide a source for seeds and are not anthropogenically disturbed or highly polluted
- Successive operation of the quarry and open pit exploitation to ensure the continuous presence of some succession areas that deliver sufficient seeds – locally collected seed maintain the local integrity of the terrain to be rehabilitated
- Preference for seed mixtures containing not only the pioneer species but also species of later successional stages which will dominate after the earlier colonisers have died out – the seeds are spread manually, by helicopter or by agricultural seed spreader
- If humic topsoil is used on certain sub-areas, it should be noted that the topsoil to be backfilled stores viable seeds, but unfortunately not only desirable seeds – in particular, in relation to the ecological restoration projects, the spread of unwanted plant species can be counterproductive
- Complete habitat transfer using wheel loaders can be carried out on a small scale but is not normally applied
- Cost-intensive biotope service, particularly regarding the xeroplant grassland and semi-dry grassland types.

The creation of fully functioning biotopes is only completed by simultaneous establishment of the fauna. Usually, the fauna return occurs automatically when habitat requirements are existent. Nevertheless, some constructive activities can be carried out to initiate and to improve fauna re-colonisation such as construction of nest boxes for birds and deposition of timber to establish shelter and breeding areas for different animals like reptiles and invertebrate species.

In the context of reclamation for forestry purposes different requirements should be kept in mind. With regard to the humic topsoil management the exact characterisation of topsoil and overburden should be integrated into the planning process of the mining operation as early as possible. Attention should be paid to the following aspects (AG 2006):

- Ploughing-in of humic topsoil to the weathered stony material to improve root penetration and development is recommended
- In the case of a lack of topsoil it can be spread into rip-lines where it provides sufficient water availability for the seedlings – in general, irrigation is rarely carried out.

With reference to the plantation the following items become important:

- Plantation occurs directly according to a quick initiation of a shrub and tree vegetation after humic topsoil has been backfilled
- Use of shrub and tree seedlings is preferred to seeding because they have a more accurate planting density and mostly a higher survival rate
- Tree transplanting using excavators and trucks can help to re-vegetate areas with extremely slow growth characteristics but it is too cost-intensive and thus usually not applied
- Protection from animals must be provided (e.g. fencing-in)
- Inoculation with symbiotic microorganisms should be an additional option.

Apart from material segregation and the possible re-use of stockpiled humic topsoil, the placement of the different substrates should be planned exactly. Materials which are used for landform construction should not interfere with future excavation areas or areas where valuable raw materials are still supposed. Irrespective of the type of open pit mining, the entire area should be made safe to reduce erosion and to prevent accidents. This goal can be achieved by installation of a bund wall around the pit and by fencing-in.

3.3.1 Uncontaminated Quarries and Mining Pits

Uncontaminated quarries and mining pits are usually associated with non-metallic substances. We can differentiate between solid rock (e.g. sandstone, limestone, marble, basalt, granite), which is related to the building sector in addition to cement factories, unconsolidated material (e.g. sand, gravel, clay, marble), which is mostly extracted in the context of the building sector, as well as minerals technically separated from the gangue (e.g. barite, fluorspar, kaolin, bentonite, gypsum, phosphate rock), which are predominantly used in the chemical industry. The arrangement of quarries and unconsolidated rock pits (e.g. gravel pits, sand pits, clay pits) generally causes a strong impact on the environmental subjects of protection, namely plants, animals, water, soil, climate, landscape and the human being and its cultural and physical goods. For this reason, rehabilitation might be required, particularly at extraction sites exceeding 10 ha, exploring more than 3,000 t day^{-1} and leading to large-scale groundwater lowering. The impact is not only focused on the environmental factors. Exploitation of raw materials influencing a wide area also means consequences for the current and planned residential areas, water reserves, nature reserves and nationwide infrastructure facilities such as traffic routes, utility services and disposal systems. Accordingly, the demand for a proper planning and subsequent rehabilitation of the impacted land is surely essential (BRGM 2001).

The climate, particularly the seasonal rainfall, impacts the success of the rehabilitation strategy enormously because it influences landform stability (e.g. erosion potential) and vegetation performance. For this reason, the rehabilitation approaches should be adapted to the climatic conditions. Accordingly, in this book indications relating to the climatic differences are provided.

3.3.1.1 Quarries

Two types of quarry arrangement are distinguished. Firstly, the quarry stones are used as large ashlars, resulting in structured and carefully treated quarry walls and, secondly, they are used for road construction and technical application purposes, resulting in large and unstructured walls interrupted by berms and transport routes. The latter always contain quarry facilities such as crusher and sorting equipment.

Quarries are ugly landscape appearances after exploitation has been finished. Therefore, rehabilitation is required and should be generally applied. Before exploitation the soil has been removed, so that the quarry landscape leaves behind preferentially denuded bedrock bodies. In Fig. 3.21 a quarry landscape is presented. The main constituents of the quarry are the bottom of the quarry, some berms that have interrupted the steep slopes at the sides, eroded cone-shaped material, heaps consisting of former excavated topsoil, subsoil and overburden and the extremely steep geological walls surrounding the quarry. The geological walls are of essential interest for geologists motivated to investigate the quarry in detail (Fig. 3.22). The walls can serve as nesting sites for birds.

In contrast to open pits of unconsolidated rock, in quarries the erosion potential is much lower, since the rocky, large-sized material with a high density appears to be more resistant to erosion. Nevertheless, steep piles tend towards reduced stability in line with erosion. Erosion can only be limited, if steep landforms in the quarry landscape are minimised (AG 2006).

In this context drainage devices should probably be taken into account. Runoff from the top of the piles within the quarry terrain is discharged using stable lines with concentrated flow which can transport the high amount of runoff to the ground level. Otherwise, the runoff is retained on the top or on the berms of the piles, leading

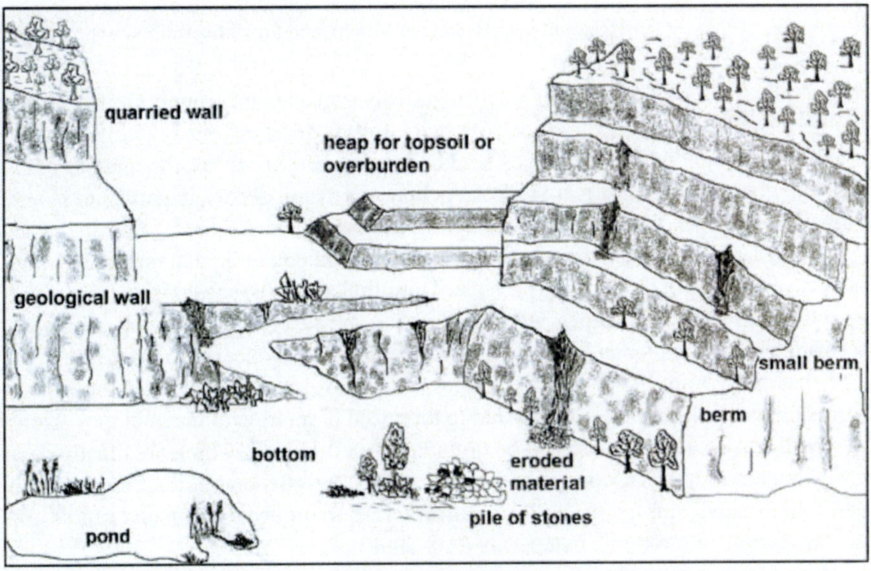

Fig. 3.21 Schematic illustration of a quarry landscape

Fig. 3.22 Geological wall at the side of a sandstone quarry

to prolonged water ponding which will negatively impact plant growth. Occasionally, the berms have to be piped in order to reduce runoff disasters.

Normally, the outer batter slopes are constructed in a linear fashion and are interrupted by berms which intercept runoff. As long as erosion occurs, in particular in arid climates with single hard rainfalls, maintenance appears to be required, since fine material will accumulate on the berms as a result of continuous sedimentation, and subsequent overtopping of the berms is possible. This problem can be solved by incorporation of rocks into the batter slopes. A better solution, however, appears to be alternative formation of concave outer batter slopes without berms to concentrate the flow.

It has been found that concave slopes, which are reminiscent of natural landforms, may reduce the erosion twofold to threefold in relation to linear slopes. Thus, the landform construction should be oriented to natural hills, which are mostly concave-shaped and which consist of diverse angles, lengths and surface textures. In addition, a rough surface of the slopes is preferred to reduce erosion and runoff and to establish vegetation in a better way (AG 2006).

While forming the slopes, a progressive technique is the more effective option. By rehabilitating separate shorter slopes in the course of time longer and more stable slopes are created without damage resulting from erosion. This gradual performance reduces the erosion risk in comparison to the construction of the whole slope at the same time (AG 2006).

The amount of overburden depends on the type of quarry and the subsequent material utilisation provided. For example, sandstone quarries whose material is used for road construction generates up to 40% fine-grained residues, while limestone quarries might generate only 10–15% unusable remains (BRGM 2001).

Just after finishing exploration natural pedogenesis starts slowly. The material consists of parent material slightly weathered in the upper part. The weathering depth depends on the rock present and it differs from a few centimetres (e.g. granite, basalt) to more than 20 cm (e.g. shale, limestone, marblestone) within about the first decade. At the flat areas of the quarry, namely the bottom and the berms, the humic A horizon can hardly be discovered but residues of the pioneer vegetation and sedimented dust allow fine material to be detected in the uppermost layer. Natural succession can be observed combined with the initial development of an A horizon (Fig. 3.23).

In slope situations the percentage of fine material usually increases at the lower slope and footslope due to erosion, while coarser material accumulates in the upper parts of the slope. This differentiation is often superimposed by weathering processes which are also responsible for material dislocation. In association with the impact of weathering stones with a shale structure, particularly mudstone and marble stone, are strongly affected, in contrast to stones resistant to weathering such as granite and sandstone containing iron.

The dust deposition as mentioned above plays a major role, in particular in association with the exploitation and transport of the extracted material. In the course of the quarry operation stones are broken and crushed continuously, causing dust development and consequently dust deposition at the bottom. Thus, the bottom of the quarry frequently reveals finely grained texture independent of the type of stone quarried.

Fig. 3.23 Basalt quarry unused for one decade revealing succession of vegetation at the bottom of the upper and lower berm

The nutrient capacity is assessed to be meagre, since the nutrients stem from the virgin stones only. Nitrogen, phosphorus and sulphur are mostly deficient due to the lack of organic matter. The native contents of the upper continental crust were, on average, only 0.0083% N, 0.067% P and 0.007% S but the values varied between the different geological formations. For example, the results for phosphorus concentrations exhibited 0.079% (sedimentary rocks), 0.044% (granites), 0.11% (gabbros), and 0.097% (metamorphic rocks) respectively (Wedepohl 1995). The values indicated total concentrations and consequently do not allow inferences to be drawn about the plant-available concentrations.

Examples of several quarries (clay covered by Quaternary deposit, limestone, rock phosphate) provided for rehabilitation in the catchments of Novgorod and St. Petersburg, Russia, exhibited low nitrogen content in the underlying bedrock varying between 0.02 and 0.2%. Simultaneously, the carbon content showed low concentrations as well, ranging from 0.11 to 1.27%. Higher values were associated with older abandoned quarries where root development caused subsurface carbon increase. In contrast, higher C and N concentrations, which increased with age, were found in the uppermost A horizons. For instance, the rock phosphate quarries indicated 2.6% C and 0.21% N at 8-year-old sites under pines, 4.3% C and 0.30% N at 15-year-old sites under larches and 7.2% C and 0.69% N at 20-year-old sites under firs respectively. A 190-year-old limestone quarry exhibiting alkaline reaction and related higher biodegradation potential showed 4.89% C and 0.36% N in the uppermost horizon (Abakumov 2000).

The concentrations of the cationic elements calcium, potassium and magnesium depend on the rock type and vary from beneficial to more or less absent. On average, in the upper continental crust values of 2.95% (Ca), 1.35% (Mg) and 2.87% (K) were reported. Again, differences between the rock types were ascertained, as the results from calcium (sedimentary rocks: 6.39%, granites: 1.07%, gabbros: 3.25%, metamorphic rocks: 2.07%), magnesium (sedimentary rocks: 1.98%, granites: 0.34%, gabbros: 4.56%, metamorphic rocks: 1.75%) and potassium (sedimentary rocks: 1.77%, granites: 3.61%, gabbros: 0.75%, metamorphic rocks: 2.66%) showed (Wedepohl 1995). The pH value also varied depending on the rock type but regarding the re-vegetation of the bare soils an optimum pH of 5.5–8.0 should be aspired to (AG 2006). With reference to re-vegetation the addition of nutrients can be of importance since the rehabilitation procedures (soil stripping, soil replacement) generally lead to a degraded state.

Some chemical compounds of originally uncontaminated rock lead to subsequent pollution, which must be taken into consideration. For instance, rocks containing sulfides tend towards conversion to sulphates and ultimately to acidification, which, in turn, can accelerate the release of heavy metals. In most rocks the percentage of toxic metals is definitely restricted but in some the metals such as the rock type serpentinite have accumulated, revealing enhanced chromium and nickel concentrations which have been geologically determined (Meuser 2010). Moreover, regarding the minerals to be exploited some minerals are frequently associated with toxic substances, as the example of barite (combined with lead and zinc) demonstrates (BRGM 2001).

The rubble zone containing sulfidic rock which remains after exploitation at the surface should be covered to prevent the ingress of oxygen, which is responsible for the acidification process. The layers above the rubble zone may consist of alternating coarse-grained and fine-grained materials which are additionally compacted by heavy trucks. Moreover, deposited waste material (piles) within the open pit area, which is susceptible to oxidation, is encapsulated (deposit at the sides and on the top) by more fine-grained material for the same reason. With reference to the cover a comparable process can also take place on tailings, which tend towards drying and subsequent sulfide oxidation (see Sect. 3.3.3). It is definitely possible to encapsulate the piles despite the general aim to design the landscape in a landform that may mimic natural environments (AG 2006).

Quarries offer manifold vegetation communities in which species are frequently able to survive, in contrast to the adjacent areas. In temperate climates the calcareous quarry sites showing limited rock weathering and root penetration are adapted to communities such as xeroplant grassland and semi-dry grassland that accept extremely dry soil conditions and a lack of nutrients. At sites exhibiting skeleton-enriched weathered material mixed with subsoil less valuable but intensively

Fig. 3.24 Limestone quarry during reclamation period. *Zone A*: lower berms just after soil management has been implemented. *Zone B*: upper berms covered with grass-legume mixture for soil conditioning. *Zone C*: upper berms already afforested

blossoming plant families such as ruderal semi-dry grassland and ruderal grassland exist. In more acid conditions (e.g. on china clay, on sandstone) heathland species such as *Calluna vulgaris* and *Erica* sp. were able to invade the quarry terrain, as shown in mining areas in south-west England. In semi-arid and arid climates a relatively fast invasion of grasses and even tree seeds has been observed on limestone and Jurassic shales, as the example of the coastal region in Kenya proved. 26 tree species centered on the nitrogen fixing *Casuarina equisetifolia* were discovered after a short period of time (Cooke and Johnson 2002). The example proves the general advantage of symbiotic and free-living nitrogen fixers in relation to the rehabilitation of mined land (AG 2006). In the USA it was observed several times that the low nitrogen supply in addition to the low water retention potential was an advantage for the restoration of prairie communities in arid climates (Monaghan 2007). The initiation of desired planned communities can be improved by addition of seeds containing hay and mulch which stems from natural sites of other locations.

The reclaimed areas of unused quarries are usually planted with tree species for afforestation purposes. In Fig. 3.24 a former limestone quarry is shown that does not exhibit steep slopes between the berms in the upper part of the quarry. Hence, the soil consisting mainly of the weathered rock was mixed with subsoil from adjacent areas as well as with topsoil. Afterwards the soil was conditioned by seeding legumes and grasses for a short period of 2 years and finally planted with trees. The initial establishment of a grassy community has an advantage over direct shrub and tree planting due to the production of a high rate of soil cover that controls erosion. Disadvantageously, the grasses can hinder the later growth of shrubs and trees, which forestry organisations in charge of quarry re-vegetation must encourage. Apart from repeated cutting, the application of selective herbicides can be taken into consideration.

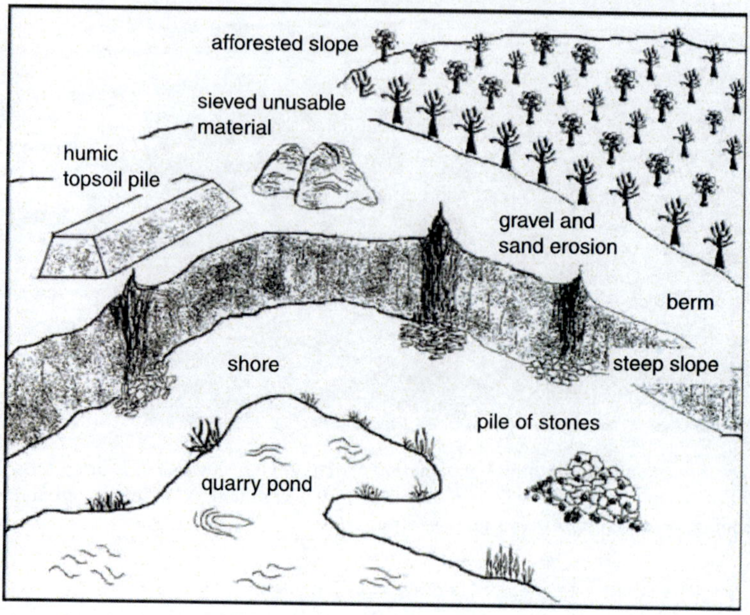

Fig. 3.25 Schematic illustration of an unconsolidated rock pit landscape

3.3.1.2 Unconsolidated Rock Pits

Unconsolidated rock is extracted, for example in gravel, sand, loam and clay pits. In Fig. 3.25 the schematic design of these pits is displayed. The exploitation occurs mainly in alluvial floodplains and terraced landscapes in soils with a relatively low distance to the aquifers or even directly influenced by groundwater. Thus, after quarrying of, in particular, sand and gravel pits lakes stemming from groundwater remain, covering most of the area of concern.

Nevertheless, in areas where uncontaminated unconsolidated rock and minerals are exploited terrestrial terrains remain. For these areas most of the information mentioned in the context of quarry handling can be transferred. Apart from the water body some deposits are present such as piles of sieved stones and piles of humic topsoil which have previously been excavated. The humic topsoil is usually removed and stockpiled for later backfilling purposes. Moreover, eroded and sedimented areas due to the unstable slopes are typical features in areas used for unconsolidated rock pits. Furthermore, parts of the catchment affected by dredge mining become dewatered terrains after mining operations.

As an alternative to a proper rehabilitation management, it is also possible to include remaining pioneer habitats from the ecological point of view. It is well-known that abandoned gravel and sand pits offer a multitude of ecologically relevant sites, namely steep faces with dens for bank swallows, piles of stones where reptiles live, dry piles of sand where insects live and temporary pools used as spawning grounds for toads. Therefore, it is ecologically worthwhile that perhaps some terrestrial areas remain generally untouched.

A possible operating sequence dealing with the terrestrial part of the mining terrain is introduced here in conjunction with the dredge mining of the heavy minerals in Zululand, South Africa. The minerals containing dune sand were firstly mined by means of a floating dredger, which pumped the sand into a tailing pond where the minerals were gravimetrically separated from the sand. Afterwards, the tailings were dewatered and the dry material deposited on the former mined land. In this example stockpiled topsoil was spread over the sandy matrix to a depth of 10 cm to establish indigenous vegetation. In fact, after seeding of fast-germinating species the vegetation cover increased quickly. Consequently, natural succession led to the growth of some trees, in particular the nitrogen-fixing *Acacia karoo*, which is well-known as a tree pioneer. This tree achieved a height of 1.5–3 m after a maximum period of 8 years, of 3–8 m after 11 years and of 9–12 m after 16 years corresponding to natural reference sites where the plant height of *Acacia karoo* amounted to 12–15 m. Nitrogen-fixing tree species that are responsible for favoured nitrogen content in the soil contribute to rapid succession of the vegetation and subsequent species richness. Gaps between dying pioneer trees may enable colonisation of secondary tree species, so that the succession can probably be facilitated. Simultaneously to the plant development the number of edaphon species showed an increasing tendency with time (Cooke and Johnson 2002).

The shores of remaining or man-made lakes reveal different slope gradients, including cliff-like shores as well as gently inclined shores. The natural slope formation depends on the cohesiveness of the material. Apart from the compactness, the presence of cementing substances such as $CaCO_3$ and iron oxides influence the slope formation.

Before the exact design of the lake is finalised the planning process should consider the possible approaches of the water body. In particular, the shore design in the context of the inclination of the shores depends predominantly on the exploitation, since size and depth of the lake might be mainly determined with reference to the previous exploitation. Consequently, the exploitation restricts the future use to a great extent. Nevertheless, in the majority of cases several solutions can be found but they clearly demand decision-making at an early stage. In mining areas where the mining occurred after dewatering the re-sloping operations should be done prior to water filling. If the groundwater table is not too deep, the creation of artificial ponds in addition to temporary pools should always be taken into consideration to contribute to the site diversity.

Swimming lakes should meet some conditions such as a minimum size of 5 ha, a Secchi depth of at least 1 m, an oligotrophic to mesotrophic state, compliance with hygiene and inclination requirements. Lakes used for sports need a minimum size of 20–40 ha, a water depth of 5–10 m and an acceptable boat density of <5 ha^{-1} (sailing boats) to <15 ha^{-1} (canoes). Motorboats are not desirable. Lakes used for fishing should consist of deep and flat shorelines and must be continuously monitored with regard to their nutrient state.

Ponds used as wildlife habitats should be irregular at shores and at the bottom to enable a high variety of habitat possibilities. Large areas of shallow water and steep slopes are included. Whereas steep banks provide shaded and cooler areas, the long shallow shore shelves are of importance for the wetland vegetation. Long slopes

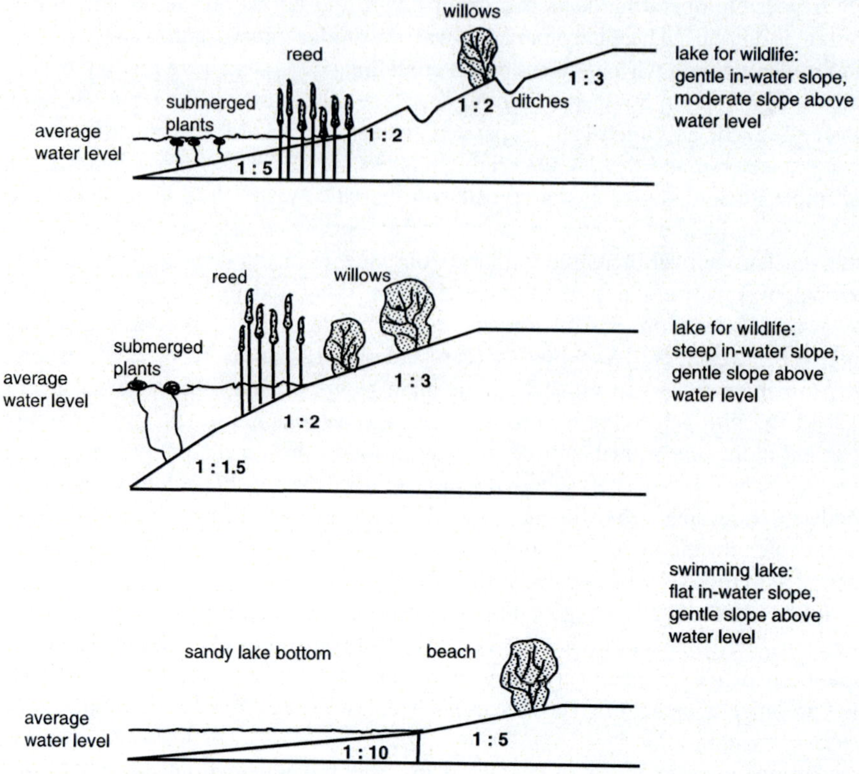

Fig. 3.26 Slope configurations of ponds

should be integrated because it is difficult to predict accurately the final water level. Most in-water slopes are gentle (>1:5) and variable to enhance plant diversity. Moreover, parts with steep slopes underwater (up to 1:1.5) are involved. Above average lake water level an inclination of 1:2–1:3 is mostly preferred but the transition from wetland to upland should occur gradually. For this reason, the upland bank slope gradient above the zone of water fluctuation is formed with a gradient of 1:5 or flatter. This inclination allows a person to escape from the water body, should they fall in. At sandy beach areas of ponds used for swimming the slope gradient is 1:10 in order to prevent accidents with drowning children (Norman et al. 1997). In Fig. 3.26 slope configurations are illustrated.

In ponds for wildlife habitats islands can be constructed by depositing and compacting sand or placing of boulders or mine waste rock at the bottom. The islands are used by e.g. turtles and birds for loafing and they increase the surviving rate of different animals because they discourage predators. The mining terrain is often located in the vicinity of rivers. Near rivers long, narrow and moderately deep post-mining ponds with irregular islands can be formed. These are connected with the river to establish a natural floodplain area. The connecting outlet channels can be used by fish such as salmon. To create more habitats for aquatic creatures tree crowns, logs, stumps, root wads, etc. can be submerged. Protruding logs are areas where turtles and other amphibians are able to bathe in the sun.

In a similar way to natural lakes, separate vegetation belts are introduced at the shallow parts of the banks. In upland direction a submerged vegetation belt, shallow marshland with reed (especially *Phragmites australis*) and wet meadow or an upland zone with wetland shrubs and trees are established. In particular, in the presence of gravelly substrate the reed contains additional species such as *Phalaris arundinacea, Glyceria maxima, Typha angustifolia, Carex* sp. and *Scripus lacustris* in temperate climates. Most of the plants show high biomass production and will vegetate the quarry pond shore rather quickly.

Willow species (*Salix* sp.) are the most useful wetland shrubs and tree. This species *Salix pupurea, Salix incana, Salix alba, Salix fragilis, Salix cinerea* and *Salix viminalis* is particularly well-known for its enormous potential to stabilise the soil. In temperate climates alders (e.g. *Alnus glutinosa*), ash trees (*Fraxinus excelsior*) and poplars (e.g. *Populus nigra, Populus alba*) are often taken as well. The species mentioned are particularly suitable since they grow rapidly and cover the soil within 2–3 years. Besides, they are adapted to the unfavourable soil fertility. Some nitrogen-fixing species such as alders, black locust (*Robinia pseudoacacia*) and buffaloberry (*Shepherdia canadensis*) are also chosen in different climates because they can ameliorate the soil. Shrubs and trees should be planted in clusters and should not be spaced over wide areas. At the shoreline the introduction of organic mud that contains seedbanks for wetland species can be an option but the use of peat should not be considered due to the expected damages to natural everglades (see Sect. 3.5.1) (Norman et al. 1997; Monaghan 2007).

In any case, stability against erosion and collapse must be assured. If the construction of acceptably inclined shores is limited, terraces (berms), which have to include ditches in which the runoff is collected and channelled, are considered. Consequently, the danger of erosion is minimised and long-term bank stabilisation can be expected.

In the shallow marsh of relatively large lakes wave erosion occurs in windy time periods. The windward lake shore can be protected from wave erosion by placing boulders on it. More cost-intensive is the construction of fascines usually applied in the field of bioengineering. The erosion and deflation potential would be also reduced, if grasses and legumes (e.g. *Lupinus* sp.) were seeded just after the shorelines have been formed. In association with the future use of the lake a temporary fencing-in might be beneficial to reduce damage to the vegetation during the initial periods. Furthermore, to some mammals like muskrats can have another detrimental effect which disturbs the development of the plants.

Moreover, chemical aspects have to be considered, since the future use requires an acceptable water quality with regard to e.g. phosphorus content and hygiene standards. The trophic state depends preferentially on the phosphorus concentration (Table 3.11) (EPA 2008). Swimming lakes and aquatic sports lakes which are used

Table 3.11 Trophic lake classes (Data from EPA 2008, modified)

Trophic class	Phosphorus concentration (mg L^{-1})	Secchi depth (m)	Oxygen saturation in summer (%)	Remarks
Oligotrophic	0–12	>8–4	>70	Clear water
Mesotrophic	>12–24	<4–2	>30–70	
Eutrophic	>24–96	<2–0.5	0–30 (deep water oxygen-free)	High plankton production, temporarily H_2S development, tendency towards sapropel formation
Hypereutrophic	>96	<0.5	0	Extreme plankton production, intensive H_2S development, intensive sapropel formation

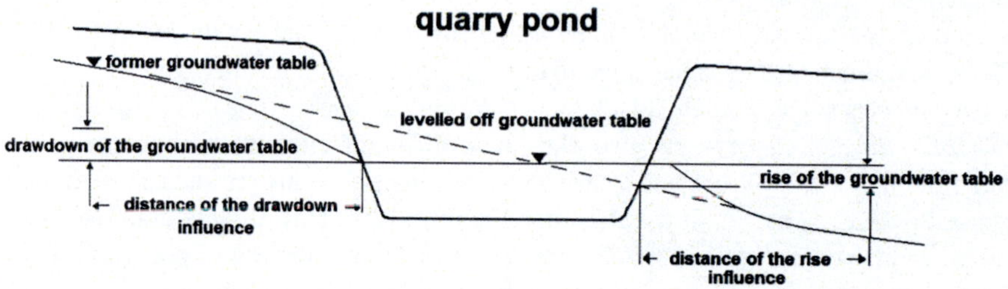

Fig. 3.27 Impact on the groundwater table by artificial lakes originating from sand and gravel exploitation

by humans and which must allow them to come into contact with the water need oligothrophic (to mesothrophic) properties. The tolerance with regard to nature reserve lakes and fishing lakes is supposed to be higher but, for instance, eutrophic and particularly hypereutrophic water bodies appear to be generally unfavourable for future use and must consequently be remediated (see Sect. 7.3.1). Eutrophic and hypereutrophic lakes result from discharges, runoff and erosion from adjacent agricultural sites, influx of contaminated groundwater, fishing use and use for recreational purposes. In the course of time a tendency towards sapropel formation at the lake bottom is detectable because of increasing phytoplankton production and subsequent sedimentation of organic material. Techniques for dealing with the remediation of eutrophic and hypereutrophic lakes are mentioned in Sect. 7.3.1.

The remaining quarry ponds cause problems to the watershed, since they are artificially created after the complete excavation of the raw material. Hence, they should be considered as groundwater that has lost the covering and protecting layers. As indicated in Fig. 3.27, the groundwater level of the watershed will respond to the new hydrological conditions caused by the establishment of the artificial lake. In consideration of the groundwater flow direction, the groundwater table will sink upstream, while downstream a table rise occurs. In particular, in areas exhibiting relatively large quarry ponds and strong groundwater incline the effects are clearly detectable.

Accordingly, in the upstream catchment there is increasing drought, leading to detrimental impacts on agricultural use and hazards to environmentally valuable wetlands. On the other hand, in the downstream catchment the opposite development happens and agricultural cropland is in danger of being exposed to too moist soil conditions. With regard to the total watershed, the appearance of large quarry ponds influences the climatic properties as well, since the waters tend towards higher evaporation in comparison with the original land surfaces. In conclusion, lake management in the context of rehabilitation measures should also pay attention to the adjacent areas.

3.3.2 Contaminated Metallic Ore Mining Pits

Shallow metallic ore bodies (<300 m) are mined in open pits or quarries (BRGM 2001). The successive stripping operations produce tons of extracted materials which are similar to the overburden and accordingly not interesting for ore extraction procedures. This material is sent directly to waste dumps formed either close to the open pit or within the extraction terrain. Consequently, smaller and relatively low ore mining waste deposits can be found in open pits. Alternatively, the solid waste is used for filling some abandoned open pits. The mass to be deposited depends on the ratio between extracted waste and the quantity of extractable ores, the so-called stripping ratio.

In mining terrain of former times dating back to the Roman period slag deposits were also present since the ore smelting occurred in the vicinity of the extraction area due to the limited transport capacity. Thus, at old ore mining sites deposits containing contaminated slag resulting from ore smelting and polluted ash produced by cleaning the furnaces can also be discovered (Meuser 2010). Moreover, because of fall-out resulting from mining processes such as crushing and metallurgical operations in metallic ore mining terrains large-scale soil pollutions must be expected, as ascertained in the area of Mansfeld, Germany. Here cupriferous schists, which offered 2.6 Mt of copper and 14,200 t of silver (BRGM 2001), were mined until 1990.

Irrespective of the source of contaminants (waste rock deposit, tailing, dust deposition, slag heap) the groundwater pathway should be preferentially taken into account. Whether, for instance, heavy metals will contaminate the aquifer, depends on different factors such as:

- Formation type and thickness on which the waste is deposited (presence of formations with low hydraulic conductivity like clay)
- Type of aquifer (confined or unconfined type) and its porosity (porous, fractured or karstic)
- Topography of the catchment (valley position or slope position).

Results from mining are steep faces and pit floors, rock dumps and tailing lagoons (see Sect. 3.3.3). In general, compared to deep underground mining (see Sect. 3.2.3), relatively little waste is left. The deposited mining substrate normally shows a very coarse texture in the rock waste piles and a fine texture in the milled tailings. Accordingly, the latter tends towards compaction after sedimentation and drying. In principle, during the mining process material segregation, in particular related to uncontaminated and contaminated portions, seems to be advisable. The segregation provides the opportunity to bury material which is toxic to plants and to separate overburden which is suitable for the rehabilitation programme, especially for the creation of the upper layers. Besides, strict segregation may avoid double-handling and unnecessary long-term stockpiling of the different materials (AG 2006).

The organic matter content exhibits very low concentrations and, furthermore, the deposited material is low in macronutrients such as nitrogen and phosphorus. Waste containing iron pyrite may generate sulphuric acid to a great extent, inducing low pH values which can reach 2.0. Moreover, the high metal content of the waste indicates toxic conditions and leachability, in particular at very low pH values (Cooke and Johnson 2002; AG 2006).

Apart from the underground, the water conductivity within the waste heaps is of decisive significance with regard to the groundwater contamination. Rainfall flows along voids and cracks. This preferential flow is minimised in the long term, if weathering occurs and subsequently finer material blocking the pathways exists. The water percolation, however, can be restricted in different ways. As soon as vegetation covers the pile, water infiltration is reduced and transpiration counteracts the downward percolation. Irrespective of the vegetation development, the water infiltration can also be restricted by bulldozer compacting of the surface, which in turn is responsible for water ponding at the surface. The latter is avoidable by sloping the top surfaces. Nonetheless, seepage from the toe takes place but its extent might decrease over time due to the decreasing hydraulic conductivity based on the weathering effects.

Old abandoned mining sites (open pit terrains, ore mining heaps – see Sect. 3.2.3) developed naturally colonised plant communities over a long period of time. These consisted of metallophytes which adapted to the extreme soil properties, in particular the high metal values. For this reason, many old ore mining sites could achieve a high conservation status, since they are refugia for rare plants and subsequent rare animals. The potential for plant species to grow under such conditions depends on the ability of young plants to survive and on the possibility of the propagules to be transported and to germinate. Metallophyte species are capable of surviving due to their high metal tolerance so that they colonise the ore mining sites without serious competition from other species. The recovery on initial bare soils involves a colonisation sequence until the achievement of a climax community.

Irrespective of the metallophytes different plant species are potentially capable of invading highly contaminated mining pits. As shown in Sudbury, United Kingdom, at a mining site indicating low pH value of 3–4.5 and subsequent high available copper and nickel concentrations as well as high aluminium content re-plantation was possible after a thin layer of dolomitic limestones was manually applied.

Table 3.12 Principles for rehabilitation of metalliferous mining sites (Data from Johnson et al. 1994)

Contamination	Techniques	Possible maintenance
Low	Liming (acid soils) Fertilizing Addition of organic matter Use of commercial or wild seed Use of indigenous species	Irrigation Inhibition of grazing Monitoring
Moderate	Liming (acid soils) Fertilizing Addition of organic matter Use of commercial seed of metal- tolerant species Use of wild seed of metallophytes	Competition with other species after fertilizer addition Inhibition of grazing Monitoring Adverse affects to wildlife animals
High	Surface treatment with 10–50 cm uncontaminated overburden Liming (acid soils) Fertilizing Use of indigenous species	Monitoring (capillary influence)
Very high	Soil cover with 30–100 cm including barrier system (see Sect. 5.1.1) Creation of a suitable rooting zone	Monitoring

Tree species with low nutrient demand such as birches, willows and trembling aspens (*Populus tremulus*) were able to grow, though the soil properties, in particular the stony material, did not afford good growth conditions. The fast improvement of growth conditions was effected by leaf litter, which transported every year bases to the soil surface by means of which the colonisation of further vegetation was initiated (Cooke and Johnson 2002).

Some metal-tolerant ecotypes of e.g. *Agrostis capillaries* and *Festuca rubra* can be sown to vegetate the mining area quickly but attention should be paid to the long-term maintenance of the area, since these species may require fertilizing and repeated seeding operations. At restoration sites thc grasses covered the bare soil satisfactorily in spite of extremely high copper, lead and zinc content. In this context, the possibility to grow so-called hyperaccumulators such as *Thlaspi caerulescens* and *Minuartia verna* at abandoned metallic ore mining sites should be discussed (see Sect. 6.4.2).

On the other hand, it is always feasible as an alternative to re-vegetate contaminated open pits by means of soil cover. The principles, suggested and published by Johnson et al. 1994, are listed in Table 3.12. For very highly contaminated sites the re-use of previously excavated and stockpiled uncontaminated topsoil is recommended. Usually, the time span between excavation and possible re-use may take a very long period and goes hand in hand with increasingly adverse soil quality. Loss of organic matter, loss of aggregate stability, increase in organic acids in line with enhanced metal availability, changes of pH, redox potential and the ammonium : nitrate ratio as well as disturbance of the edaphon and subsequent biological activity are typical negative aspects to be taken into consideration, if optimised soil management has not been integrated (see Sect. 4.2).

To prevent leaching and groundwater hazard in highly contaminated open pits different cover systems are sometimes the only method to be applied (see also Sect. 5.1.1). The necessary materials could be present within the mining terrain. Stockpiled fine material, rocks and topsoil can be re-used for cover formation purposes, if it is sure that the material is uncontaminated. Based on the experience in Australian pits an adequate cover looks as follows (from below):

- Traffic-compacted, weathered waste rock, at least 1 m thick
- Clayey, compacted sealing layer measuring 0.5 m thickness and a hydraulic conductivity of 10^{-8} m s^{-1} to prevent percolation or breakthrough of contaminants
- Rocky soil layer with a minimum thickness of 1.5 m to store water and to enable deep-reaching root growth, if required
- Topsoil indicating a minimum thickness of 0.5 m, high water storage capacity and biological activity that is probably fertilized and seeded or planted with vegetation (AG 2006).

3.3.3 *Tailings*

Sites used for open pit mining of metallic ores are normally associated with the formation of tailing ponds nearby. While the mechanical separation of valuable ore and gangue (mainly crushing and sorting) and the subsequent deposit of mining waste appears to be the less important process of underground mining from the environmental point of view, the following production of the concentrate is frequently combined with physical and chemical methods which are environmentally problematical. The physical separation is mostly based on gravity separation. In a lot of mining areas flotation by means of detergents is applied instead. Furthermore, chemical measures are involved in the solution mining process to separate and to purify the valuable metal. Consequently, toxic substances such as cyanides (gold extraction), acids (e.g. copper, nickel), etc. are taken for these purposes.

The products from ore processing are usually fluid matrices such as aqueous solutions from cyaniding and slurries of fine particles which contain additives required for purification processes. The fine particles derive from preparation steps in which the crushed ore-gangue mixture is finely ground prior to the subsequent chemical extraction procedure. All remaining fluid matrices containing high concentrations of metals have to be disposed of usually in tailing ponds. It is assumed that about 1.2 billion tons of tailings were stored all over Europe. For instance, with regard to copper mining in a similar way to the copper ore grade, the produced tailings increased from 17 to 290 Mt a^{-1} worldwide over the same period. The majority of tailing ponds serve as permanent disposal facilities but most of the tailings are in an accessible condition, so that a future reprocessing is generally possible (see Sect. 2.3.3) (BRGM 2001; Cooke and Johnson 2002).

The degree of contamination achieves enormous dimensions, as the example of the copper mine tailings in the Welsh Mynydd Parys, United Kingdom, has shown. The average total metal concentrations were 1,905 mg kg^{-1} for copper, 7,692 mg kg^{-1} for lead and 1,036 mg kg^{-1} for zinc respectively. In addition, the mobile rates analysed with DTPA extraction revealed high values, e.g. copper 7.5 mg kg^{-1}, lead 107 mg $kg^{-1}$and zinc 3.1 mg kg^{-1}. The high mobility required the consideration of remediation strategies which were investigated on the basis of alkaline amendments according to the stabilisation approach (see Sect. 5.4) (Khan and Jones 2008).

The tailings are dumped via pipelines in lagoons or settling basins, which are surrounded by dams. The embankment of the dams should be compacted to reach geotechnical stability and should be built with gentle slopes between 27 and 36% to prevent erosion and allow vegetation. The dam stability is particularly important in relation to the polluted water that would damage the surrounding landscape in the case of spills containing e.g. cyanides, acids and alkalis. In 1998 such a disaster happened in the Aznalcollar mine, Spain, which produces concentrates of copper, lead, silver and zinc from pyritical ores. The fine-grained metal compounds were treated in a flotation process. For this purpose, different agents such as sulphur dioxide, calcium hydroxide, copper sulphate pentahydrate and organic compounds were added. The fluid tailings were pumped into an artificial lagoon covering 1.5 km^2. In the Spanish catastrophe the surrounding dam failed at a length of 50 m, discharging three million m^3 of metal-contaminated sludge and four million m^3 of acidic water. About 4,500 ha of land, including parts of the Coto Donana National Park, were affected. The only way of remediation was disposal of the excavated sludge in an old mine pit adjacent to the contaminated area. Remaining contamination had to be treated because the quality criteria could not be kept, in particular regarding arsenic and some heavy metals in the context of an agricultural use. Methods of soil stabilisation had to be integrated (see Sect. 5.4) (BRGM 2001).

After the flooding in the course of time the water content of the tailing ponds tends to decrease due to evaporation and evapotranspiration in combination with the development of vegetation. Altogether, even in a temperate climate a net evaporative moisture flux persists in tailings in spite of periodic rewetting by intensive precipitation. If the tailings consist of sulphides, increasing oxidation in line with potential acidification and production of soluble metals occurs, which would endanger the environment by metal leaching and subsequent percolation downwards. Initially, high leaching rates are expected but due to the settling of the fine particles the hydraulic conductivity within the slurry diminishes. Moreover, the leaching process is strongly influenced by the hydraulic conductivity of the foundation belowground (AG 2006). Besides seepage through the dams and leaching, overflow of the dam wall caused by extreme rainfall or mismanagement might be a further source of environmental pollution.

Supplementary tailing remediation appears to be difficult. The lateral seepage can be reduced by construction of a slurry trench cut-off wall, which should be applied if the dams are primarily damaged or tend towards extreme weathering. This side barrier installation occurs in a similar way to the side barrier systems introduced in Sect. 5.2.1. The later construction of underdrains below

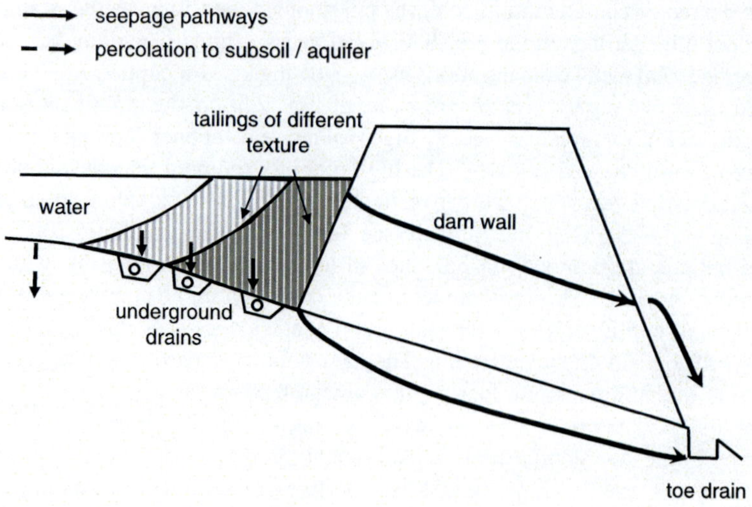

Fig. 3.28 Remediation strategies for the seepage of tailing ponds

the tailings which collect the seepage water is theoretically feasible but extremely cost-intensive. Seepage can alternatively be collected with the help of toe drains (Fig. 3.28) (BRGM 2001).

A cover system including clay liners (see Sect. 5.1.1) is possibly required in areas with high water permeability. However, because of the silty to clayey texture of the tailing material beneath, the cover system is in danger of swelling-shrinkage alterations, which, in turn, cause cracking and subsequent increased permeability by about 100-fold higher values. For this reason, based on the Australian experience with tailings the following cover system is preferred (from below):

- Fine-grained layered tailings
- Fresh waste rock with minimal fines (capillary break) to avoid transfer of contaminants into the sealing layer above and to limit root penetration
- Clayey, compacted sealing layer measuring 0.5 m in thickness and a hydraulic conductivity of 10^{-8} m s^{-1} to prevent percolation or breakthrough of contaminants
- Rocky soil layer with a minimum thickness of 1.5 m to store water and to enable deep-reaching root growth, if required
- Topsoil indicating a minimum thickness of 0.5 m, high water storage capacity and biological activity that is probably fertilized and seeded or planted with vegetation (AG 2006).

3.4 Subsided Mining Terrain

The problem of subsidence can exist in every kind of underground mining (coal, salt, ore). Large-scale land subsidence such as sinking of the ground above an underground void or depression is distinguished from collapse of usually small-scale caves, which is also termed creation of sinkholes. Deep mining includes vertical sinking of shafts to the raw material deposit (e.g. coal seams) and digging of horizontal tunnels to and through the raw material to be exploited. Regarding the underground coal mining different technologies are applied:

- For room-and-pillar mining entries are opened which provide access and ventilation already prior to the mining procedures. During mining and starting from the access ways so-called rooms are extracted whereas pillars are left out between the rooms. Approximately 50% of the coal seams remain in place and these are capable of holding the mine roof. Thus, a collapse hardly occurs as long as the remaining pillars remain untouched.
- Retreat mining normally follows, involving the removal of the pillars. Up to 75–86% of the coal seams are then exploited. Without protection measures the roof tends to collapse as soon as the pillars are extracted. Large-scale subsidence can occur, which impacts the overlying strata as well. When containing aquifers the overburden sinking simultaneously to the exploited seams will break apart and will create holes where the water can drain downwards.
- Longwall mining usually leads to large-scale subsidence because machines are used which are capable of exploiting nearly 100% of the coal seam. The seams must be protected by a set of wooden chocks which hold up the mine roof but the chocks lose their stability in the course of time so that collapsing takes place with a time delay (Durant 2011).

Apart from the removal of pillars originating from mining activity subsidence can also result from accidental coal fires which may destroy the pillars. In general, the overlying rocks are strongly affected by mine roof collapses and the affected terrain is much wider than the area where the collapse directly occurred. With decreasing distance between the seams collapsed and the soil surface and with decreasing stability of the overburden the impact may accelerate. A risk potential that quickly appears is particularly high in mining areas where the thickness of the overburden is less than 30 m. If the thickness exceeds 100 m, the subsidence effects appear with a time delay of 5–10 years.

Apart from large-scale subsidence, the spontaneous creation of small caves near the surface which tend to increase in the course of time is important. These small caves cause the most destructive events and catastrophes can even occur. They are very dangerous when occurring only a few metres below the soil surface. For example, in Bochum, Germany, such a cave measuring a depth of 40 m occurred in a residential area in 2000 and led to the destruction of some houses and garages requiring immediate remediation measures. The cave resulted from a shaft which had been incompletely filled after closing of the coal mine. Another catastrophic disaster with 11 killed miners was reported in Lassing, Austria, where in 1998 a cave 30 m deep and 50 m wide was abruptly formed. At a depth of 60 m below the surface water and mud broke into the shafts, causing instability of the shaft configuration and subsequent subsidence (Hersche and Wenker 2000).

Salt mines attract special interest, if they get into contact with water. As shown in the Les Salines Mine in Cardona Diapir, Spain, an inflow of water from a river into the mine galleries resulted in the generation of a cave system. After a flood event in 1998 the cave

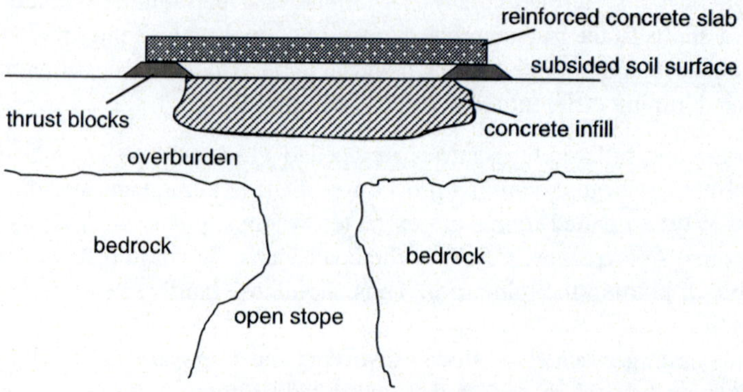

Fig. 3.29 Arrangement for concrete capping with slab (Data from Carter et al. 1995)

generation occurred abruptly and to a great extent. A number of sinkholes were created in addition to an elongated depression area measuring 100 m, which acted as the main swallow hole. 137 sinkholes with a diameter ranging from 1 to 135 m, but usually smaller than 2 m, have been identified. Many sinkholes tended towards enlargement caused by continuous inflow of fresh water originating from the river nearby. In the entire catchment infrastructure pipes and roads were strongly affected (Lucha et al. 2008).

There are different strategies for remediating subsided and collapsed mining terrain. In the case of rapidly appearing local caves which are mostly associated with sinkholes (e.g. caused by water in salt mines) and collapsed shafts the relatively small open stopes are protected by a bridge-deck cap consisting, for instance, of a reinforced concrete slab (Fig. 3.29). It is possible to use slab spans of up to 10 m in order to close the cave. The slabs should be anchored to the bedrock. Additionally, below the slab concrete is filled in to enhance the stability of the remaining overburden. If sufficient overburden depth exists, concrete infill which is compacted by rollers (Fig. 5.5) can alternatively be carried out (Fig. 3.30). The roller compacted concrete consisting of well-graded granular fill and cement allows a later utilisation of the protected site, e.g. as a car park or road. Possible openings of deeper stopes, however, must be previously closed by cover plates which are placed across the stope openings. Again, it is necessary to key possible road lanes into the underlying bedrock (Carter et al. 1995).

Larger cavities in the underground must be filled with concrete, which is pumped via the openings into the underground. The openings normally do not exist near the surface, requiring the small-scale excavation of overburden to find suitable openings where the fill can be pumped in. Since cavity areas can be far away from the sites where the infill takes place, the concrete filling operations are carried out by means of high pressure which helps to achieve tight filling and avoidance of remaining voids. Lower sections of the stopes are easier to reach pneumatically. The air compressors have the effect of blowing the fill into sections of the stopes, because a gravitational approach shows limited success. In general, at first a granular fill, which is covered by a final concrete plug, is used. Apart from concrete and granular fill, materials such as crushed slag and fly ash are alternatively used (Fig. 3.31).

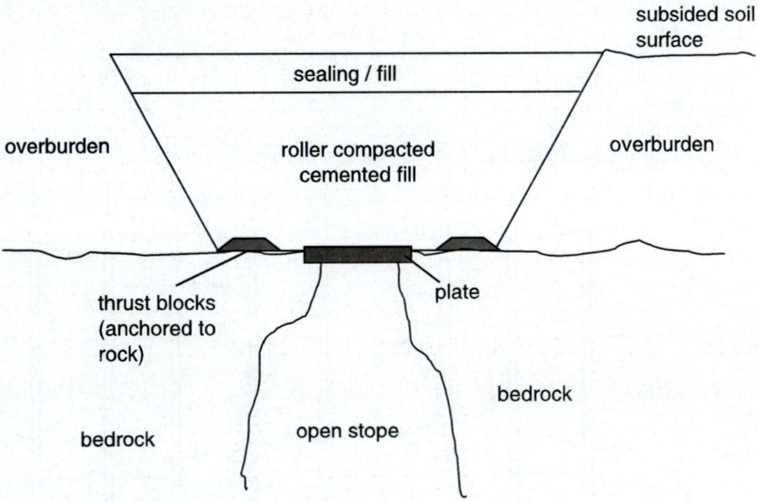

Fig. 3.30 Arrangement of roller-compacted cemented fill in addition to plates placed above open stopes (Data from Carter et al. 1995)

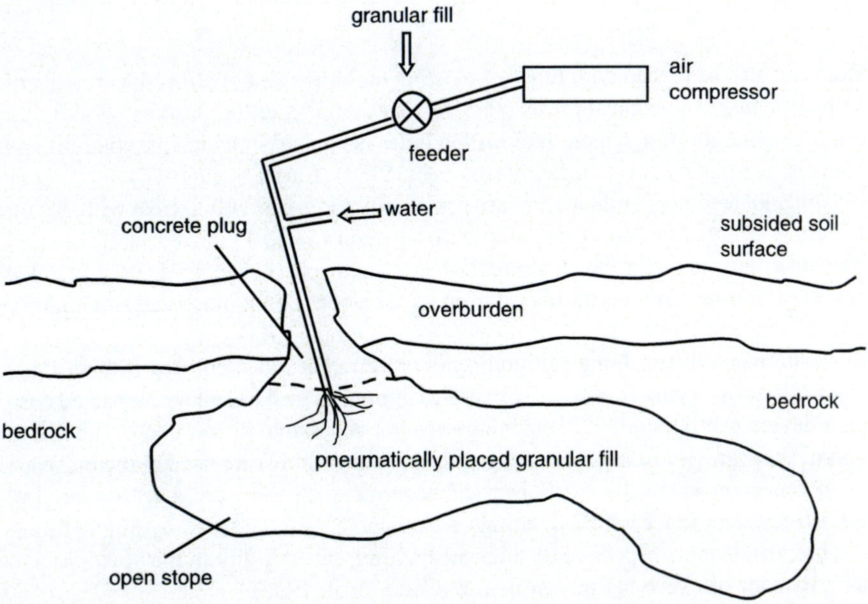

Fig. 3.31 Arrangement based on pneumatic technique for stope backfilling consisting of concrete plug above granular fill (Data from Carter et al. 1995)

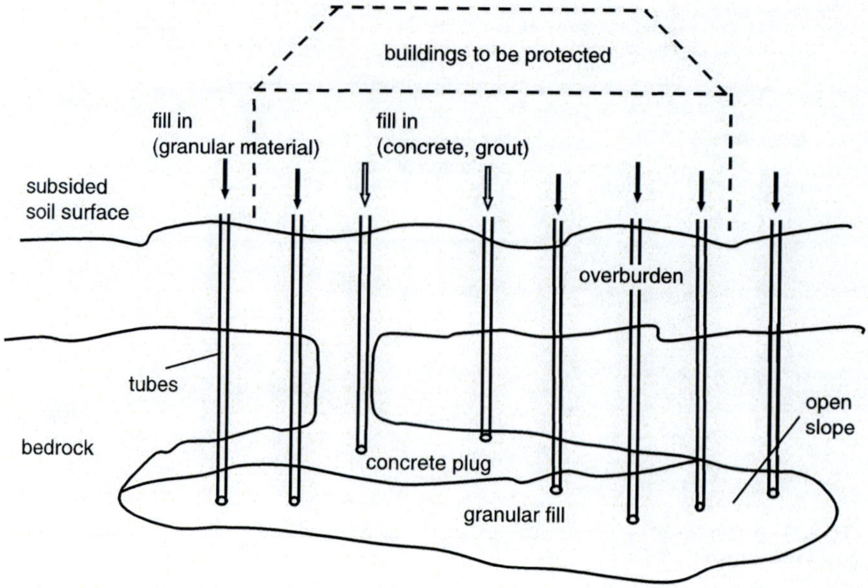

Fig. 3.32 Gravity filling arrangement consisting of concrete plug above granular fill using tubes (Data from Carter et al. 1995)

Both can also be placed with tubes, if existing buildings do not allow the creation of stope openings by excavation or other factors complicate the process (Fig. 3.32). For tube configuration a pattern of drilled holes is applied in which granular fill and concrete are placed gravitationally (see Sect. 5.2.2) (Carter et al. 1995).

Voids located deeper than 15 m are principally more difficult to treat with dry fill materials. For this reason, stopes at a depth between 15 and 45 m are usually secured by subsurface stabilisation. Columns of grout and gravel are formed at selected spacings in mine and overburden. Grout as a cement containing water-rich slurry which is injected through drilled holes is used (Durant 2011).

In summary, the bridging solution is not preferred at subsided sites with unstable sidewalls or overlying structures. Concrete capping as well as roller compacted concrete layers can be applied, if openings are not accessible or too deep. The filling approaches (gravity filling, pneumatic filling, slurry filling) are used in the presence of incompetent sidewalls. By means of pneumatic fill even areas with overlying infrastructure or inaccessible openings are treatable. Alternatively, hydraulic filling can be carried out irrespective of sidewall stability, accessibility of the openings and the existence of overlying infrastructure (Carter et al. 1995).

The techniques are applied to all kinds of mining. For instance, in Cobalt 450 km north of Toronto, Canada, the described approaches were used in areas which are affected by subsidence in the context of ore mining, in particular silver mining. The affected area was previously investigated (geophysical methods such as micro-gravity and radar, drilling technique) to identify all hazard zones which covered 15% of the inhabited town area. In total 17 problematical areas have been

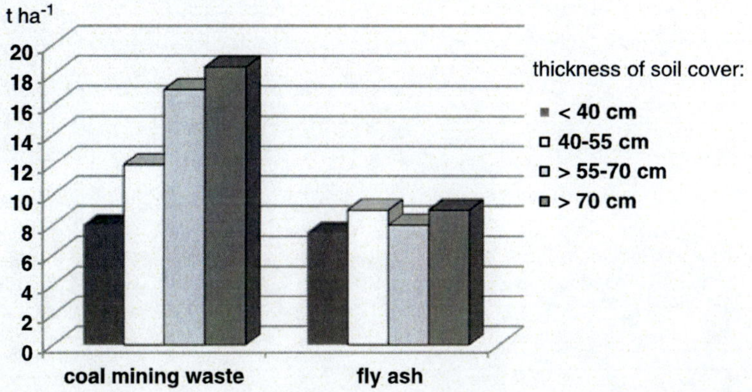

Fig. 3.33 Average wheat biomass (t ha^{-1}) depending on soil cover thickness (cm) (Data from Makowsky et al. 2010)

found where near-surface veins were exploited, where mining actually occurred close to the surface or where the stability of the overburden appeared to be weak (Carter et al. 1995).

Large-scale subsidence is remediated using more drastic approaches. They are applied in mining areas which have been influenced by subsidence over a long period. One impressive example is the Double H region around Huainan, China, with an underground mining history starting at the beginning of the twentieth century. The coal mining area contains 19 mines which produce 77 million tons of hard coal annually with an increasing tendency. As a consequence, subsidence up to 21 m covering an area of 133 km^2 occurred. In future it is expected that 1,250 km^2 will be influenced by subsidence. Since the soil surface reached deep areas, which are deeper than the groundwater table, flooding occurred after switching off water pumps, mainly affecting cropland. Consequently, a high percentage of the former farmland has been converted into fish ponds and reservoirs.

An attempt was made to compensate for the loss of agricultural land by depositing coal mining waste and fly ash on the subsided terrain (Fig. 3.33). Above these deposits loamy material originating from alluvial floodplains was deposited. In the meantime, the reclamation process was implemented over approximately 8 km^2. The coal mining waste derived from the Permian period consists of coarse conglomerate, sandstone and mudstone. Because of low pyrite concentrations acidity was not analysed but the growth conditions for wheat that is predominantly grown in that province are nevertheless detrimental due to a lack of nearly all relevant macronutrients. The fly ash deposits did not reveal any contamination due to their origin from various coal power stations. Texture (silt to fine sand) and nutrient capacity are favourable but the layered structure in association with high underground soil compaction means a barrier for root penetration. Obviously, the transition between the loamy soil material ultimately deposited and both underlying technogenic substrates resulted in a strong limitation for root growth. Accordingly, the wheat production

Fig. 3.34 Deposit of coal mining waste on subsided and flooded terrain in the Double H Region, China (From: Huainan Mining Industry (Group) CO., LTD.)

showed low yields as long as the loamy soil deposit was too thin. In a research project dealing with experimental plots of different cover classes (<40 cm, 40–55 cm, >55–70 cm and >70 cm) it has been found that there was a tendency for wheat growth to increase with increasing cover thickness. Regarding underlying coal mining waste the wheat biomass production increased significantly from <40 to >70 cm soil cover thickness. In contrast, increasing cover did not indicate higher biomass production related to the fly ash deposit (Fig. 3.34). In conclusion, it transpired that no substantial advantage was recognised if more than approx. 60 cm was applied to coal mining waste and more than approx. 45 cm to fly ash (Makowsky et al. 2010).

The deposit of suitable material to allow an agricultural use is usually not feasible due to the enormous material transportation and the quantity of soil material required. Thus, in some large-scale subsided mining areas arrangements have been made for wetland development which considers ecological demands. For instance, in the German Ruhr area where between 1800 and 1990 a volume of 9.54 billion tons of hard coal seams measuring thicknesses from 0.6 to 4.5 m in an area covering 3,044 km^2 to an average depth of 920 m below the surface was exploited, extreme subsidence occurrences were ascertained. It has been calculated that a volume of 8 km^3 was removed from the Carbonaceous bedrock. Based on longwall mining subsidence occurred, causing sanitary problems and an increase in epidemics until the beginning of the twentieth century. The maximum subsidence reached up to 24 m. Apart from damages to buildings, the relative rise of groundwater, in particular in association with long periods of rainfall, resulted in flooding and swamping of large areas. Hence, a number of pumping stations were installed with the aim of draining the land by consuming energy. In 1989 about 340 km^2 had to be continuously pumped in the central Ruhr area (Drecker et al. 1995).

After the closing of many collieries the pumping activities were reduced to a great extent. Due to the lack of wetlands in urban agglomerations like the densely populated Ruhr area the wetlands formed by mining subsidence became very important for flora and fauna. For instance, one lake in Dortmund (Lake Lanstrop), with a width of 200 m and a length of 450 m formed by subsidence of about 9 m, which meant a depth lower than the average groundwater level, served as a fishing centre and led to an enrichment of the landscape in the urbanised territory. However, the artificial creation of the mining lake caused adverse effects as well. The adjacent agriculturally used areas were strongly affected by too high groundwater levels and some residents had problems with the sewage from their homes, which did not satisfactorily flow off. Some lakes may represent secondary biotopes for wildlife, as the example of Lake Hallerey located in the same city has shown. This water body was isolated from the residential and industrial areas by highways and railway embankments and its size was similar to that of Lake Lanstrop. Subsequently, the lake was used as a nature reserve. Currently it is becoming one of the most important resting places for birds in Northwest Germany. Besides the possibility to use mining lakes for ecological purposes, the closing of the pumping stations in line with rising groundwater offered more environment-friendly opportunities. For example, in Dortmund during mining operations the streambeds of streams such as River Emscher and nearly all its tributaries had to be channelled and sealed at the bottom to prevent flooding but also downward water percolation, which would affect the underlying mining tunnels and shafts. After the closure of the mine drainage stations the removal of the concrete lining of the channelised streambeds was possible and excessive rehabilitation measures of the River Emscher were initiated (Drecker et al. 1995), which continue to the present day (see Sect. 7.3.2).

3.5 Peatlands

Peatlands (mires) are specific ecosystems exhibiting natural characteristics which lead to the categorisation of ecologically valuable areas and which are frequently designated as nature reserves. The maintenance of peatlands as an important factor for biodiversity and a sink for carbon is widely accepted. In Europe, peatlands covering about 20% of the land are frequently altered for agricultural (50%), forestry (30%) and exploitation (10%) purposes. Approximately 60% of pristine peatlands have disappeared (Vasander et al. 2003).

In the context of global warming there is a growing tendency towards awareness about changing abandoned peatlands after harvesting as carbon sources to the atmosphere into carbon sinks by restoration. Due to the removal of vegetation in cut-away peatlands only low carbon fixation occurs. Re-vegetation improves the carbon fixation significantly and discharges the greenhouse gas emission. On the other hand, in cut-away peatlands organic matter is permanently oxidised as a result of the lowered water table, which is also responsible for reduced CH_4 fluxes. After rewetting the CH_4 flux from the anoxic peat layers below to the oxic peat layers above and subsequently into the atmosphere increases, leading to a new greenhouse gas problem. However, as found in rewetted Finnish restoration areas, the CH_4 emission will fall below the quantities of pristine mires of the same region (Vasander et al. 2003).

The following natural peatland types must be differentiated:

- Raised bogs, whose nutrients are derived from rainwater (ombrotrophic system). They form a dome and the surface is raised above the adjacent areas and accordingly not influenced by groundwater. They reveal high water retention potential combined with high evapotranspiration and consequently relatively low groundwater replenishment. The evapotranspiration even exceeds the evaporation of surface water bodies. The bogs are dominated by *Sphagnum* mosses as peat-forming plants requiring environments which are moist and poor in nutrients. In particular, nitrogen exhibits rather low concentrations. In principle, the pH value displays acid conditions varying mostly between 2.5 and 3.5. The annual rainfall should reach at least 700–750 mm and the number of rainy days should not fall below approx. 175 days (Wheeler et al. 1995; Bord na Mona 2001).
- Fens, whose nutrients are derived from mineral-rich groundwater sources (minerotrophic system). Usually, nutrient capacity and acid neutralising capacity are higher than in raised bogs and consequently a more diverse flora with sedge species can be found. The nutrient status does not automatically imply eutrophic conditions and mesotrophic or even oligotrophic characteristics can also exist. In contrast to the raised bogs, fens are often located in groundwater-influenced landscape depressions which are lower than the surrounding areas. Consequently, the species of fen prefer higher pH values (levels >5.0) than the bog vegetation (Wheeler et al. 1995; Lamers et al. 2002).

Based upon the classification system in Ireland a distinction is made between blanket bogs on gentle slopes with moderately deep peat accumulation (on average 2.6 m) and raised bogs on flat plains with a peat accumulation of 9 m on average . In Germany one differentiates between nutrient-rich and groundwater-influenced low moors on alluvial sediments, precipitation-dependent high moors and transitional moors which reveal both characteristics. Another classification is based on the water quality (marine and freshwater swamps) or on the climatic conditions (peatlands of boreal, temperate and tropical climate). Moreover, regarding the peat origin distinctions between moss peat, sedge peat and wood peat are made (Andriesse 1988).

Topographically, most of the bogs in boreal, temperate and tropical climates consist of a hummock and hollow system as integral parts of the ecosystem. The wet hollows either support a high concentration of *Sphagnum* mosses or are completely filled with water (small pools up to 50 cm deep), whereas the hummocks are predominantly refugial areas for other typical bog species such as *Eriophorum vaginatum* (cotton grass) and *Drosera rotundifolia* (sundew). The peat mosses grow 3.5–20 cm a^{-1} and are biodegraded after dying. The growth of the organic residues amounts to a few millimetres per year (on average 4.3) (Brady and Weil 2008). With regard to the carbon content in Finnish bogs, for example, the average long-term carbon accumulation is 24.0 g C m^{-2} (Vasander et al. 2003). Accordingly, the bog surface rises slowly but continuously. Bogs reveal an extensively treeless surface and only some small water bodies (Wheeler et al. 1995). In Fig. 3.35 a bog landscape is visualised.

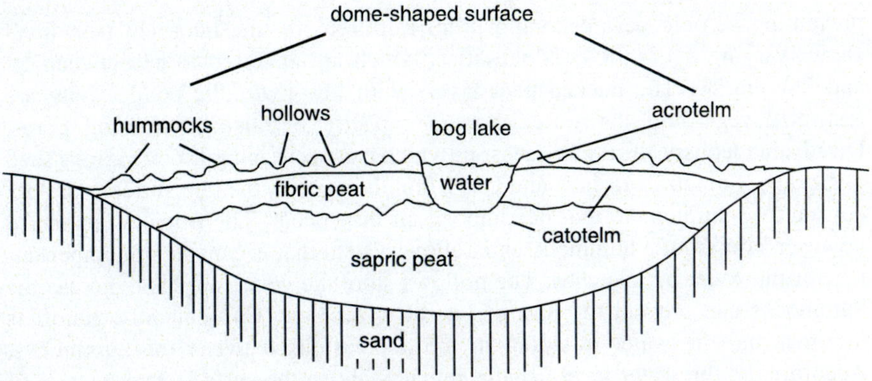

Fig. 3.35 Schematic illustration of a typical bog landscape

Table 3.13 Characteristics of fibric and sapric peat (Data from Andriesse 1988, added)

	Fibric peat	Sapric peat
Colour	Light yellowish brown to reddish brown	Very dark grey to black
Degradation	Low to moderate	High
Wet bulk density (g L^{-1})	<100	>200
Pore volume (vol%)	91–98	85–93
Saturated water content (% of dried material)	850–>3,000	<450
Organic matter (%)	94–99	94–99
pH value ($CaCl_2$)	2.5–3.5	2.5–3.5

Under boreal and temperate climatic conditions undamaged natural bogs, which can achieve a thickness of 10 m or more, show a dual layered constitution with an upper acrotelm in which the peat-forming *Sphagnum* mosses grow and an underlying catotelm which is permanently waterlogged. The acrotelm can usually be differentiated into a light brown fibric peat layer and a dark grey to black sapric peat layer underneath. The fibric peat is less decomposed according to the presence of more fragments of plant tissue than the sapric one. Properties of the two peat types are mentioned in Table 3.13.

In tropical areas, for instance in south-east Asia, peat swamps are also landscaped as dome-shaped, raised bogs provided by rainwater (ombotrophic bogs). The main difference is that the peat is derived from trees and shrubs (e.g. *Dactylocladus, Gonystylus, Madhuca, Palaquium, Parastemon, Shorea, Swintonia*) instead of mosses. Obviously, the same self-regulating forces as present in the temperate zone are decisive. The water percolation into the mineral soil underneath is minimal due to the impermeable subsoils in tropical areas (e.g. mangrove clay, podsols with hardpan layers) and the low rate of disappearing water is substituted by rainfall. Therefore, the bogs remain permanently wet, preventing aerobic peat decomposition. Furthermore, the humified peat itself shows varying hydraulic conductivities, which are analysed to be between 1^{-8} and 2.3^{-3} m s^{-1}. The packed peat layers with low hydraulic conductivity are restricted regarding the water storage capacity because of missing pores. Disadvantageously, during dry seasons which can occur in monsoon regions such as Borneo, Sumatra and Java and last 1 month or more, the bog surfaces are not constantly saturated, so that decomposition may occur. The tropical peatlands, however, consist of a hummock and hollow system that is capable of compensating for the water deficiencies. The hollows store the water surplus, whereas the hummocks can prevent the loss of lateral water flow. Consequently, runoff is minimal and the water is stored in the depressions between the hummocks. Accordingly, the water level usually remains above the surface. Irrespective of the slope gradients erosion is widely excluded, since the hummocks and aboveground buttress roots hinder water flow and the development of erosion channels (Dommain et al. 2010).

3.5.1 Peat Harvesting

In particular, bogs are often used on a large scale for raw material extraction purposes and are consequently considerably damaged. After peat extraction it is estimated as necessary to restore the used areas and to re-initiate the establishment of peat bogs, if an agricultural use is not provided for.

Also, in former times, a small-scale use of peat took place, e.g. to provide for fuel. These mostly peasant peat cuttings re-colonised spontaneously and consequently disappeared in the course of time. In Sweden, for instance, one restored large-scale ombotrophic bog was compared with 11 shallow hand-dug bogs, which remained unused for the last 50 years on average. The small-scale bogs showed a higher percentage of bare peat surface (17%, ombotrophic bog: 2%), a lower *Spaghnum* cover (54%; ombotrophic bog: 74%), but a higher diversity of *Sphagnum* species (11–21, ombotrophic bog: 9) and some *Carex* species (sedges), which are usually more common in fens and which had previously been absent in the ombotrophic bog (Soro et al. 1999).

During the peat extraction deep ditches are cut to drain the land (Fig. 3.36). Afterwards, the peat is exploited in different ways. With regard to the fibric peat above, cut-away bogs prevail peat cutting machinery (Fig. 3.37). After the cutting process the sods are dried in the open air (Fig. 3.38) and transported to the peat processing facilities. In this way layer by layer is extracted and removed until a residual sapric peat remains. One adequate method applied to sapric peat is based on milled surfaces in the autumn which are then exposed to freeze-thaw processes in wintertime. It is possible to excavate the peat treated in this way or to extract it by suction. The sapric peat can also be excavated in moist conditions by means of grab dredgers.

Fig. 3.36 Drainage ditches for drying purposes before the bog is exploited

Fig. 3.37 Machinery for cutting peat sods in operating state

Fig. 3.38 Peat sods piled for drying purposes

3.5.2 Sphagnum Bog Restoration

3.5.2.1 Principles

In many countries such as Canada, Estonia, Finland, Germany, Ireland, Sweden and the United Kingdom (Northern Ireland, England) bog restoration projects have existed for several decades after the environmental awareness in relation to exploited and damaged peatlands gained in importance. In the meantime, restoration guidelines have been published (e.g. Canadian Peatland Restoration Guide – http://www.peatmoss.com/pdf/Englishbook.pdf, Global Peatland Restoration Manual – http://www.imcg.net/docum/prm/prm.htm) which are the basis for the restoration approaches described below. In most countries with a high peatland cover (e.g. Estonia 22.5%, Finland 32.1% and Sweden 23.1% of land area) the percentage of pristine peatlands in relation to dried (agriculture, forestry) and cut-away areas has been reduced dramatically (e.g. percentage in Estonia 32%, in Finland 40%, in Sweden 47%) (Vasander et al. 2003).

The decisive objectives to be achieved are the generation of a plant cover dominated by peatland species as well as long-term water management which enable a surface-near water table. Measures are aimed at the restoration of the ecological functions, particularly the interaction between vegetation and hydrology, and the self-regulating mechanisms which lead to peat accumulation over a long period of time. Theoretically, if the hydrological conditions can be maintained, the targeted vegetation is very similar to what was there before (Quinty and Rochefort 2003).

With regard to peatlands the term restoration is used for the general re-establishment of a functional self-regulating and peat accumulating ecosystem (see definitions in Sect. 3.3). In this context the restoration process consists of three stages:

- Rewetting, which requires the technical removal of the drainage system in addition to land-forming processes and clearance of undesired vegetation such as *Molinia caerulescens* (purple moor-grass) and *Betula pubescens* (birch) hindering the growth of peat-forming species. This stage usually takes 3–5 years
- The development of appropriate vegetation which can last decades (this stage is also termed re-naturation)
- The long-term accumulation of the peaty surface which is expected to take up to some centuries (this stage is also called regeneration) (Wheeler et al. 1995).

Worked bogs which cannot fulfill the start conditions for restoration, in particular in the presence of a remaining peat depth of less than 50 cm, should also be treated in an ecologically acceptable way, e.g. by means of a restoration to heathland. With regard to an environmentally friendly approach that is not focused on the peatland restoration alternative uses are possible:

- Plantation of reed canary grass (*Phalaris arundinacea*) for the large-scale production of bio-energy or chemical pulp makes sense on cut-away peatlands because they are large-sized, flat, weed-free and supplied with infrastructure. The moist soil conditions can be achieved by ditch blocking (Leinonen et al. 1998; Finell 2003).
- Similar conditions prevail in association with the establishment of energy forest using fast-growing tree species capable of coppicing such as *Salix viminalis, Salix dasyclados*, alders (*Alnus*) and poplars (*Populus*). The cut-away areas must be fertilized prior to planting. The water table is recommended to be at a depth of 30–40 cm, which might be realistic due to the high evapotranspiration rates of the trees. As long as frost does not occur in the long term energy forest is possible but climatic limitations in countries such as Finland and Sweden must be taken into account (Hörnsten 1992).
- The cultivation of cranberries (*Vaccinium oxycoccus*) is also an option for use of the cut-away peatlands. The closing of the drainage system, phosphate fertilizing and the presence of decomposed fibric peat are important pre-requisites. Peatland vegetation growing parallel (e.g. peat mosses, cotton grass) might even allow peat accumulation simultaneously (Chiasson and Chiasson 2000). Furthermore, other berry species can be grown for the same purposes like cloudberries (*Rubus chamaemorus*), currants (*Ribes rubrum, Ribes nigrum*), gooseberries (*Ribes oxycanthoides*), strawberries (*Fragaria vesca*), etc.

In Table 3.14 advantageous and disadvantageous factors influencing the restoration success for bogs are summarised.

3.5.2.2 Water Management

In order to raise the water table as close as possible to the land surface the drainage ditches are blocked every 50–100 m by filling them with peat that is afterwards compacted (Quinty and Rochefort 2003). Water led into the ditches will penetrate

Table 3.14 Advantageous and disadvantageous factors influencing the restoration success for raised bogs (Data from Gorham and Rochefort 2003, modified and added)

Type of factor	Factor	Positive influence	Negative influence
Peat cutting process	Use of heavy machinery (soil compaction)	No	Yes
	Temporary storage of the acrotelm	Conducted	Not conducted
	Remaining peat layer thickness	>50 cm	<50 cm
	Time period between peat exploitation and restoration	Short	Long
	Connection to other still existing peatlands	Present	Absent
Geographical conditions	Topography	Flat	Sloping
	General hydrology	Simple	Complex
	Groundwater table	Deep	Shallow
	Water quality (groundwater, surface water)	Poor in nutrients, acid	Rich in nutrients, alkaline
Climatic conditions	Climate	Wet and stable	Dry, cyclical (floods, drought)
	Annual precipitation	>750 mm	<750 mm
	Frost-heaving	No	Yes
	Permafrost	Absent	Present
	Long-term dry periods after cut-away (subsidence)	No	Yes
Vegetation	Remaining bog vegetation	Present	Absent
	Seed bank of bog vegetation (propagules)	Present	Absent
	Distance to bog propagules	Near	Far
	Invasive species	Few	Many
	Fen species (e.g. sedges)	Few	Many
	Intensive shrub and tree development after cut-away	No	Yes
Anthropogenic influence	Nutrient input	Low	High
	Air pollution (NO_x, SO_2)	Low	High
	Agricultural land use after cut-away	No	Yes
	Forestry after cut-away	No	Yes
	Catchment	Densely populated	Sparsely populated
		Strongly affected by agriculture	Less affected by agriculture

into the peat body and spread throughout the area. Combined with drainage and subsequent subsidence of the peatlands it can be detrimental that the peat was strongly compacted during the exploitation period, exacerbating the water penetration into the peaty body (Vasander et al. 2003).

With regard to the interruption of the drainage system there are some problematical aspects that must be taken into account. Firstly, sometimes it is difficult to find all ditches because in the course of time vegetation was able to grow between the last peat extraction and the rehabilitation, overgrowing and concealing the ditches. Secondly, the rewetting process can only occur, if the precipitation is sufficient, exceeding, for example, 750 mm a^{-1} in the northern part of Europe. A combination between the rarity of peat-forming mosses and climatic conditions which are different from those in the past might make the restoration process impossible, as already found in many lowland bogs in England. Another problem to be solved refers to the possible subsidence in some parts of the rewetted area, leading to damage to the closed drainage ditches and subsequent new leakages.

The rewetting would be easier if the peat surface lay closer to the groundwater but in this case it must be expected that the establishment of fen vegetation prevails. Influenced by the underlying mineral soil, tall sedge species are favoured. These species are also capable of forming peat that is different from bog peat, e.g. more decomposed and with lower water storage potential. Alternatively, water can be added using pipelines. This water, however, must achieve the quality standards for bog water, which is principally poor in nutrients and acid.

Nevertheless, as a result of compression and decomposition during peat harvesting the water retention potential of the restored peatlands is reduced, while the chemical properties mostly do not show any significant difference. The growth of *Sphagnum* mosses is only guaranteed if some important physical features are acceptable in the whole season, e.g. a high water table (mean: 30 cm below surface) and a high moisture content (>50 vol%) (Price and Whitehead 2001).

3.5.2.3 Land-Forming

Worked areas adjacent to still existing peatland habitats appear to be the best option for restoration purposes. Existing habitats nearby may provide sufficient diaspores and besides they can serve as yardsticks for the restoration project. Thus, it is of immense importance that some sections of the entire peatland area are preserved in a natural state while exploiting.

In general, before exploitation the uppermost approximately 20–30 cm of the bog (acrotelm) should be removed and stored in mounds at the sides because this material contains the majority of the viable seeds of the bog-like vegetation. In the acrotelm, where aerobic and anaerobic conditions exist, the plant growth and consequently the peat formation take place, whereas in the underlying catotelm permanent anaerobic conditions prevail. Later on, the stored overburden, which must be kept fully wetted within an area with a high water table, can be backfilled on the remaining peat. Up to 30 cm of the material is deposited. It is necessary initially to compact this material and to remove wood, branches and other disturbing artefacts which make the compaction difficult. The use of the active, intact and peat-forming layer of the peatland increases the success of the restoration because the kick-starting of a new acrotelm would take up much more time (Quinty and Rochefort 2003).

In the absence of acrotelm material at least the peat surface should be broken up because in dry conditions which can exist prior to the restoration measure the surface tends towards surface crusts. It is also possible that the dried surface appears powder-like and susceptible to deflation. A lot of time is sometimes taken up

Fig. 3.39 Failed bog rehabilitation area where large lakes are visible causing wave action that inhibits the peat moss development

between drainage and subsequent cut-away and the restoration. Thus, the original bog reveals some changes caused by the long-term drying process. The organic matter tends towards biodegradation, which in turn provokes subsidence and nutrient release. Under the dry conditions an independent growth of hummocks cannot occur. Due to these reasons the degradation processes argue for a rapid initiation of the restoration (Gorham and Rochefort 2003).

The soil preparation includes different technical approaches. To achieve the bog restoration at least 50 cm of sapric peat must be left to impede drainage, since the hydraulic conductivity of the sapric peat layer appears to be very low (Quinty and Rochefort 2003). Fibric peat is not recommended. If the residual peat layer cannot guarantee the minimum thickness, the areas can be ensured with bentonite mats or plastic liners (see Sect. 5.1.1). Thin remaining peat layers additionally influenced by groundwater might only allow the establishment of a fen. Moreover, too thin peat layers in combination with missing waterlogging due to sandy material underneath will create heathland habitats, which are also of ecological interest. Establishment of heathland might be the better solution, since otherwise a complete translocation of peat from one area to another would be required. Thin peat layers underlain by permafrost are also not capable of being restored successfully.

The rewetting process should make sure that the rehabilitated bog is continuously moist but the average water level must not be higher than 30 cm below surface, because otherwise the presence of large water bodies (lakes) cannot be excluded (Fig. 3.39). Large water bodies are generally undesirable, since wave action that can sweep away peat mosses and other desired plants in windy periods must be avoided. For this reason, land-forming procedures are required. The entirely rehabilitated bog should be split into individual polders similar to paddyfields (Fig. 3.40). While forming the surface peat should be pushed upslope, leading to the creation of relatively flat terraces. The re-profiling of a dome-shaped surface, which is typical for untouched bogs, might be difficult to execute. It is worthy of note that

Fig. 3.40 Construction of polders interrupted by dams to prepare bog rehabilitation

natural peatlands are not completely flat. Because they consist of hummocks and hollows the previously compacted surface should be roughed up to implement the complex hummock and hollow system. The microtopography produces a complex network of small open waters instead of larger lagoons, which are not desired. Within the individual polders compacted peat should be predominantly placed in the deepest and wettest part, since there the *Sphagnum* growth and maintenance might be more beneficial (Price et al. 2003; Quinty and Rochefort 2003).

The polders are connected to each other by means of an overflow system. This system discharges the surplus water continuously. The polders are surrounded by dams or bunds constructed from highly decomposed and compacted peat. The dams have a width between 50 and 75 cm and are raised 30 cm above the bog surface. Wide dams are preferred to high dams due to their resistance to pressure from the water body. The dams reduce the wind speed that is responsible for wave action in open water areas and for the spread of applied amendments such as mulch. The polders should be constructed for each 10 cm drop in level of the drain so that the maximum water level across the polder is kept within 10 cm. Consequently, polders showing the generation of lakes are rapidly emptied. The bog area is slightly inclined so that it is possible that the water flows from polder to polder, until it is ultimately collected in an artificial water basin (Fig. 3.41). The sides of the rehabilitated bog should be taken into consideration to minimise water losses. By constructing surrounding berms the creation of a restored peatland can occur while quarrying is in progress in other parts of the large-scale peatland terrain. The berms are required as long as the main ditches are still functioning. The water levels should be controlled by a monitoring system in order to check the effectiveness of the established infrastructure (IPCC 2008).

3.5.2.4 Plant Management

Although all conditions are adequate, *Sphagnum* growth can be limited because of a lack of propagules to be dispersed. To improve the *Sphagnum* development the mosses can be harvested from donor sites often located in the proximity of the

Fig. 3.41 Water basin for final collection of the overflowing water originating from plenty of polders

restoration area. The upper 10 cm with plant fragments are removed with a rotovator, transported to the restoration area and spread with a manure spreader. *Sphagnum* species can be directly harvested and consequently spread in thin layers over the bare peat. The ratio harvest area to spread area should amount to 1:10. In this way the cut-away surface can be successfully inoculated with the original moss vegetation (IPCC 2008; Rochefort et al. 2003; Quinty and Rochefort 2003).

Moreover, plugs with plant fragments consisting of typical bog vegetation such as *Eriophorum angustifolium* (cotton grass) can be taken from donor areas as well. In addition, in particular in relation to the hummocks, some species such as *Calluna vulgaris* and *Erica cinerea* (heather), *Empetrum nigrum* (crowberry), *Vaccinium myrtillus* (bilberry), *Vaccinium vitis-idaea* (cowberry) can be introduced by collecting and spreading ripe seeds. The plant development should be monitored continuously (Anonymous 2010).

With the help of *Spaghnum* spreading and plug relocation it is feasible to re-establish the bog vegetation within a relatively short period of 3–5 years, while the implementation of a stable watertable might take more time (up to one decade) and ultimately a functioning ecosystem that starts accumulating peat might even need much more time. Irrespective of the main object to regenerate the bog, which can be difficult to achieve due to the previous strong bog degradation, or to the climatic deficiencies in central Europe, for example, a minimum target of the restoration process can be the re-establishing of rare species and the creation of new habitats for different kinds of wildlife such as birds (Gorham and Rochefort 2003).

After the plant material has been spread straw mulch (3,000 kg ha^{-1}) is used to protect it from desiccation. The straw mulch protects the diaspores, since the micro-climatic conditions remain wet and temperate (Rochefort et al. 2003). The stabilisation of the vegetated surface can alternatively be achieved by using biodegradable jute matting (Wheeler et al. 1995).

Attention must be paid to the maintenance of the rehabilitated bog. Problems are associated with some disturbing plant species which are also adapted to the soil conditions but which suppress the development of the moor-like vegetation to a great extent. It must be avoided that exotic and native weedy species such as *Rumex acetosella* (sheep sorrel) and *Epilobium angustifolium* (rosebay willow-herb) colonise the area and prevent the re-growth of *Sphagnum*. In particular, *Molinia caerulea* (purple moor-grass) must be permanently eliminated (Wheeler et al. 1995). This can be done by mowing and by sheep grazing using e.g. White Polled Heath (*Ovis aries (ammon) f. aries*).

In addition, grazing animals help to keep the shrub level down. Birches and pines are tree species which are adapted to the specific soil conditions and which grow rapidly, resulting in an increased transpiration of the peatland and an additional throwing of shadow on the area of concern. Both results are detrimental regarding the natural bog vegetation and consequently it appears to be necessary to fell periodically most of the trees and to pull smaller shrubs. Apart from the root system, wood and branches should be completely cleared to prevent additional nutrient accumulation. The shrub and tree control should be done regularly, e.g. annually in the first decade (Quinty and Rochefort 2003). It should be noted that few pines and birches, however, are acceptable due to the advantages for birds, which are able to use the plants in the context of the predator-prey relationship.

Some species usually typical for bog vegetation such as *Eriophorum vaginatum* (cotton grass) can also cause reduced peat growth, since they develop without acrotelm forming, which is required for the *Spaghnum* development. For this reason, the dispersion of *Eriophorum* must be observed carefully. Cotton grass was found to be the plant species that is able to accumulate the highest percentage of nitrogen. Fortunately, this species grows rapidly and dominantly in restored peatlands (Gorham and Rochefort 2003; Silvan 2004).

3.5.2.5 Nutrient Management

Attention should also be paid to the nutrient household. A reason for nutrient input is related to the precipitation. In particular, in areas influenced by air pollution originating from vehicle traffic and industry or from agriculture with intensified animal husbandry nitric oxides (NO_x) and sulphur dioxide (SO_2) accumulate in the air (Meuser 2010). The subsequent acid rain containing H_2SO_4 and HNO_3 can be assessed as beneficial with reference to the pH value but extremely negative with regard to the nutrient contamination. Accordingly, nutrient-enriched precipitation reduces the optimistic calculations relating to the nutrient household of the rehabilitated bog significantly. In contrast, restored fens, which normally indicate higher pH values, are endangered by acid rain, because they are vulnerable to acid deposition (Gorham and Rochefort 2003).

In northern Germany, for example, where large areas are intended for bog restoration, the nitrogen input based upon the rainfall amounted to 27 kg ha^{-1} a^{-1} N on average, while the nitrogen leaching exhibited only 2 kg ha^{-1} a^{-1}. Hence, nitrogen accumulation happens over a long period of time. With regard to the uppermost 20 cm of the peatland nitrogen was available as NH_4-N (3–33 kg ha^{-1}), while the NO_3-N content fell below 5 kg ha^{-1}. Obviously, the nitrification was inhibited due to the high moisture content of the soil. The nitrogen accumulation might be impacted by the low N uptake of the moor-like vegetation. Furthermore, it was possible to observe the same effects with regard to sulphur. The sulphur input stemming from the precipitation indicated 73 kg ha^{-1} a^{-1} and the sulphur leaching was reduced to approximately 10 kg ha^{-1} a^{-1}, causing sulphur accumulation in the peat as well. In contrast, the phosphorus input as well as output showed negligibly low values (unpublished data).

In relation to the side effects of agriculturally used areas the input of fertilizers must be strongly reduced due to the demand for low nutrient concentrations in the bog. Hence, between agriculturally used areas and the bog that is rehabilitated a protecting wood or heathland vegetation (recommended width 30–150 m) must be created. This should consist mainly of coniferous trees such as pines and spruces, which prevent fall of leaves. In addition, the adjacent agriculture should be extensive, e.g. in association with organic farming.

On the other hand, a light phosphorus fertilizing can improve the peat accumulation through stimulation of the vascular plant growth. In Canadian studies it has been found that an application of 150 kg ha^{-1} granulated rock phosphate promotes the growth of some bog mosses (Quinty and Rochefort 2003).

Nevertheless, nutrient accumulation resulting from tourism should be restricted as well. Therefore, in regions located in proximity to urbanised areas many citizens want to visit the generally beautiful countryside at the weekends. The areas are suitable for walking as well as bird and wildlife watching and can serve as educational trails. In principle, overcrowding appears to be harmful to the bog requiring necessary limitations on the visitors. Accordingly, visitor management might be helpful in order to minimise damages to the vegetation and nutrient contamination caused by waste which has been thrown away. For instance, plank roadways (Fig. 3.42) are constructed to achieve this object. Sometimes the designation of fenced-in zones cannot be avoided.

3.5.3 *Wooded Peatland Restoration*

The general prerequisites for enabling a successful restoration are in a poor state after long-term agricultural use of remaining peat layers. Organic material has been mineralised, leading to peat shrinkage and peat loss.

Fig. 3.42 Plank roadway to manage the visitors walking in a restored bog in Roztocze National Park, Poland

Densely wooded peatlands are also restored because they are threatened in relation to both the peatland ground and forest species. Areas which have been drained for forestry in the long term are difficult to restore and the achieved natural state is sometimes different from the original peatland state but nevertheless ecologically precious. The object of the restoration can be the re-establishment of spruce mires, which in Scandinavian countries are often located between drained forest and natural peatland ecosystems. The state of originally pristine spruce mires is easier to realise if the tree cover remains intact. Otherwise, after tree removal, a natural tree stand structure can mostly be achieved, since pioneer tree species colonise the area rapidly, e.g. *Betula pubescens* (birch).

Restored wooded areas can show accelerated nutrient concentrations in the outflow, because the areas have been fertilized prior to restoration and the felling residues were left on the site. For this reason, buffer zones should be integrated which reduce the nutrient loading of adjacent watercourses. The buffer zones, irrespective of cutting procedures of the forest stands, are rewetted sections of the peatland previously drained. They are more capable of efficient nutrient retention than sedimentation ponds. Research projects in Finland considering the throughflow of buffer zones ascertained that the nitrogen retention reached 38–74% and phosphorus

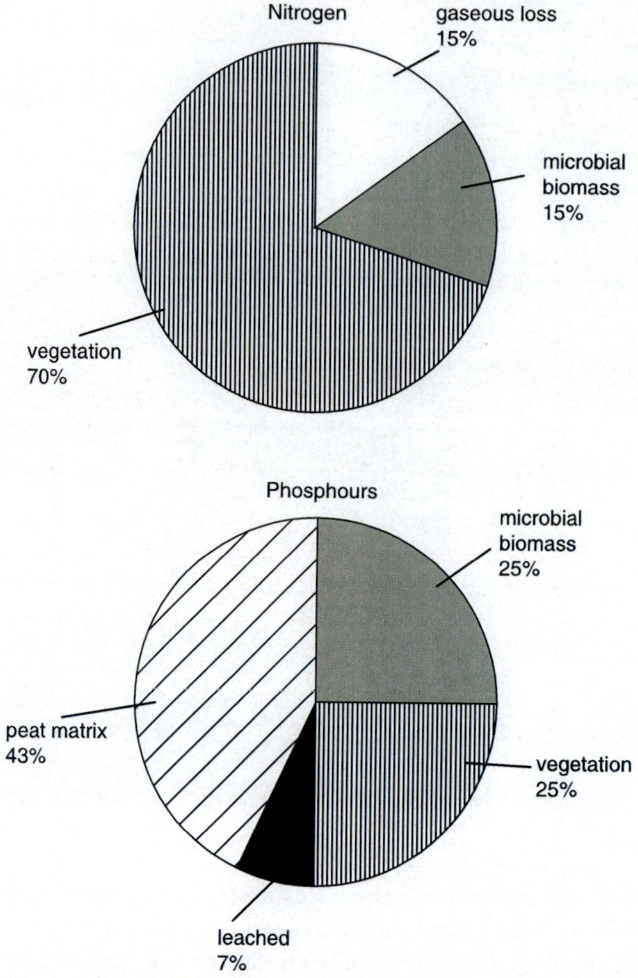

Fig. 3.43 Retention of N and P (%) in a restored peatland buffer during the first year after artificial nutrient fertilizing (Data from Silvan 2004)

retention 37–68%. In Canada, domestic water was filtered in comparable buffer strips up to 93% (N) and 98% (P). In an experimental study in Southern Finland the buffer zone was artificially loaded by nitrogen and phosphorus fertilizers. The added nitrogen and phosphorus were restrained by the rewetted buffer area (Fig. 3.43). In particular for nitrogen, the retention of the vegetation was dominant in the growing season but it must be noted that it is short-term without annual harvesting.

Retention was also associated with the peat matrix itself in the case of phosphorus, while nitrogen can only be restrained by adsorption of NH_4^+ onto the organic matter. Phosphorus retention normally also results from aluminium and iron phosphate formation. Both elements are very common in acid peats. Particularly, iron phosphates such as the blue colored viviante can immobilise the nutrient. However, in anaerobic conditions the redistribution of phosphorus after reduction of iron compounds caused by rewetting must be taken into account (Silvan 2004).

Furthermore, according to nitrogen gaseous losses are of importance as well. The areas under investigation show high N_2O losses caused by microbial de-nitrification, which can contribute to the greenhouse effect. This negative result should be kept in mind. Whether the de-nitrification process will produce mainly N_2O or will react quickly to the harmless dinitrogen (N_2) depends on the rewetting process and might be different from time to time (Silvan 2004).

In tropical regions the landscaping requires particularly accurate handling. Blocking of ditches and canals as well as placement of dams surrounding the area of concern are also important prerequisites. It should be noted that the dams must be installed in a network with the possibility of allowing surplus water to leave the terrain. In deforested areas the runoff is strong, so that the water is drained too quickly in wet seasons. Consequently, surface peat tends to dry and to decompose. Additionally, hummocks must be created to re-plant buttressed trees which are able to grow beneficially on hummocks. The tree growth is particularly important at the margins of a tropical peat dome, because there trees may grow significantly better than in the centre of the dome due to larger water fluctuations and subsequent enhanced peat decomposition in line with higher nutrient availability. Consequently, the established vegetation varies between centre and margins in a way that buttresses and hummocks producing tall trees are more located at the edges, whereas in the dome centre species adapted to the more frequent flooding are concentrated. The dispersion of peatland seedlings should take these circumstances into consideration (Dommain et al. 2010).

In general, cut-away areas in tropical regions are problematical with regard to damage by fire. For this reason, at least a rapid rewetting should take place to prevent further peat degradation. However, rewetting procedures are often endangered, since oil palm and *Acacia* cultivation, which need a lowered water table, are located in the proximity of the peatland to be restored (Dommain et al. 2010).

3.5.4 Fen Restoration

Fen restoration projects occur in lowlands usually influenced by groundwater. The most important factor for restoration purposes is the raising of the groundwater table, which is, for instance, carried out by turning off the pumps or changing of the groundwater flow performance. To compensate for the lowered groundwater table surface water is occasionally used that should be rich in iron in order to bind phosphates. The used water must not be rich in nutrients or natural helophyte filters are involved which serve as buffer zones for the inflowing water (Lamers et al. 2002). On the other hand, too excessive input of (acid) rainwater might also be negative due to increased growth of *Sphagnum* mosses, which are not typical vegetation for restored fens. In principle, acidification can play an adverse role with reference to fen restoration. Infiltration of base-rich water by flooding or carrying out liming are helpful as long as the generation of high nutrient concentrations can be avoided.

During summertime permanent water flooding, however, is detrimental with regard to the aerobic conditions in the surface layers of fen meadows, which are required to supply predominantly the growing vegetation with NO_3^- instead of NH_4^+, because nitrate seems to be the preferred nitrogen source of the fen vegetation. In conclusion, the water regime should allow fluctuating water tables, with a tendency towards lower water tables in summer and higher ones in winter (Lamers et al. 2002).

There are controversial discussions about the simultaneously occurring re-establishment of the vegetation in damaged fens. Because most fens to be restored are influenced by groundwater with mineral enrichment and high buffering capacity against acidification, an abundance of high biomass-producing species such as monotonous reed plants (e.g. *Phragmites australis*) or willows (*Salix*) appears to be the result of restoration. The latter are less valuable from the perspective of botanists but probably very important for bird populations. As reported for restored fens in the catchment of Utrecht in the Netherlands, the nesting facilities improved. Nevertheless, to leave the area to natural succession will not foster the fen-adapted plant communities. In the central European lowlands, for instance, tall graminoid species tend to grow intensively, suppressing the fen-like vegetation. To preserve the desired vegetation control and mowing appear to be required (Wheeler et al. 1995; Lamers et al. 2002).

In restored fen landscapes the presence of some lakes is common. Lake creation leads to the establishment of marsh vegetation and is beneficial for wetland animals, fish and waterfowl. The lake bottoms should retain at least 30–60 cm of peaty material to support the invasion of aquatic plants (see Sect. 7.3.1). Fishing would require deeper lakes, so that the peat must be excavated as deeply as possible. Other utilisations such as swimming and water sports are not normally recommended in the restored terrains.

With regard to the lake creation in fen areas for which there are restoration plans, natural terrestrialisation should take place. This is possible, if enough diaspores of plants capable of terrestrialisation (e.g. *Calla palustris, Stratiotes aloides*) are present. Thus, the re-introduction of these diaspores should be taken into consideration. Moreover, to establish rare species re-introduction of seeds by man is necessary.

In addition to atmospheric deposition, the presence of arable land in the proximity of the restored fens may contribute to nutrient accumulation caused by runoff. In countries where intensively used agricultural land is located close to fens intended for restoration like the Netherlands it is difficult to restore fen areas. In particular, raising the groundwater table is undesired in the surrounding arable land. The nutrient input based on surface runoff and a detrimental groundwater quality should be minimised to prevent eutrophication of the fen area. In particular, in restoration areas with surface water bodies (see Sect. 7.3.1), the eutrophication must be avoided, e.g. by establishment of submerged vegetation, which provides habitats for zooplankton, reduces the re-suspension of peaty sediments caused by wind action and competes with algae – all processes which make a clearwater quality possible. Manipulation of the fish stocks in association with only moderate phosphate accumulation is another adequate measure. In this context, sufficient re-migration of fish using the interconnecting ditches might be the major problem. If lakes within the restored terrain show high phosphate loading in the sediment, sometimes dredging is required in order to prevent eutrophication. The addition of iron to bind soluble phosphate might be an alternative solution, whilst calcium addition creates the negative effect of peat mineralisation. In restored semi-terrestrial fens which are dominated by tall gramnoids due to eutrophication the cutting of the aboveground sods (depth 5–10 cm) is an option to create the desired diaspore bank (Lamers et al. 2002).

References

Abakumov, E. V. (2000). *Soils of recultivated quarries of North-west of Russia.* 1st international conference on soils of urban, industrial, traffic and mining areas. Proceedings, Vol. III, pp. 1821–825, Essen.

AG 2006 – Australian Government. (2006). *Mine development program for mining industry.* Commonwealth of Australia. Rehabilitation. Leading Practice Sustainable.

Andriesse, J. P. (1988). *Nature and management of tropical peat soils.* FAO – Food and Agriculture Organization of the United Nations. http://www.fao.org/docrep/x5872e/x5872e07.htm. Accessed 15 Feb 2012.

Anonymous. (2010). *Guidelines for peatland restoration.* Peatlands and Uplands Biodiversity Delivery Group. http://www.qpani.org/documents/PeatlandRestorationguidelinesfinal.pdf. Accessed 18 Feb 2012.

Boles, P. (1983). *Shrub cover and density on Western rangelands in relation to reclamation success standards for surface mined lands in Wyoming.* Lexington: OES Publications, University of Kentucky.

Bord na Mona. (2001). Raised bog information leaflet. http://www.bnm.ie/files/20061124040538raisedbogs.pdf. Accessed 3 Mar 2012.

Brady, N. C., & Weil, R. R. (2008). *The nature and properties of soils.* Upper Saddle River: Pearson Education.

BRGM – DG Environment. (2001). *Management of mining, quarrying and ore-processing waste in the European Union 2001.* European Commission, BRGM/RP-50319-FR.

Carter, T. G., Mackasey, W. O., & Steed, C. M. (1995, June). *Coordinated approach to remediation of abandoned mine hazards.* Proceedings conference on mining and the environment, Sudbury, Ontario, Canada.

Chiasson, L., & Chiasson, G. (2000). Cranberry culture as a reclamation option. In L. Rochefort & J.-Y. Daigle (Eds.), *Proceedings of the 11th international peat congress, 2000, Quebec, Canada.* Shippagan: International Peat Society.

Cooke, J. A., & Johnson, M. S. (2002). Ecological restoration of land with particular reference to the mining of metals and industrial minerals: A review of theory and practice. *Environmental Reviews, 10,* 41–71.

Dommain, R., Couwenberg, J., & Joosten, H. (2010). Hydrological self-regulation of domed peatlands in south-east Asia and consequences for conservation and restoration. *Mires and Peat, 6,* 1–17.

Drecker, P., Genske, D. D., Heinrich, K., & Noll, H. P. (1995, October). Subsidence and wetland development in the Ruhr district of Germany. In *Land Subsidence, Proceedings 5th International Symposium on Land Subsidence*, The Hague, the Netherlands.

Durant, W. (2011). *Land subsidence in Western Pennsylvania.* The Pittsburgh Geological Society. www.pittsburghgeologicalsociety.org/subsidence. Accessed 1 Dec 2011.

EPA – Environmental Protection Agency (of the United States). (2008). *Carlson's trophic state index. Aquatic Biodiversity*. www.epa.gov/bioindicators/aquatic/carlson.html. Accessed 15 Oct 2011.

Finell, M., (2003). *The use of reed canary grass (Phalaris arundinaceae) as a short fibre rawmaterial for the pulp and paper industry*. Dissertation Acta Universitatis Agriculturae Sueciae, Uppsala, Sweden.

GOA – Government of Australia. (1995a). *Revegetation methods*. Guideline for the revegetation of mined areas. www.derm.qld.gov.au/register/p01206ba.pdf. Accessed 19 Mar 2012.

GOA – Government of Australia. (1995b). Assessment *and management of saline/sodic wastes*. www.derm.qld.gov.au/register/p01206an.pdf. Accessed 30 Mar 2012.

Gorham, E., & Rochefort, L. (2003). Peatland restoration: A brief assessment with special reference to *Sphagnum* bogs. *Wetlands Ecology and Management, 11*, 109–119.

Grandt, A. F. (1988). Productivity of reclaimed lands – Cropland. In *Reclamation of surface mined lands* (Vol. I). Boca Raton: CRC Press Inc.

Grzesik, K., & Mikolajczak, J. (2009). *The extractive waste management from hard coal mines in Poland*. Institute of Environmental Protection and Management, AGH-UST University of Science and Technology, Krakow.

Heiden, S., Erb, R., Liebmann, H., & Kahle, K. (2001). *Potash mining – Environmental impact and chance*. Berlin: Erich Schmidt (in German).

Hersche, B., & Wenker, O. (2000). Case report: Lassing mining accident. *The Internet Journal of Rescue and Disaster Medicine, 1*(2). www.ispub.com/journal/the-internet-journal-of-rescue-and-disaster-medicine/volume-2-number-1/case-report-lassing-mining-accident.html. Accessed 31 Oct 2012.

Hörnsten, L., (1992). *Treatments of peat bogs harvested by deep digging technique*. Vattenfall Research, Vällingby, Sweden. (in Swedish)

IPCC – Irish Peatland Conservation Council. (2008). *Restoration of drained peatlands*. http://www.ipcc.ie/diymandrained.html. Accessed 1 Feb 2012.

Jedrczak, A. (1992). *Chemical composition of water of anthropogenic lake district in the Luk Muzakowski*. Zielona Gora, WSI, 54, 144 p. (in Polish)

Jochimsen, M. E. (2001). Vegetation development and species assemblages in a long-term reclamation project on mine spoil. *Ecological Engineering, 17*, 187–198.

Johnson, M. S., Cooke, J. A., & Stevenson, J. K. (1994). Revegetation of metalliferous wastes and land after metal mining. In R. E. Hester & R. M. Harison (Eds.), *Mining and its environmental impact* (Issues environmental science and technology). Letchworth: Royal Society of Chemistry.

Kahl, L., Kluge, R., & Thomas, S. (2000). Evaluation of an attempt to cultivate shrubs and trees on a heap of a potash mine. *Landscape and Urban Planning, 51*, 109–112.

Khan, M. J., & Jones, D. L. (2008). Chemical and organic immobilization treatments for reducing phytoavailability of heavy metals in copper-mine tailings. *Journal of Plant Nutrient and Soil Science, 171*, 908–916.

Klapper, H. (2003). Technologies for lake restoration. *Journal of Limnology, 62*, 73–90.

Komnitsas, K., Paspaliaris, I., Zilberchmidt, M., & Groudev, S. (2001). Environmental impacts of coal waste disposal sites – Efficiency of desulfurization technologies. *Global Nest: The International Journal, 3*, 109–116.

Lamers, L. P. M., Smolders, A. J. P., & Roelofs, J. G. M. (2002). The restoration of fens in the Netherlands. *Hydrobiologia, 478*, 107–130.

Leinonen, A., Lindh, T., Paappanen, T., Kallio, E., Flyktman, M., Hakkarainen, J., Käyhkö, V., Peronius, P., Puuronen, M., & Mikkonen, T., (1998). Cultivation and production of reed canary grass for mixed fuel as a method for reclamation of a peat production area. In: T. Malterer, K. Johnson, & J. Stewart (Eds.), *Peatland restoration and reclamation-techniques and regulatory considerations*. Proceedings international peat symposium, Duluth, Minnesota. Jyväsklyä: International Peat Society.

Lucha, P., Cardona, F., Gutierrez, F., & Guerrero, J. (2008). Natural and human-induced dissolution and subsidence processes in the salt outcrop of Cardona Diapir (NE Spain). *Environmental Geology, 53*, 1023–1035.

Lücke, M. (1997). *Investigations of the rehabilitation of potash mining waste heaps*. Dissertation University of Kassel-Witzenhausen. (in German)

Luscar Ltd, Cardinal River Coals Ltd, Pisces Environmental Consulting Services Ltd, & Bighorn Environmental Design Ltd. (1994). *Development of sports fisheries in lakes created by coal mining operations in the Eastern slopes*. Alberta: Edmonton.

Makineci, E., Gungor, B. S., & Kumbasli, M. (2011). Natural plant revegetation on reclaimed coal mine landscapes in Agacli-Istanbul. *African Journal of Biotechnology, 10*, 3248–3259.

Makowsky, L, Nix, K. Meuser, H., Yan, J., & Xu, L. (2010, September). Towards a sustainable use of cover soil for agricultural reclamation of coal mining waste and fly ash deposits. In *Proceedings International Workshop on Diffuse Pollution-Management Measures and Control Technique*, Huainan, China.

McCarthy, D. F. (2007). *Essentials of soil mechanics and foundations*. Upper Saddle River: Pearson.

Meuser, H. (2010). *Contaminated urban soils*. Dordrecht: Springer.

Monaghan, M. M. (2007). *Design, signage and maintenance guidelines: Waterfront trail*. www.waterfronttrail.org/library-publications.html. Accessed 10 Jan 2012.

Norman, D. K., Wampler, P. J., Throop, A. H., Schnitzer, E. F., & Roloff, J. M. (1997). *Best management practices for reclaiming surface mines in Washington and Oregon*. Open file report Washington Division of Geology and Earth Resources and Oregon Department of Geology and Mineral Industries.

Pflug, W. (Ed.). (1998). *Lignite coal opencast mines and rehabilitation*. Berlin: Springer (in German).

Price, J. S., Heathwaite, A. L., & Baird, A. (2003). Hydrological processes in abandoned and restored peatlands: An overview of management approaches. *Wetlands Ecology and Management, 11*, 65–83.

Price, J. S., & Whitehead, G. S. (2001). Developing hydrological thresholds for *Sphagnum* recolonization on an abandoned cutover bog. *Wetlands, 21*, 32–40.

Quinty, F., & Rochefort, L. (2003). Peatland restoration guide. Canadian Sphagnum Peat Moss Association and New Brunswick Department of Natural Resources and Energy. Quebec, Canada. http://www.peatmoss.com/pdf/Englishbook.pdf. Accessed 28 Jan 2012.

Resulovic, H. (2000). *Specific processes in minesoils in postrecultivated period* . Proceedings Vol. 3. Paper presented at the 1st international conference on soils of urban, industrial, traffic and mining areas, pp. 993–997, Essen, Germany.

Rochefort, L., Quinty, F., Campeau, S., Johnson, K., & Malterer, T. (2003). North American approach to the restoration of *Sphagnum* dominated peatlands. *Wetland Ecology Management, 11*, 3–20.

Ruzek, L., Vorisek, K., Vrablikova, J., Strnadova, S., & Vrablik, P. (2003). Chemical and biological characteristics of reclaimed soils in the most region (Czech Republic). *Plant Soil Environment, 49*, 346–351.

Salazar, M., Bosch, A. D., & Poch, R. M. (2000). *Restoration of soils of coal mining areas using pig slurry in semiarid Mediterranean environments*. 1st international conference on soils of urban, industrial, traffic and mining areas. Proceedings, Vol. III, pp. 1001–1006, Essen.

Schaaf, W., Gast, M., Wilden, R., & Hüttl, R. F. (2000). *Development of element cycles at post-mining sites*. 1st international conference on soils of urban, industrial, traffic and mining areas. Proceedings, Vol. III, pp. 1015–1021, Essen.

Schraps, W. G., Schneider, S., & Baumgarten, H. (2000). *Soil investigations for reclamation of spoil heaps*. Proceedings Vol. 3. Paper presented at the 1st international conference on soils of urban, industrial, traffic and mining areas, Essen, Germany.

Schulz, D. (1996). Recultivation of mining waste dumps in the Ruhr area, Germany. *Water, Air, and Soil Pollution, 91*, 89–98.

Scullion, J. (2000). *Long-term studies of soil improvement after opencast coal mining*. 1st international conference on soils of urban, industrial, traffic and mining areas. Proceedings, Vol. III, pp. 1059–1064, Essen.

Silvan, N. (2004). *Nutrient retention in a restored peatland buffer*. Academic dissertation, Faculty of Agriculture and Forestry, University of Helsinki, Finland.

Sokolovska, M., Bech, J., & Delkov, A. (2000). *Forest vegetation peculiarities of waste banks from underground coal production*. Proceedings Vol. 3. Paper presented at the 1st international conference on soils of urban, industrial, traffic and mining areas, Essen, Germany.

Soro, A., Sundberg, S., & Rydin, H. (1999). Species diversity, niche metrics and species associations in harvested and undisturbed bogs. *Journal of Vegetation Sciences, 10*, 549–560.

Svoboda, I. (1993). Land restoration in the North Czech Brown Coal Basin: The new challenges and opportunities. *International Journal of Environmental Issues in Minerals and Energy Industry, 118*, 151–154.

Thornburg, A. A. (1982). *Plant material for use on surface-mined lands in arid and semi-arid regions* (SCS-TP-157). Washington, DC: U.S. Department of Agriculture

UFZ – Centre for Environmental Research. (1999). *Remediation of abandoned surface coal mining sites* (Report No. 11). Leipzig-Halle: UFZ – Centre for Environmental Research.

Vasander, H., Tuittila, E. S., Lode, E., Lundin, L., Ilomets, M., Sallantaus, T., Heikkilä, R., Pitkänen, M.-L., & Laine, J. (2003). Status and restoration of peatlands in northern Europe. *Wetlands Ecology and Management, 11*, 51–63.

Wedepohl, K. H. (1995). The composition of the continental crust. *Geochimica et Cosmochimica Acta, 59*, 1217–1232.

Wheeler, B. D., Shaw, S. C., Fojt, W. J., & Robertson, R. A. (Eds.). (1995). *Restoration of temperate wetlands*. Chichester: Wiley.

Chapter 4
Treatment of Contaminated Land

Abstract Apart from a general overview of soil remediation measures, this chapter focuses on the site and soil management requirements. The site management involves the zoning strategy in contaminated land as well as the placement of different soil materials which are excavated and used. Processes such as dewatering and technical devices for excavation procedures are introduced in detail. Special attention is paid to the opportunities to reuse excavated materials. Transportation, stockpiling and backfilling are further important aspects which are mentioned. Since most of the soil remediation projects must include site clearance operations, in particular the demolition of buildings and underground installations, this chapter gives information about the dismantling of buildings, selective deconstruction and material separation. In this context, exposure to hazardous materials such as asbestos and contaminated building debris is discussed. Regarding hazardous waste, safety measures to protect workers and engineers who come into contact with the materials during the site clearance are also presented. The content of this chapter is a decisive prerequisite for understanding remediation approaches related to soil containment and soil decontamination.

Keywords Building demolition • Excavation • Hazardous material • Material reuse • Soil management • Technical devices • Working safety

4.1 Overview of Soil Remediation Measures

4.1.1 Containment and Decontamination Approaches

The soil remediation measures can be distinguished according to the location where the treatment occurs:

- *In situ* means an application without any excavation and transportation of contaminated material. In some examples (e.g. barriers installation, encapsulation)

H. Meuser, *Soil Remediation and Rehabilitation: Treatment of Contaminated and Disturbed Land*, Environmental Pollution 23, DOI 10.1007/978-94-007-5751-6_4,

Table 4.1 Site-related soil remediation measures

Treatment	*in situ*	on site	off site
Excavation followed by landfill disposal or by recycling	–	–	+
Surface cover (geotextile-based and bentonite-based)	+	–	–
Sealing	+	–	–
Barriers installation at the sides	+	–	–
Encapsulation	+	–	–
Solidification (cement-based)	+	+	(+)
Asphalt batching	–	+	(+)
Vitrification	+	(+)	(+)
Stabilisation	+	–	–
Soil washing/soil extraction	(+)	+	+
Bioremediation (biopile, bioreactor, *in situ*, landfarming)	+	+	+
Phytoremediation	+	–	–
Thermal treatment	(+)	–	+
Electro-remediation	+	–	(+)

+=Usual application
(+)=Exceptional application
–=Unusual application/application excluded

only a very small portion of the contaminated soil needs to be excavated in order to enable the remediation technique

- On site measures are combined with previous excavation and transportation over short distances within the defined contaminated land. The equipment is usually transported and constructed near the contaminated site
- In contrast, off site measures occur in specialised decontamination facilities that are mostly not located in the vicinity of the contaminated site. Accordingly, transport over long distances must be included.

Another approach describes *in situ* as a measure in unexcavated and relatively undisturbed soil and *ex situ* as an application which includes excavation of soil or extraction of groundwater. Accordingly, on site techniques can take place on the contaminated site in both applications *in situ* and *ex situ* (Nathanail and Bardos 2004).

Remediation can be defined as decontamination that destroys, removes or detoxifies the contaminated material and as a containment-based approach that leaves the contaminants in place, but attempts to prevent migration.

In Table 4.1 the soil remediation measures are listed. In the case of the containment approaches soil cover, side barrier installation, encapsulation as well as stabilisation the contaminated soil remains untouched to a large extent. Thus, these strategies are termed *in situ* measures. Some decontamination treatments, however, can also be assigned to the *in situ* methods, namely phytoremediation and electro-remediation. The most important decontamination measures soil washing, bioremediation and thermal treatment reveal all kinds of relations to the site, but soil washing applications on site and incineration techniques *in situ,* apart from hot steam injections, are rarely applied. Finally, off site application of solidification is also carried out as an exception.

It should be noted that in the past in a number of case studies the containment approaches were preferred to decontamination measures due to the insufficient development and standardisation of decontamination techniques. Up to the middle of the first decade of the twentieth century, despite the decontamination opportunities, containment appeared to remain the most widely used technique in contaminated land management in terms of the number of sites. The main reasons are that an environmentally cost-effective treatment-based remedy is not available and the containment may generally provide a comparably rapid risk management (Nathanail and Bardos 2004).

4.1.2 Immediate Response Action

If the public authority recognizes an immediate danger to human health, the response action must be initiated without any time delay. Apart from informing the affected residents immediately and implementing keep off-rules (fencing-in, information boards) different quick response actions can be taken into account:

- Evacuation from the contaminated sites (people evacuation, closing of industrial or construction sites)
- Covering of the soil surface, leading to prevented ingestion, wind erosion, and water infiltration
- Regulations prohibiting activities such as gardening, vegetable consumption, and groundwater extraction
- Direct intervention in the water management including collection of percolating water (drainage) and the closing of waterworks in charge of the public drinking water supply.

Immediate response actions might cause an attraction of interest at short notice. In particular, the media will possibly react and make sure that there are detailed reports about the contaminated area of concern in the form of newspaper articles, television and radio reports as well as internet publications. The public interest will ensure that there is a more careful and scientifically adequate handling of the contaminated land.

On the other hand, the public interest that is attracted may also lead to different reactions of the affected inhabitants. Sometimes, the overreaction of some people living in the contaminated area causes the public authority to order and implement inadequate and disproportionate measures. The authority should react appropriately to the contamination situation and it ought not to fall for the line of some alarmists or of people who are ignorant about soil pollution. The problem is also associated with long-term remediation approaches.

4.2 Site and Soil Management

4.2.1 *Site Management*

In the course of the remediation the differentiation of contaminated and uncontaminated soil is of immense interest. For this reason, a zoning appears to be sensible, preventing mixing of substrates that must be kept apart (Fig. 4.1). Usually, a black zone is defined in which the contaminated buildings and the soil that must be treated are located and a white zone that is generally not touched by contaminated materials. Between both borders fences are put up and these are interrupted by some gates consisting of decontamination facilities like tyre-wash plants for trucks (Figs. 4.2 and 4.3) and boot-wash plants for the workers and engineers. Trucks going from the black zone into the white zone must be decontaminated. The stringent differentiation on site can be more visible, if trucks, excavators and bulldozers are flagged white (clean) and black (contaminated).

Sensitive installations such as office containers, the first aid ward and the canteen must in any case be sited in the white zone. Since the soil excavation and handling progresses continuously, the border fences must be put up temporarily, requiring exact soil management at the construction site. Permanent fences are only installed for the bordering of the complete area of concern. After passing a checkpoint trucks are allowed to leave the area. It should bc noted that in spite of integrated tyre-wash plants it is mostly necessary to provide for continuous street-cleaning outside of the fenced-in area depending on the frequency of truck journeys.

With reference to the realization of soil clean-up in the first instance a construction site setup plan is developed as generally occurs in the context of construction processes. This plan should take into account all site-specific operations and handlings provided. In detail, the following areas and devices are differentiated:

- Excavated areas, divided into sites exhibiting excavated topsoil and excavated subsoil
- Interim storage facilities, divided into uncontaminated topsoil heaps, uncontaminated subsoil heaps and distinct types of polluted soil heaps
- Storage facilities for demolition and construction materials
- On site clean-up facilities
- Temporary construction roads built during dry weather conditions and consisting of construction debris levelled above geotextiles.

4.2.2 *Excavation and Stockpiling*

Up to now excavation involving removal from the contaminated site, transportation and reuse or treatment has been the most common remediation measure and has subsequently been classified as the standard remediation approach. Nearly all remediation processes need some form of excavation and material handling apart from the in-situ treatment approaches.

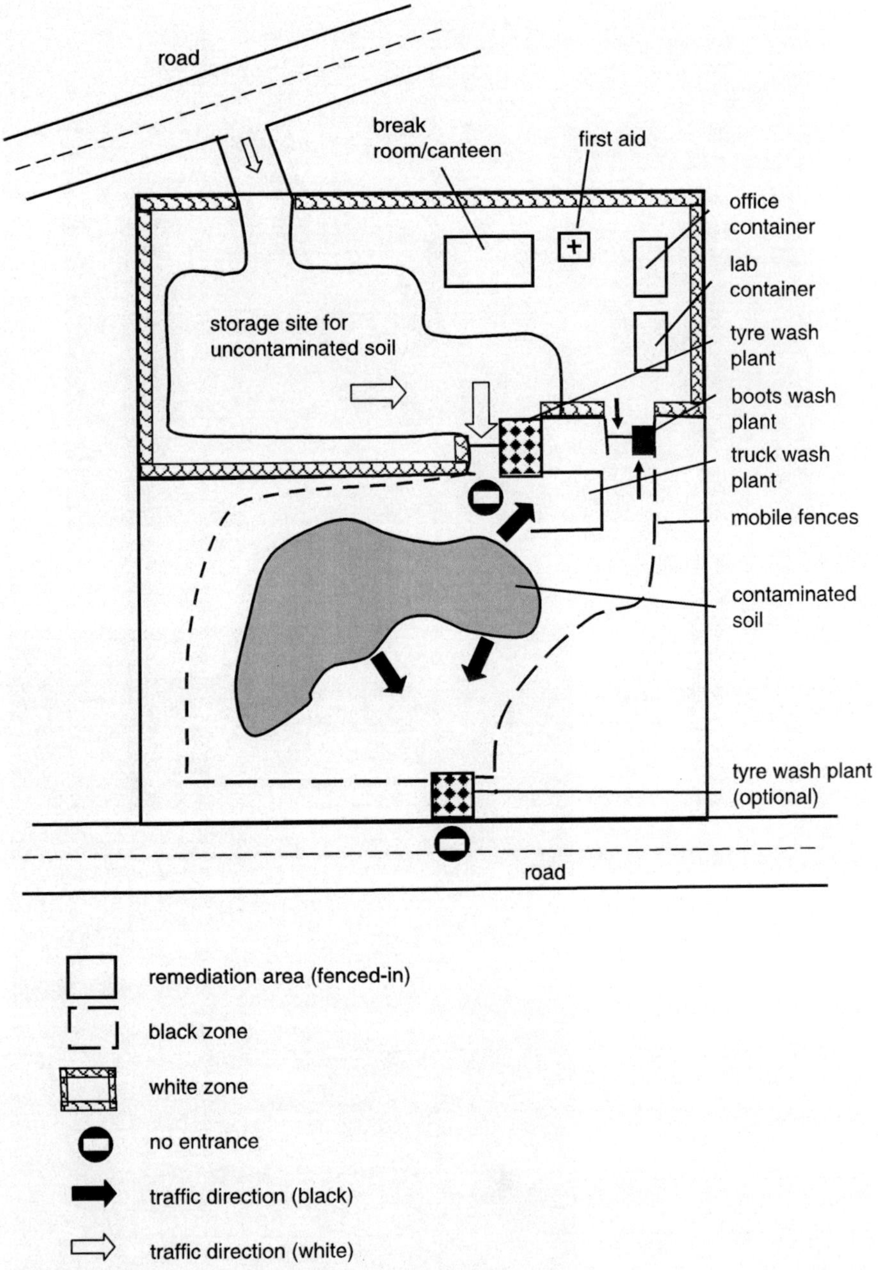

Fig. 4.1 Principle of the zoning of a remediation site

Fig. 4.2 Tyre-wash plant

Fig. 4.3 Truck-wash plant operated by a worker wearing protective clothes and breather

Besides the complete excavation to the maximum contaminated depth and replacement by clean soil, there are further reasons for excavation. For instance, the excavation only becomes necessary to lower the surface in order to superimpose an optimum cover system. Excavation can also be focused on the removal of so-called hot spots, meaning selected highly contaminated areas. Ultimately, excavation is often the prelude for other decontamination techniques.

The costs of excavation do not include some necessary operations such as transportation, pre-treatment, fees and regulatory restrictions and subsequently these costs must be included in the entire calculation. Furthermore, some excavation processes require addition of water to suppress dust development, leading to additional costs for periodical irrigation. As long as the soil is exposed and disturbed contaminants become airborne and subsequently problematical. On the other hand, some examples dealing with contaminated soil below the groundwater table demand dewatering procedures after excavation or dredging and this also incurs costs. Section 6.1 provides information about measures relating to the pre-treatment after the excavation process has been carried out.

During the excavation dewatering must take place, if the groundwater is in contact with the contaminated soil or if groundwater moves into the excavated area in spite of side walls which have been constructed. To enable continuous excavation and to prevent damage to the structures adjacent to the excavation groundwater must be pumped off. Moreover, rainwater can cause problems to the excavation process particularly after stormwater occurrences. Wet soil (saturated conditions) or ponds in the open excavation holes may increase the weight to be transported. Thus, sheeting with waterproofed material captures the precipitation and the water afterwards pumped off does not need to be treated, because there was no contact between the contaminated soil below and the captured water. In addition, damming that surrounds the entire area of concern helps to solve the problem. In contrast, the construction of ditches makes less sense, since the ditches become contaminated due to the uncontrolled runoff.

In association with contaminated sediment removal requiring dredging the contamination of the adjacent water body is designated as a problematical result, since suspension of the sediment can occur after the sediment has been disturbed (see Sect. 7.2.1). The implementation of temporary side barriers might counteract difficulties.

Not only do contaminated sediments to be excavated reduce the practicability of excavation, but also boundaries are met in terrestrial soils. Constraints are possible in association with neighbouring highly sensitive land-use types that must not be contaminated during the soil handling.

Excavation procedures become generally more problematical with increasing depth of the contaminated soil. Some situations make excavation processes difficult and must be considered as limiting factors, e.g.:

- Mining shafts are identified
- Absent access routes or working platforms prevent the soil excavation from using heavy technical equipment
- Massive foundations are present in the soil
- Contaminants have already migrated into the natural bedrock.

Before starting the remediation action at the contaminated land (abandoned, derelict land) bulky refuse and waste visible at the soil surface should be removed. This material may be the source of polluted hot spots, but it is easy to collect. Apart from residues of the ruined buildings, problematical waste constituents are important such as wires, tubes, bitumen, asphalt, asbestos, lacquers and paints. The bulky refuse frequently discovered may include furniture, rusty vehicles, refrigerators, electronic equipment, etc. Collection and treatment should be carried out selectively (see Sect. 4.3).

Moreover, information about the topography based on a topography survey should be available. The existing contours should be checked to make sure that the subsequently excavated depths coincide with the depths of the contaminated soil, in particular in the case of hot spots to be removed. Survey monuments, markers and control points can help to provide orientation during the excavation procedure.

During excavation the methods used must ensure that the walls of the excavation do not slump into the holes or trenches. Physical support is already needed when working in a depth of > 1.2 m (Nathanail and Bardos 2004). With regard to the protection of workers and engineers, the protection of adjacent structures such as buildings and basements as well as the prevention of groundwater infiltration to the areas which are excavated, the performance of the excavation must not endanger the stability of the walls (see also Sect. 4.4). For this reason, sheeting piles (e.g. steel sheets) that are usually driven into the ground at construction sites using a vibratory pile driver must be installed to stabilise the walls and to impede water infiltration. The technique can be applied to depths as great as 75 m. Excavated slopes that are common with reference to large-scale excavation procedures should be at a 1.5:1 ratio (horizontal to vertical) to prevent abrupt erosion. In areas with limited space slopes can be protected with shoring and shielding systems combined with dewatering conducted by sheets at the slope surface. Moreover, the freeze wall technology as explained in Sect. 5.2.1 can also be an adequate solution for temporary wall stabilisation.

Afterwards, excavation voids remain which must be filled back with clean material. Hence, the calculated mass to be excavated mostly complies with the clean material required. However, the backfilled material decreases in cases where subsequent construction design does not necessitate soil. Examples are basements, cellars, underground car parks and subway tunnels.

The most frequent substrates of contaminated land are:

- The contaminated humic topsoil
- The contaminated subsoil
- Construction rubble originated from foundations and other underground facilities
- Mixtures of both soil and constructions debris.

The soil movement machinery which is usually used comprises crawler excavator (Fig. 4.4), wheel excavator (Fig.4.5), wheel loader (Fig. 4.6), crawler-type loader and bulldozer (Fig. 4.7). Technical data of the excavators are listed in Table 4.2. These data refer to the equipment normally used in soil remediation but some

Fig. 4.4 Crawler excavator

Fig. 4.5 Wheel excavator

Fig. 4.6 Wheel loader that loads up a screening machine

Fig. 4.7 Bulldozer

Table 4.2 Technical data (ranges) for soil movement applied to soil remediation projects on site (Data from different vendors)

	Weight (t)	Horsepower	Speed ($km\ h^{-1}$)	Content (m^3)
Crawler excavator	14–35	100–200	0–11	0.2–2.5 (bucket)
Wheel loader	5–17	60–270	0–40	0.8–8.5 (bucket)
Crawler-type loader	11–23	100–170	0–10	1.2–2.4 (bucket)
Bulldozer	17–32	160–250	0–11	3.2–7.2 (blade)

specialist companies need larger machines such as companies dealing with quarries, mines and landfill sites. It should be noted that the access to the site plays a major role in selection of the equipment. Too large equipment is impractical in the presence of buildings, utilities and vegetation (trees). Another aspect is related to the weight of the machine used, which can also be of importance with reference to ground collapse. In general, the technical equipment requires supplied air and other special techniques where the material to be excavated is highly contaminated, explosive or otherwise dangerous (McCarthy 2007).

The crawler excavator is equipped with a bucket in the front for excavating soil and then loading it onto transport vehicles or discharging it on a stockyard. Thus, the excavating machinery is usually applied for the excavating operations of ditches and trenches. While there is a preference for using crawler excavators at loamy to clayey sites, wheel excavators are suitable for more sandy and gravelly soils.

A wheel loader has a front mounted square wide bucket connected to the end of two booms (arms) to scoop up loose material from the ground and move it from one place to another without pushing the material across the ground. A loader is commonly used to move a stockpiled material from ground level and deposit it into a truck or into an open trench. The crawler-type loader functions in a similar way to the wheel loader but it moves on tracks (tracked machines), which are required at sites with loamy to clayey soils and in moist conditions.

A bulldozer is a tracked machine equipped with a substantial metal plate (blade) used to push large quantities of soil. Bulldozers are mainly used to perform excavation and to push the soil to areas where it is spread. Alternatively, scrapers and graders are used to keep the soil uniformly spread. Unfortunately, the term bulldozer is often used to mean any heavy equipment (loaders, excavators), but, to be precise, the term refers only to a machine fitted with a dozer blade.

Moreover, the backhoe excavator consists of a shovel or bucket attachment at the front (front loader) for moving the material. In addition, on the back side of the machine there is an articulating arm with a small backhoe at the end serving as a digging tool. The seat of a backhoe excavator can be moved so that the operator can use the controls for both attachments as necessary. Full-sized backhoe excavator machines are designed for the larger tasks, where tons of material need to be moved to make deep and extensive holes and trenches.

Sometimes cable excavators are used as an alternative. The cable excavator carries different attachments such as a hoe and a shovel. The cable excavator uses its arm to lower the bucket and extract soil from the earth. What makes it different is that it operates with a series of cables or wire ropes that are pulled and hoisted in the direction where it is moved.

In some cases which necessarily involve tree protection on site the usual machinery cannot be applied. If the environmental protection agency does not accept cutting down old trees such as oaks and beeches which have grown on the contaminated site, a specialised technique termed vacuum excavator technique can help to solve this problem. The technique allows the removal of contaminated soil while protecting the tree at the same time (Figs. 4.8 and 4.9). The soil is withdrawn by suction in the vicinity of the trunks, while the roots are damaged to a less extent, thus guaranteeing

Fig. 4.8 Vacuum excavator in use

survival of the trees. The technique, also available for subsoils containing infrastructure pipes, is more effective in gravelly and sandy soils and generally more effective in dry soils than in moist soils. The skeleton content should not exceed a grain size of 25 cm. Afterwards, uncontaminated soil will be backfilled.

This cost-intensive and relatively long-term procedure can be the only way to combine the removal of contaminated soil and the protection of valuable trees, and it is consequently rarely applied in highly sensitive areas such as kindergartens and playgrounds which are located in, for instance, ancient parklands.

After excavation immediate transportation does not always occur. Temporary storage must therefore take place. Soil containing hazardous waste with extremely high contaminant concentration (e.g. values >z2 based on the German regulations, see Table 4.5 in Sect. 4.2.4) should be staged in a sealed container or sealed drum. Non-hazardous soil indicating elevated concentration (e.g. values from z1 to z2 based on the German regulations, see Table 4.5) must be stored on an impervious surface (e.g. concrete, asphalt, plastic sheets) and must be covered with plastic sheeting. Volatilisation, runoff, leaching as well as dust emission must be entirely excluded and the location for stockpiling should not be accessible to the public.

Stockpiling should generally be based upon a strict segregation of the contaminated masses according to the guidance of the public authority. The segregation improves the reuse opportunities and minimises the need for decontamination.

Fig. 4.9 Surviving tree roots after the vacuum excavator has removed the soil

Stockpiling for the uncontaminated material which has been delivered may also become necessary in some cases. Long-term piling lasting more than a few months requires additional rules:

- The storage facility must not be located at sites where there are symptoms of stagnant water
- The gradient of the piles should exceed 4%, trapeze is the preferred shape and the piles are surrounded by drainage ditches, if necessary
- The maximum height is 2 m (topsoil) and 4 m (subsoil) respectively and an immediate cover with plants and grass is recommended to avoid erosion
- The deposit should occur in dry weather periods.

Theoretically, mixing of piled contaminated soil with uncontaminated soil (e.g. sand) would reduce the contamination level significantly. Apart from a lot of governmental regulations prohibiting this option, the mixing might not be an acceptable and environmentally friendly treatment option and should not find acceptance, because it will not remove the toxic substances.

Table 4.3 Examples for the specific gravity of different substrates

Substrate	Specific gravity g cm^{-3}
Soil	
Humic topsoil	1.4
Sand/gravel	1.6
Loam/clay (dry)	1.8
Natural cobbles	
Limestone	2.6
Sandstone	2.5
Shale	1.7
Granite	2.7
Basalt	3.0
Technogenic substrates	
Coke	0.9
Coal	1.6
Ashes	0.9
Slag	2.9

Stockpiling is also used in circumstances where the excavated soil has not been characterised and analysed in detail. In this case each pile is inspected, sampled and analysed. The sampling strategies usually differ according to the regulations. The frequency of sampling and analysis changes when it is not possible to apply field analytical methods or it is suspected that hot spots exist in the pile. For instance, the guidance in New Jersey, USA (NJDEP 1998) suggests:

- Dividing of the pile into a grid of 18 m lots
- Minimum of two borings in each 18 m lot at 60 cm-deep intervals
- Lab analysis for a minimum of two samples per 76.500 m^3 of piled material.

4.2.3 *Transportation*

Attention should be paid to the specific gravity of the substrates intended for excavation. It is difficult to calculate the transport capacity, if it is estimated that the excavated material indicates a specific gravity of only 1.5 g cm^{-3}, the mean value for soils normally defined. Based on a detailed soil survey including a substrate differentiation the masses to be excavated should be calculated for each substrate separately, because the specific gravity of the substrates might differ enormously (Table 4.3).

Differences can already be discovered between distinct soil matrices. In the presence of a high percentage of natural cobbles like iron containing basalt and granite an increase in the specific gravity is to be expected, leading to an increase in truck journeys required. In contrast, the presence of technogenic substrates revealing a lower specific gravity such as coke and ashes might also alter the number of truck journeys. Technogenic substrates causing soil pollution are typical components of deposited soils (heaps, fills).

Fig. 4.10 Bell tipper truck

A detailed field investigation which includes the recognition, description and estimation of substrates eliminates the risk of a problematical miscalculation. The example of the slag types with varying specific gravities and maximum values up to 4.0 g cm^{-3} (lead slag) shows impressively the necessity to pay attention to the specific gravity before calculating the amount of truck journeys. An initial separation of the coarse fraction (cobbles, boulders) and the fine fraction consisting of gravel and fine earth on site might be helpful for the calculation of the transport capacities. Moreover, the amounts of transportation required depend on the truck type. Effective loading capacity, for instance, have bell tipper trucks (Fig. 4.10).

Apart from the density, attention should also be paid to the swelling features of disturbed and excavated material. The undisturbed soil material shows fewer voids than it does after the excavation. Hence, swelling processes may start just after excavation, in particular if the material is influenced by precipitation. Clays for instance, may swell up to 40% and sand and gravel up to 10–15% (NJDEP 1998). Accordingly, the volume to be transported increases significantly.

During the material transport some regulations should be taken into account. The trucks must be lined with plastic or covered to prevent spreading of fugitive material and gas emission. Furthermore, oxidative reactions must be considered, if possible. For instance, during the transport over a long distance in a dry soil contaminated with chromium the chemical reaction to the toxic CrVI can deteriorate the toxicity of the material. Another example is material containing pyrite, which is usually present in mining areas. During excavation and transport oxidation processes occur, causing pyrite oxidation and subsequently acidification of the material (H_2SO_4 formation). In the case of a simultaneous heavy metal contamination the oxidation process might be responsible for an enhanced mobility of the metals after dumping (see Sect. 3.2.1). The same reaction

Table 4.4 Possible sites appropriate for substrate reuse (Based on regulations associated with the German Waste Avoidance and Management Act (GOG 2011))

	Type of reuse	Sites
↓	Unlimited reuse	All sites
↓	Limited reuse	All sites except for sensitive land-use types (playgrounds, sports fields, gardens, agricultural sites, water reserves)
↓	Reuse without surface cover	Mining reclamation Construction of sealed roads and industrial areas (e.g. road-beds, embankments) Parks with complete vegetation cover and missing bare soil Landfills (levelling course) (exceptions: see above, in addition nature reserves and flood areas)
Increasing contamination	Reuse with clay caps and geotextiles	Construction of sealed roads and industrial areas (e.g. road-beds, embankments) Sound-insulating walls Landfills (levelling course) (exceptions: see above, in addition drainage layers, pipe shafts, and hydrological zones with high water conductivity)

might occur during shipment of contaminated soil. Hence, shipment of hazardous material must be containerised prior to leaving the site in a similar way to transport by truck.

4.2.4 *Reuse of the Excavated Material and Backfilling*

In the context of excavation the primary aim should be to reuse substrates. Most of the city, county or governmental districts provide applications or guidance for reuse purposes. The opportunity to reuse the former excavated material depends on the level of contamination. With increasing contamination the opportunities tend to be more limited or finally excluded. Table 4.4 provides information about the recycling opportunities that are principally feasible. On the basis of regulations associated with the German Waste Avoidance and Management Act (GOG 2011) four classes are defined. Land-use types with high sensitivity tend to be more restricted with regard to recycling purposes. The reuse of contaminated material will improve, if technical protective measures are combined with the deposit of the material. Accordingly, contaminated material can possibly be used again but subsequently different types of soil cover must be added (see Sect. 5.1.1). The reuse also includes the area of concern, so that contaminated material (<z2) can be used elsewhere in the planned clean-up area as long as the restrictions and rules are followed. Accordingly, the movement of contaminated material from a highly sensitive land-use type such as a kindergarten to a less sensitive one such as a landscaped area is feasible.

Table 4.5 shows the threshold values dependent on the distinct recycling classes the planned German Ordinance will define in the near future. Already one parameter exceeding the defined class leads to a grouping to the next higher class. It should be noted that the reuse is only feasible, if exact knowledge about the material, the area of excavation and the area of depositing is present, e.g. results of chemical analysis, classification of the material, quantity of the material, hydrogeology of the site for depositing, etc.

In any case the reuse of excavated material will be limited with regard to general well-known highly contaminated sites such as industrially influenced floodplains, sewage sludge treatment farms, fields filled with dredged harbour sludge, excavated urban soils containing a high percentage of potentially contaminated technogenic substrates like ashes and slag and sites which reveal geologically determined contamination (ore bodies). Furthermore, the reuse appears to be problematical for material derived from treatment plants that cannot decontaminate all analysed pollutants, e.g. heavy metals in bioremediation pits (see Sect. 6.3.4).

Highly polluted material that exceeds the values of the class z2 cannot be reused any more, so that a final disposal in e.g. a landfill will be the only opportunity apart from decontamination measures that is preferred. Generally speaking, the transfer of contaminated soil to a landfill appears to be more the transfer of an environmental problem than a good solution. In general, five kinds of locations of emplacement should be distinguished:

- Deposits of uncontaminated mineral soils and rocks (valuable raw material)
- Deposits of inert construction rubble and materials similar to its leaching behaviour
- Deposits of household waste and industrial waste with a low degree of contamination (landfills)
- Deposits of highly contaminated industrial waste including hazardous waste
- Deposits of hazardous waste with high toxicity, which are only adequate to mine shafts (e.g. radioactive substances, explosive waste from military sites, fly ashes).

If an off-site disposal is planned, it will be of importance whether the receiving facility is permitted to accept the contaminated material and whether the disposal is in compliance with all permissions needed. The opportunities to dispose of contaminated material will be more restricted in future and consequently the costs of disposal are going to increase because the capacity of permitted disposal facilities will become more and more limited. Nowadays, service providers offer a number of treatment-based solutions leading to a more appropriate minimisation of risk than excavation and landfilling alone.

Excavation and soil cover operations should take the planned land-use types into consideration. One obvious problem in urban areas, which are affected by a multitude of contamination sources and which cover a high percentage of contaminated land, is the lack of clean material to be used for remediation purposes. In particular, in the urban agglomerations the supply of sufficient clean soil is not always certain. Thus, in some agglomerations the public authority decided to create storage facilities for uncontaminated material. As can be seen in the photographs of Figs. 4.11 and 4.12, clean subsoil originating from construction sites and certified as being

Table 4.5 Reuse opportunities for excavated humus-poor soil (Based on regulations associated with the German Waste Avoidance and Management Act (GOG 2011))

Reuse	Unlimited reuse			Backfilling of excavated holes and trenches	Limited open backfilling technically structured	Limited backfilling technically structured in an impermeable way	Landfilling and transportation to a soil remediation centre
Class	<z0[a]			z0–z0*	z0*–z1	z1–z2	>z2
	Clay	Loam/Silt	Sand[b]				
Cd	1.5[c]	1[d]	0.4	1[e]	3	10	>10
Cr	100	60	30	120	180	600	>600
Cu	60	40	20	80	120	400 (600)[f]	>400 (600)
Hg	1	0.5	0.1	1	1.5	5	>5
Ni	70[c]	50[d]	15	100	150	500	>500
Pb	100[g]	70[h]	40	140	210	700 (1,000)[f]	>700 (1,000)
Zn	200[c]	150[d]	60	300	450	1.500	>1.500
As	20	15	10	15[i]	45	150	>150
Tl	1	0.7	0.4	0.7[j]	2.1	70	>70
TPH[k]	100	100	100	200	300	1,000	>1,000
ΣVCHC	1	1	1	1	3	10	>10
ΣBTEX	1	1	1	1	3	10	>10

EOX	1	1	1	1[l]	3[l]	10[l]	>10
ΣPCB	0.05	0.05	0.05	0.1	0.15	0.5	>0.5
Benzo(a)pyrene	0.3	0.3	0.3	0.6	0.9	3	>3
ΣPAH-EPA	1	1	1	6	9	30	>30
Cyanides	1	1	1	1	3	10	>10

[a]Related to humus content <8%
[b]Silty sand related to loam/silt
[c]Use texture loam/silt at pH < 6.0
[d]Use texture sand for pH < 6.0
[e]1.5 mg kg^{-1} for clayey soils
[f]In discussion
[g]Use texture loam/silt for pH < 5.0
[h]Use texture sand for pH < 5.0
[i]20 mg kg^{-1} for clayey soils
[j]1.0 mg kg^{-1} for clayey soils
[k]TPH chain length C10–C22
[l]Further investigations necessary

Fig. 4.11 Process of soil mixing for the supply of clean soil: piles of uncontaminated material divided into different texture classes (sand, loess, compost at the back)

Fig. 4.12 Process of soil mixing for the supply of clean soil: drum-sieve for mixing as well as for organic matter addition purposes

unpolluted is collected. The heaps consisting of distinct texture classes are deposited separately. The material is mostly poor-quality material with regard to the nutrient status which needs to be managed, e.g. by the addition of organic matter. Afterwards, it is mixed and augmented with organic matter like compost and bark mulch by drum-sieves according to the intended use. The rehabilitation requirements for the soil used for backfilling or soil cover are introduced in Sect. 5.1.1.

Nevertheless, the transportation of contaminated and uncontaminated material should be limited as far as possible and a proper soil management can avoid an unnecessary intensive movement of soil. In Fig. 4.13 the soil contamination of a

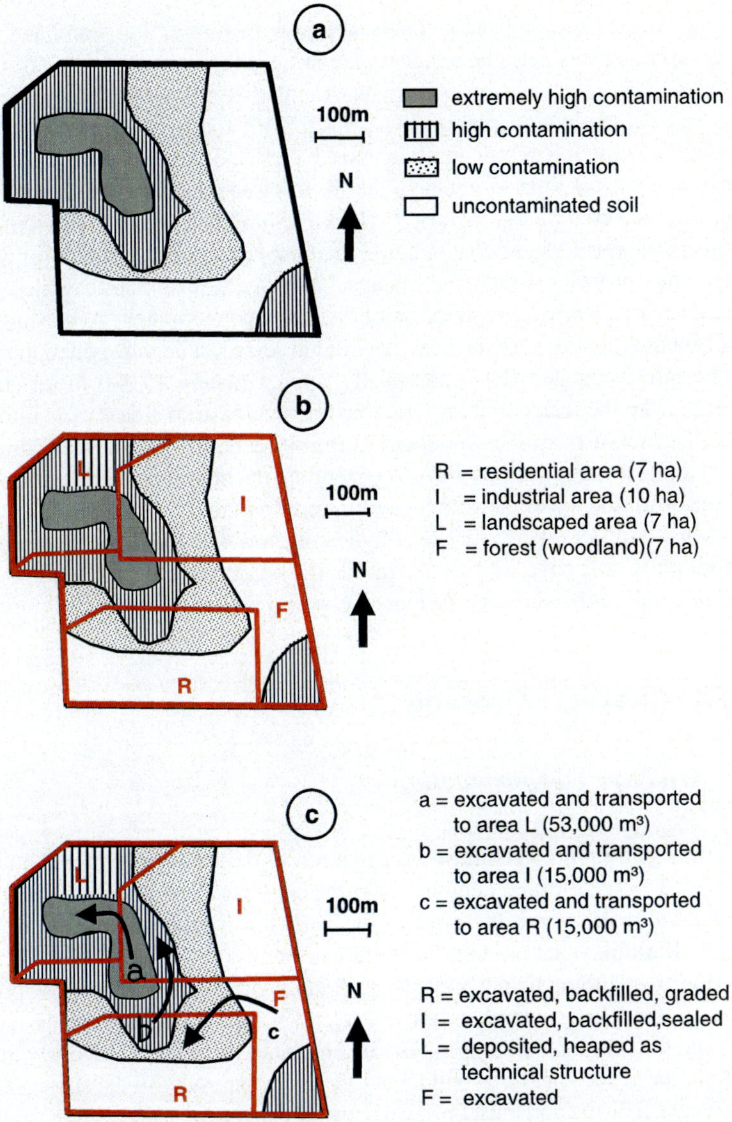

Fig. 4.13 Redevelopment of a former industrial plot. (**a**) Soil contamination. (**b**) Planned land-use types. (**c**) Mass flow of the excavated material

former industrial area (metal processing industry) of approximately 31 ha is illustrated. The enhanced values refer mainly to the upper 3 m. With reference to the highest contaminated area more than 80,000 m^3 material was identified. There are some areas indicating an extremely high level of soil contamination, while other areas show moderate and low contamination or are entirely unpolluted. The redevelopment plan of the public authority provided for the introduction of a residential area (7 ha), an industrial area with predominantly sealed surfaces (10 ha), a landscaped area (7 ha), which will be used as landscape park in future, and woodland (7 ha), which is supposed to shield the residential area from the industrial area.

Figure 4.13 shows the redevelopment plan involving the land-use types mentioned and the soil contamination present. The position of the land-use types took into account that ideally only a little material must be transported. Because of the limited budget it was not possible to contemplate *ex situ* treatment. The extremely highly contaminated material (heavy metals, PAH, PCB) should be piled up with foundation fragments of the former industrial buildings and used as a landscape park for walking purposes after properly conducted encapsulation (see Sect. 5.2.2), which avoids direct contact and downward migration of dissolved contaminants. According to the plan, only about 52,500 m^3 of the heavily polluted soil must be transported to the area where the pile is situated (L). Additionally, approximately 15,000 m^3 of soil was transported from the residential area into the future industrial area (I) and ultimately unpolluted material from the woodland to the residential area for backfilling purposes (R). In this way it was possible to concentrate the highly contaminated material in the areas with the lowest sensitivity and in turn to plan the land-use type with the highest sensitivity (residents) in areas indicating low soil pollution and requiring some additional soil cover measures. In the industrial area the soil handling was reduced to a minimum, since the surface was sealed on a large-scale.

4.3 Site Clearance Operation

4.3.1 Selective Deconstruction

Most of the clean-up projects must pay attention to the frequently ruined buildings formerly used which must be demolished in the context of site clearance operations. Hence, the soil experts are involved in the demolition processes which occurred previously. Buildings that are kept in use are normally not demolished in a certain remediation area, although they may place restraints on the soil remediation process, because e.g. the placement of groundwater and vapour extraction wells is made difficult and the scope for excavation of contaminated material is obviously limited. In addition, for *in situ* measures buildings form a discontinuity in the subsurface.

In principle, a distinction must be made between conventional demolition neglecting any separation and recycling of the material that must be broken down and innovative, selective demolition called deconstruction involving material-specific listing and disposal. An uncontrolled dismantling of buildings leads to the production of a great amount of unsorted waste to be disposed of. For this reason, an organised demolition strategy which is focused on the reuse of the demolition material (buildings and installations) is preferable to landfilling, which takes up a high amount of unnecessary space.

Deconstruction is an environmentally friendly approach that allows the reuse and recycling of building components but it requires a longer period of time compared with fast demolition. Deconstruction lowers the need for virgin raw materials, since there is a new life cycle for the building materials. Apart from the safety regulations to protect the workers from danger to their health (see Sect. 4.4), many aspects must be taken into consideration. In any case, it is important to take an integrated approach together with soil remediation and other environmental restoration after the decommissioning has taken place.

Beforehand, the following points should be taken into consideration and clarified:

- For installations and machinery potentially interested buyers should be searched for and found. Moreover, vendors must be found dealing with the reuse or recycle of the installations to be dismantled, in particular with the cleaning, sorting and storage equipment.
- The buildings need to be checked with regard to archaeological importance, which requires cooperation with preservationists.
- The building stability needs to be checked with regard to the preferred sequence of demolition. This requires experts who deal with the structural details, such as architects and construction engineers. A survey of the stability of buildings to be demolished must also be carried out with regard to the actual demolition programme (sequence of breaking down) and the choice of the demolition technique.
- Vendors must be chosen who are able to handle the selected demolition technique, e.g. ball demolition, bulldozers, excavators, explosives, etc.
- A detailed report on the monitoring of environmental hazards caused by demolition work must be written. The monitoring programme includes all negative impacts of the demolition such as the soil and water contamination, air pollution (dust, gases), vibration and noise emission. Fire hoses are used to support wet demolition suppressing dust and fire during the process.
- The present use but also the use in former times with particular reference to the location of installations and accident sites with a hazard potential must be investigated (Sect. 4.3.2). The present use is established by a visual inspection and the historical development can be ascertained by means of maps, production records and waste handling policies.
- Moreover, a detailed financial plan including sale of re-used materials must be drawn up.

A schedule for organised demolition and site clearance is shown in Fig. 4.14. In the first instance, a catalogue of machinery and installations to be sold is drawn up. Non-structural components such as intact doors, windows and appliances can be reclaimed and sold. Buildings or parts of buildings to be protected for archaeological reasons and buildings to be protected because they will be used after the demolition work are described. Special attention must be paid to the description of decontamination tasks involving drainage by means of tanks and pipes, the evacuation of solid waste, the excavation of polluted soil and the removal of asbestos and other toxic waste. Permits and submitted notifications are obtained for handling the asbestos or for the removal of other hazardous materials such as lead painted walls and site-specific safety plans must be developed.

Afterwards, materials that can be reused after stipulation, sorting and cleaning of the storage facility are registered as marketable components. Equally, materials that cannot be reused must be named and registered as well. The demolition starts at the

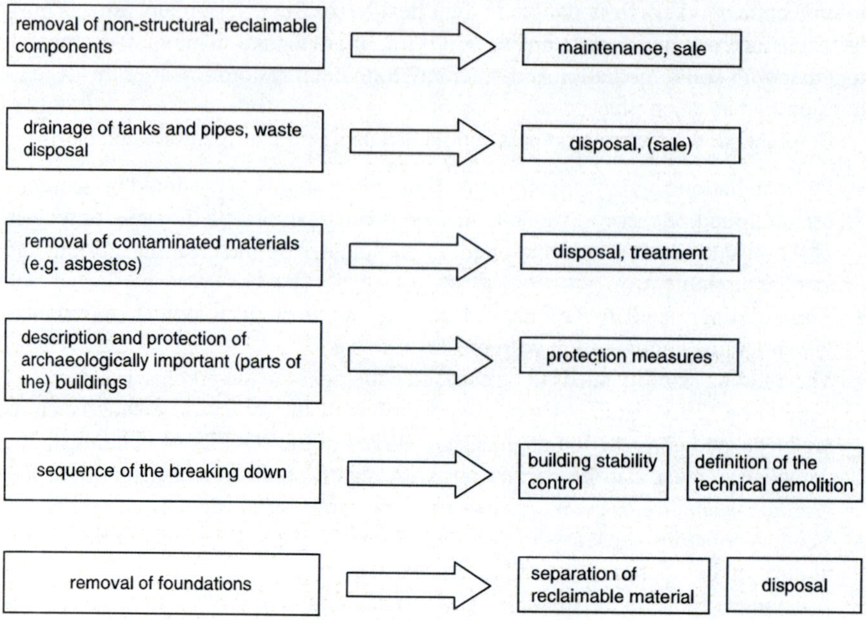

Fig. 4.14 Course of action for the selective demolition of buildings

roof and works down to the foundation. During the dismantling of the structural building components bricks and dimension stones are separated and stored in a secure and dry location. In principle, it is possible to reuse used bricks and dimension stones (e.g. limestone, sandstone) due to their durability. However, as a result of damages and colour changes over time it is preferable to reuse these bricks and stones in the road construction sector where the material can be used for the construction of the base course. Furthermore, important materials to be reused might be construction wood and metals (e.g. steel, copper; see also Sect. 2.3.1).

The subsequent removal of the foundation is closely linked to the soil excavation procedure. The simultaneous removal of foundation structures and surrounding soil often leads to a mixture of soil and debris, which is pre-treated (Sect. 6.1) and ultimately treated and decontaminated in different ways.

Usually, the sequence of the demolition leaves a framework consisting mainly of concrete and construction steel that must be broken down. Various types of technology are used for this:

- Hydraulic excavators (Fig. 4.15) undermine and pull the residual building, which then falls in the desired direction. The walls are undermined at the base of the building. Concrete decks are additionally removed by hoe rams and hydraulic shears remove structural steel, if required.
- Ball cranes (Fig. 4.16) are used in the case of higher buildings to demolish the structure. Because of the uncontrollable swinging wrecking ball special safety regulations must be observed.

Fig. 4.15 Hydraulic excavator demolishing an industrial building; an employee tries to minimize dust development using a water hose (With kind permission of IGfAU company, Melle, Germany)

Fig. 4.16 Ball crane demolishing an industrial building (With kind permission of IGfAU company, Melle, Germany)

Fig. 4.17 Demolishing excavator with shear attachment (*left*), hoe ram and crawler excavator (*centre*) as well as bulldozer (*right*) in full operation (With kind permission of IGfAU company, Melle, Germany)

- Demolition excavators which reach the higher floors of high-rise buildings (Fig. 4.17) are also used but they need a shear attachment for the reinforced steel structures (Fig. 4.18) and hydraulic hammers to crush the concrete. Alternatively, wheel loaders and bulldozers are used to ram smaller buildings.

4.3.2 Separation and Treatment of Hazardous Material

Just like the soil, the material of the buildings to be pulled down might be contaminated, since buildings and facilities of industrial and commercial complexes that are of particular interest for the clean-up projects can, generally speaking, be characterised as contaminated. Hazardous residues might be present in industrially used buildings where toxic materials have been used, stored and manufactured. Furthermore, accidents at an earlier stage and uncontrolled handling during the post-closure period might cause the release of hazardous substances. With regard to this post-closure period clean-up activities should begin as soon as possible, since areas abandoned for a long time might invite use of the area for illegal waste deposit purposes.

Fig. 4.18 Hydraulic crushing hammer

Consequently, the different installations should be previously investigated and analysed. It becomes necessary to make an inventory of all hazardous materials and liquids. This inventory including sampling and analysis of expected contaminated materials is required to carry out the demolition work in a safe manner.

Figures 4.19, 4.20, and 4.21 reveal typical potentially contaminated details of buildings which are usually sampled and assessed. The sample documented in Fig. 4.19 shows flush-mounted insulating material, Fig. 4.20 tar paper on the top of the roof and Fig. 4.21 the inlet of a heating pipe containing problematical insulation material. Using the waste catalogue of the European Union (EU) (EPA 2002) in Table 4.6 as a basis, examples of toxic substances often combined with building material are listed.

Special attention will be paid to the asbestos problem. Asbestos is the term for a group of fibrous silicate minerals with a crystalline structure either bound within the matrix and consequently with a low degree of airborne fibres or present as friable material that generates airborne fibres. Damaged, originally non-friable materials (e.g. pipe laggings) and already friable materials are the cause of asbestosis and lung cancer after inhalation by human beings. For this reason, the asbestos containing

Fig. 4.19 Flush-mounted insulating material which may be contaminated (With kind permission of IGfAU company, Melle, Germany)

Fig. 4.20 PAH contaminated tar paper on the roof of a building (With kind permission of Wessling company, Altenberge, Germany)

Fig. 4.21 Heating pipe with contaminated insulation material (With kind permission of IGfAU company, Melle, Germany)

Table 4.6 Contaminants in building material (examples) (Data from EPA 2002)

Waste	Catalogue No. (EU)	Examples
Insulating material made of artificial fibres	170603	Glass wool, rock wool, slag wool
Insulating material containing asbestos	170601	Fillers, fire doors, light building boards
Asbestos building material	170605	Facing tile, roof tile, asbestos cement
Equipment containing asbestos	160212	Night-storage heater, electrical device, boiler
Building material containing lead	170903	Insulating material made of lead sheet, filler (lead wool), roof tile, lead glazing, cable coatings, water tubes
Building material containing cadmium	170903	Pigments, corrosion protection, batteries, cable coatings, photovoltaic
Building material containing chromium	170903	Concrete and cement additives, pigments, wood impregnation, electrical heater
Building material containing mercury	170901	Paints, wood impregnation
Fluorescent device containing mercury	200121	Lamps, gauges
Building material containing PAH	170303	Tar paper
	170301	Tar asphalt
	170204	Parquet floor (filler containing tar)
Building material containing PCB	160209	Transformer, condenser
	170902	Filler (containing PCB)

construction debris must be kept wet during destruction and must be safely stored. The treatment opportunities are limited and are mostly focused on solidification and volume reduction by washing (Nathanail and Bardos 2004). As long as asbestos-containing material is neither damaged nor disturbed the exposure might be relatively low. In contrast, damaged and disturbed installations already releasing fibres cause danger to human health and the environment. Hence, handling must include enclosure and removal in an authorised way. The example of asbestos fibres reveals the general problem of areas to be remediated in association with the adjacent areas which can be affected. For this reason, demolishing contaminated buildings simultaneously to soil handling (e.g. backfilling of uncontaminated soil, construction of side barriers) requires care (Fig. 4.22). The release of contaminants during the demolition is caused by effluents generated by wet demolition, erosion, dust dispersion, vehicle movement and the dynamic impact of the machines used, such as ball crane and explosives.

Moreover, apart from the contaminants listed biological contaminants are often present in ruined and long-term abandoned buildings as follows:

- Pigeon excrement (causing infections, e.g. salmonella, fungi)
- Timber worms
- Mould (e.g. due to pet excrement, potting soil residues, damp indoor climate).

Fig. 4.22 Demolition of a contaminated building (in the *background*) simultaneously to the construction of a side barrier (arranged in the *foreground*) (With kind permission of IGfAU company, Melle, Germany)

The contamination of the demolished material can be treated in a washing plant comparable to the soil (see Sect. 6.2.1). If the contamination (e.g. TPH, PAH) is limited to the surface of the construction debris, the material should be conditioned and broken into smaller pieces and afterwards washed for a short time. In the case of deep-reaching contamination of the substances the material should be broken up intensively and afterwards treated for a long time with water and extraction agents. The use of agents is sometimes important in order to clean the material to an acceptable standard. For example, highly contaminated interior walls of demolished chimneys showing enhanced sulphur values at a low pH value are treated with lime and foundation walls containing PAH and conditioned with tar are treated with tensides. The separation process including the light fraction removal is carried out as described in the context of soil washing.

Before dismantling it is often not feasible to decontaminate the structures, because the contaminants are inaccessible or the contaminated dust has penetrated into the mechanical and electrical equipment. Nevertheless, in a variety of cases it is possible, prior to the breaking down, to eliminate the surface contamination of the buildings intended for demolition. Various measures are feasible:

- Cleaning with high pressure water jet: impact to a depth of approx. 10 mm, treatment of contaminated masonry, floor pavement and concrete
- Sand and grit (sharp-edged metallic grains) blasting: impact to a depth of approx. 3–10 mm, treatment of contaminated masonry, floor pavement and concrete

- Milling: deep-reaching impact after reapplication, removal of contaminated dust and plaster
- Peeling using hydraulic breaker: deep-reaching impact after reapplication, removal of contaminated foundations and asphalt
- Suction cleaning with industrial vacuum cleaner: withdrawal of dust, liquids and solid waste
- Flame treatment with acetylene-oxygen-flame (temperature up to 3,200°C): impact to a depth of approx. 30 mm, removal of oily and fatty surfaces
- Cleaning with water and detergents: treatment of oil contaminated surfaces made of concrete.

4.4 Working Safety

4.4.1 Sources of Danger to Human Health

In the context of soil rehabilitation and remediation measures a lot of dangers to human health should be taken into consideration. In general, a distinction is made between mechanical hazards, chemical hazards and other disadvantageous occurrences.

Hazard to human health is related to skin absorption and penetration, oral ingestion and inhalation, the impact of toxic gases (asphyxiation), fire and explosion and the risk of injury. Regarding the latter, brownfields frequently have residual aboveground facilities such as ruined buildings which are prone to collapse and foundations belowground whose correct location and size is unknown. If in-door operations are necessary, as is typical for the demolition of damaged buildings (Fig. 4.23), material can fall down, endangering the experts and workers. Moreover, damaged facilities and equipment contain sharp-edged objects, which may injure humans in the form of cuts and bruises to hands and feet. In and on the ground sharp materials are sometimes present, which may be a danger for human beings after they enter the buildings. If soil handling occurs, the stability of the soil to be treated is definitely important. Pits and slopes can collapse and especially underground facilities are in danger of subsiding or completely collapsing.

Detailed investigations are often carried out regarding the planning and operation of soil treatment because of insufficient information about the contaminated site. In this context, drilling for instance can cause damage to underground installations such as gas pipelines and electric cables. This can endanger humans at work on the site. In a similar way, explosives cause problems to the people dealing with the contaminated site, in particular at former explosive factories, magazines and military sites as well as sites which were influenced by wartime activities where unexploded bombs and mines remained in the soil.

In general, persons involved in the clean-up process should beware of the danger that toxic gases may volatilize. In relation to the volatile components that are released from the contaminated soil as well as in-door air the inhalation exposure causes dangers to human health. In rooms including basements, manholes and

Fig. 4.23 Damaged contaminated industrial building in Pernik, Bulgaria, before demolition occurs

trenches an explosive and reductive atmosphere can be generated and, moreover, some inhaled toxic gases will be directly responsible for people becoming unconscious. The development of gases is particularly significant, if the person must go into pits, tanks or underground installations without any information about the atmosphere. Apart from gases, dust can also be a hazard for the people. In this context, even heavy metals which are substantial components of the soil matrix as localised at hot spots with extremely high concentrations, can be inhaled. In conclusion, the inhalation of toxic substances appears to be the most important reason for adversely affected persons, whereas the oral and dermal pathway might be of less importance. Nevertheless, a direct ingestion of contaminated material cannot be entirely excluded.

Apart from the dangers associated with the handling of soil and ruined buildings, common harmful effects on physical health such as noise associated with the drilling and demolition works and the electrical safety of the machinery installed or used should be itemized.

4.4.2 *Safety Measures*

During the last few decades regulations and recommendations about working safety have been published in many countries dealing with contaminated sites (e.g. BDA 1992; Steeds et al. 1996; OSHA 2011). The most important advices are summarized and discussed in the following section.

The best way to minimize risky situations is to develop a health and safety plan prior to the remediation procedures. This plan should include and exactly describe all adequate protection measures, e.g. the necessity to wear protective clothing. Advice about the clothes in relation to the site conditions, e.g. the expected presence of acid, basic, toxic, etc. substances, should also be followed. The development of the plan is usually organized by an occupational hygienist who is adept at dealing with contaminated sites. He or she should advise on the risk, explain the work to be used, assist in training of the engineers and employees, analyze soil and waste and carry out a monitoring system. The hygienist should monitor the fence off to prevent unauthorised access and should implement zoning (dirty area, clean area). The fences should be at least 2 m high and very difficult to climb. Furthermore, his or her field of operation includes the control of dust, for example sheeting of open vehicles which transport contaminated materials and use of water sprays (see Sect. 4.2.1). Moreover, information, instructions and sometimes training should be organised by the hygienist to ensure that safe working is possible in spite of the danger present.

The hygienist is also in charge of the installation of the hygiene facilities such as washing facilities with a high standard, storage capacity for contaminated clothing including footwear, existence of restrooms or transportable toilets which are cleaned daily, boot wash and wheel wash facilities where tyres and underbody of the trucks are cleaned before travelling on public roads. Clean cabs using positive pressure or vacuum should be controlled as well.

A standing and inflexible rule is that nobody is allowed to work alone, if they are exposed to different hazards endangering human health. Of course, if toxic substances cannot be excluded and accordingly the persons concerned can come into contact with them, eating, drinking or smoking is undesirable.

If unexploded bombs are found, work must be stopped to identify the objects more exactly. Further disturbance of the findings should be avoided and an immediate evacuation may be necessary. Police and accredited units dealing with unexploded bombs, shells and rockets must be informed as soon as possible. Nowadays, the specialists possess well-established procedures to deactivate the problematical articles.

A feasible way to stabilise pits and prevent them from collapsing is, for instance, to sheet the walls, if the persons must move into the excavated area. In general, if the excavated soil reaches a depth of more than approximately 1.50 m, stabilisation of the surrounding walls must be carried out in order to avoid a very sudden collapse of the excavated area. Insecure ladders and scaffolds are also causes of hazard to humans (OSHA 2011).

There are general safety measures to which attention should be paid. Employees and engineers must wear a personal safety outfit such as safety boots with a metal cap and sole, helmets which are disposed of in the case of damages, impervious gloves as well as protective overalls. With reference to the clothes used it is advantageous to wear tight clothes to reduce the danger of being caught by rotating machines occasionally used in association with soil investigation, soil management and, in particular, building demolition. Disposable overalls are preferred to washable ones.

Fig. 4.24 Workers with full protective clothing who dismantle cement asbestos panels (With kind permission of IGfAU company, Melle, Germany)

The overalls should be pulled down on the outside of the boots. Heavy duty plastic bags should be used to transport the contaminated clothing to a garbage incinerator or other disposal systems.

Furthermore, depending on the exact activities of the affected persons an additional outfit can be of importance in relation to noise protection and dangers to the eyes and the respiratory system. Thus, it appears to make sense to use splash guard for the helmets, ear protectors or protective glasses. The persons involved are requested to wear appropriate respiratory protective equipment (Fig. 4.3) that is aerated from the outside in the presence of supposed toxic or choking atmospheres. However, it is vitally important to position somebody outside the underground installations or deep pits who is able to manage rescue measures, if necessary. Furthermore, in the case of an accident caused by toxic gases persons should never try to rescue others without taking any effective safety measures to protect themselves.

As explained in Sect. 4.3.2, special attention must be paid to asbestos and other toxic fibres that is buried on site or installed in aboveground structures at many sites. Its friable fibres can be released into the atmosphere, particularly in dry conditions. Measurable levels of airborne fibres are generated during incorrect handling of soil and buildings to be demolished. For humans the prevention of uncontrolled emission is the most important aim. Water spraying to suppress the dust development is always helpful. In general, full protective clothing is required and masks or pressure respirators must be used (Fig. 4.24). Furthermore, personal washing facilities and wheel wash plants for vehicles must be taken into consideration. Besides asbestos, other toxic and dangerous substances such as radioactive substances or anthrax can also necessitate special precautions. In relation to anthrax, which is occasionally present, for example, at tanneries, wool sorting companies and industrial sites producing gelatin, the spores can enter the skin or be inhaled. Hence, use of adequate protective clothes and gloves in addition to maximum hygiene will be required (HSE 1991).

Some technical field instruments help to detect toxic substances at once. They are helpful tools for identifying potentially dangerous substances and it is generally recommended that they are used during soil remediation:

- Photo-Ionisations-Detectors (PID) that can qualitatively detect volatile aliphatic, aromatic and halogenated components; the sum concentration of the parameters of concern is indicated and the results are quickly displayed by warning signals.
- Test tubes which are specific to a number of organic and inorganic compounds; even the concentration is displayed and the results are visible after few minutes.
- Explosimeters which inform about explosive and choking atmospheres; again, immediate results are indicated by warning signals.
- Mobile X-ray fluorescence device that is preferentially important with reference to the metals; the results are produced immediately.

If the quick detection does not result in reliable data, the only opportunity to experience the contamination level might be the sampling of soil and soil vapour by drilling technique as well as dust by means of samples in order to get information about the dispersion of toxic gases and dust particles. It is absolutely essential to wait for the analytical results of the laboratory before continuing the working process on site.

Regarding the oxygen deficiency in confined spaces such as basements, trenches, etc. it is recommended to test oxygen concentration, toxic gases (e.g. carbon dioxide, hydrogen sulfide) as well as explosive gases continuously. Depending on the results, breathing apparatus is provided or even ventilation may be necessary.

Nevertheless, in spite of the many safety measures applied it can never be avoided that accidents occur. For this reason, engineers and employees should be trained in first aid. First aiders should periodically receive training in order to be able to deal with the site-specific pollutants. The most important rule to be obeyed is to keep calm and to phone the emergency services as soon as possible. People involved in clean-up processes follow the safety plan and know the principles of first aid such as securing the scene and location of the accident. They are able to provide help to the injured people, assist police, fire brigade and ambulance and help to disperse curious onlookers.

References

BDA – British Drilling Association. (1992). *Guidance notes for the safe drilling of landfills and contaminated land*. www.britishdrillingassociation.co.uk. Accessed 9 Jan 2012.

EPA. (2002). *European waste catalogue and hazardous waste list of the European Union*. Johnstown: Environmental Protection Agency.

GOG – Government of Germany. (2011). *German Waste Avoidance and Management Act*. BGBl. I No. 51, 1986.

HSE – Health and Safety Executive. (1991). *Protection of workers and the general public during development of contaminated land*. London: Library and Information Services, Baynards House.

McCarthy, D. F. (2007). *Essentials of soil mechanics and foundations*. Upper Saddle River: Pearson.

Nathanail, C. P., & Bardos, R. P. (2004). *Reclamation of contaminated land*. Chichester: Wiley.
NJDEP. (1998). *Revised guidance document for the remediation of contaminated soils*. Department of Environmental Protection, State of New Jersey, USA.
OSHA – Occupational Safety and Health Administration. (2011). *OSHA fact sheet*. United States Department of Labor. www.osha.gov/about.html. Accessed 13 Jan 2012.
Steeds, J. E., Shepherd, E., & Barry, D. L. (1996). *A guide for safe working on contaminated sites*. London: Construction Industry Research and Information Association (CIRIA) Publications.

Chapter 5
Soil Containment

Abstract Alternatively to decontamination strategies, soil containment can help to reduce contaminant migration and subsequent endangering of the environment and human health. This chapter discusses all relevant methods of soil containment starting with surface cover systems, including use of geotextiles and bentonite, which prevent direct contact to the contaminated depths. Regarding lateral migration, side barriers are introduced and the encapsulation approach, which interrupts migration completely, is also described. Furthermore, the field of solidification is discussed, taking cement-based solidification, asphalt batching as well as vitrification into consideration. The technical construction of all applications is characterised exactly and discussed in the context of soil properties and the presence of different contaminants. Finally, the soil-protective stabilisation methods are presented. The chapter gives a comprehensive overview of relevant measures beyond the decontamination approaches.

Keywords Containment • Contaminant migration • Cover system • Encapsulation • Solidification • Stabilisation

5.1 Surface Cover

5.1.1 Geotextile-Based and Bentonite-Based Cover Systems

The purpose of cover systems is to prevent contaminant migration and to interrupt the pathways of contaminant linkages. Furthermore, they are constructed to prevent dust blowing off a contaminated surface.

The surface cover is also applied in cases where the bottom of the contaminated zone touches the groundwater table, since the part of the contaminated zone above the groundwater table cannot be leached out anymore. Therefore, the cover helps to diminish the contamination of the aquifer to an acceptable level (Genske 2003).

H. Meuser, *Soil Remediation and Rehabilitation: Treatment of Contaminated and Disturbed Land*, Environmental Pollution 23, DOI 10.1007/978-94-007-5751-6_5,

The complexity of the surface cover depends on the level of contamination analysed. The simplest solution, if the contaminants exceed the defined thresholds to a small extent, is the intensification of the vegetation in order to avoid bare soil that can be directly touched or dispersed by deflation. With increasing contamination more technical input must be applied. A single barrier cap can be acceptable, if the main functions of the soil cover, the reduction of surface infiltration, prevention of direct contact and limitation of gas emission and erosion control are feasible.

The aim of a field study near Aachen, Germany, was to gather knowledge about the minimum thickness of uncontaminated soil needed to deposit contaminated material related to the pathway soil – plant. In the field experiment soil the heavy metal concentration was high (Cd 10.3 mg kg^{-1}, Pb 610 mg kg^{-1} and Zn 1,830 mg kg^{-1}) and the pH value neutral. A number of crops such as lettuce, carrots, mangold, celery and endive were grown. Different cover configurations were introduced including systems which avoid technical barriers such as geotextiles. It was found that a thickness of 40 cm uncontaminated material was sufficient to yield plant concentrations below or in the range of thresholds usually applied to vegetable consumption. Only the element zinc was markedly reduced by exceeding a thickness of 40 cm. In any case, 70 cm exclude the danger of plant transfers causing the thresholds to be exceeded. Compared with the experiment plot without soil cover the metal reduction of the edible part of the vegetables reached 80% for Cd, 70% for Pb and 60% for Zn by a covering of at least 40 cm uncontaminated soil. Subsequently, the additional installation of a technical barrier such as geotextiles did not appear to be necessary. However, it should be noted that plants developing an intensive and deep-reaching root system may require a thicker soil cover in the absence of geotextiles. Figure 5.1 shows the requirements with reference to wheat grown in contaminated soil. It was possible to achieve acceptable cadmium and zinc concentrations by depositing at least 70 cm. This thickness was sufficient to diminish the transfer soil to plant. Generally speaking, a soil layer of 40 cm was obviously not able to eliminate completely contact between the roots and the polluted soil. With increasing density and development of the root system that is typical for agricultural crops with a long vegetation time the thickness should be increased simultaneously in order to interrupt the transfer of contaminants into the root tissue (Delschen 2000).

It has been observed that in some cases only a relatively thin cover of clean material (e.g. 30 cm) was deposited on the remaining contaminated soil and afterwards ploughed in. Consequently, clean and contaminated material had been mixed, resulting in reduction of the contamination. According to the Waste Management Laws this idea is prohibited in a number of countries and generally not recommended.

Soil cover involving the construction of geotextiles, bentonite layers and drainage layers means considerable additional work. In Fig. 5.2 an optimised surface cover is illustrated. Above the contaminated soil a combined lining is visible, consisting of a two-course bentonite layer (thickness 15–30 cm) and a water-impermeable geotextile with a thickness of a few cm. Generally speaking, a single barrier cap is appropriate for insoluble contaminants, while composite caps are recommended for more soluble contaminants such as hexavalent chromium. These layers are overlain by a gravelly drainage layer (thickness 10–20 cm), a protective layer (45–60 cm)

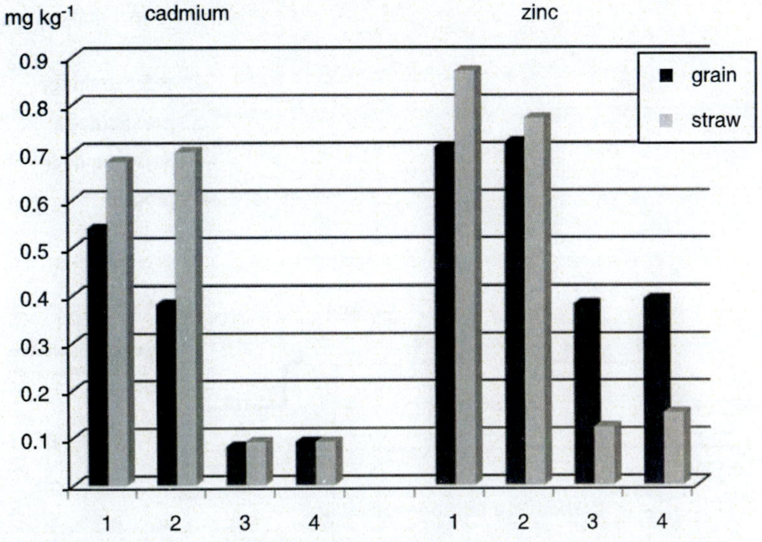

Fig. 5.1 Cd and Zn concentration in wheat (mg kg^{-1} FM) related to different cover systems installed on highly contaminated soil (Cd: 10.3 mg kg^{-1}, Pb: 601 mg kg^{-1}, Zn: 1,830 mg kg^{-1}). Cover thicknesses: *1* = none, *2* = 40 cm topsoil, *3* = 40 cm topsoil + 30 cm subsoil, *4* = 40 cm topsoil + 65 cm subsoil (Data from Delschen 2000)

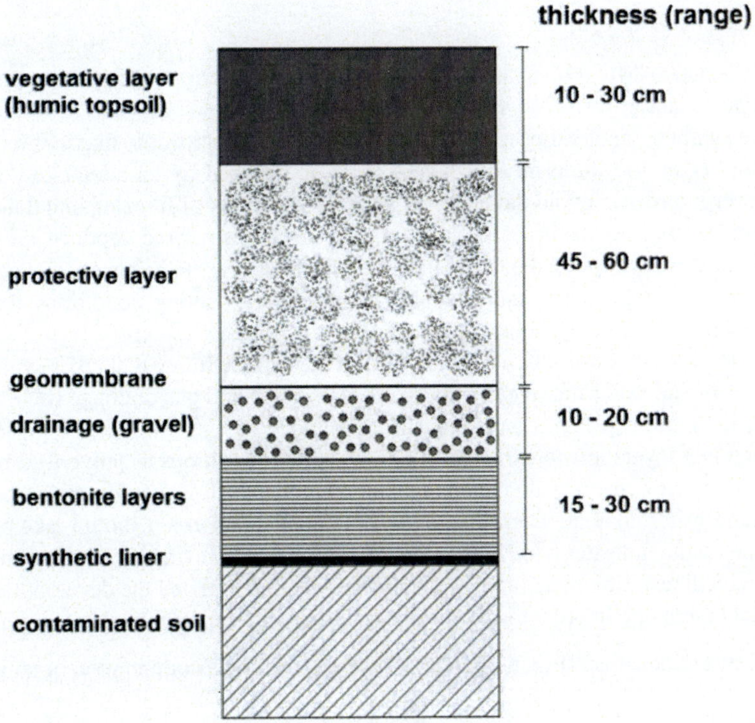

Fig. 5.2 Standardised surface cover of a residential area with gardens

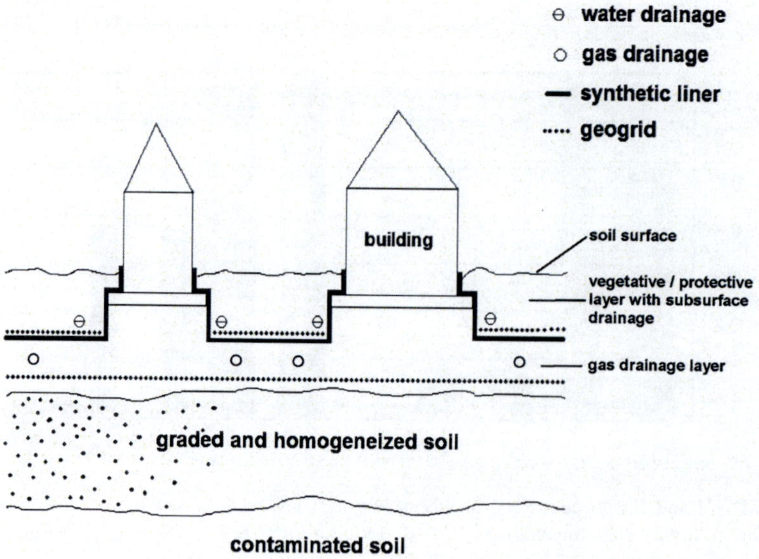

Fig. 5.3 Remediation concept considering water and gas drainage, use of impermeable synthetic liners and geogrids for liner protection and reinforcement of the ground

and a vegetative layer (humic topsoil varying from 10 to 30 cm). The latter is necessary to support plant growth (NJDEP 1998; Genske 2003).

The synthetic geotextiles must be principally located below the maximum depth of frost. Thus, the impermeable layers must be covered by an adequately thick soil layer in order to avoid damages caused by the actions of freezing and thawing. The construction of the drainage layer should also consider the depth of the frost penetration. In any case, a depth of more than 80 cm appears to be beneficial because of higher frost resistance and concomitant optimised rooting conditions for the vegetation.

A remediation concept based on the idea of the surface lining that takes synthetic liners, geonets, water drainage (above the geotextiles) and gas drainage (below the geotextiles) into consideration is introduced in Fig. 5.3.

Bentonite layers are constructed in relatively dry conditions as pulverised material. The water content which exists during the construction should not fall below the future water content. Therefore, a pre-drying of the material should take place. If the bentonite indicates a too high water content, a rapid gravimetric drainage has to be expected, resulting in a volume reduction of the material and the development of permeable fissures. In order to compact the material different techniques are applied:

- Sheep's-foot rollers (Fig. 5.4) compact the material by a combination of tamping and kneading. The technique is suitable for clayey materials. Three to five passages can compact a layer thickness of 30 cm. Similar results can be reached using pneumatic tyre rollers as long as they weigh a lot (approx. 40–50 t).

Fig. 5.4 Sheep's-foot roller

- A few smooth-drum roller (Fig. 5.5) or smooth-drum vibrator passages can follow, although this equipment is more successful in compacting gravelly and sandy materials. For this reason, the only use of conventional smooth-drum rollers or vibrators is not recommended.

The compaction equipment consists of rollers which are filled with water or sand to increase their weight. Its principal function is to reduce the water content and to homogenise the material (McCarthy 2007). Figure 5.6 shows a large-scale project in Osnabrück, Germany, where a contaminated landfill was covered using different equipment such as crawler excavator (left), bell tipper truck (centre) and sheep's-foot roller (right)(see Sect. 4.2.2). Alternatively, impermeable layers can also be constructed using bentonite mats, which overlap one another by 0.5–1 m.

Normally, in the case of a composite cap the compacted clay layer is overlain by the synthetic liner (NJDEP 1998). However, this construction demonstrates some disadvantages which should be taken into account. For this reason, the construction of the synthetic liner under the bentonite layer might be an alternative possibility.

Fig. 5.5 Smooth roller

Fig. 5.6 Construction of a landfill cover system in Osnabrück, Germany

Table 5.1 Influence of earthworms and moles on soil transport potential (Data from Hartmann 2003)

	Earthworms (average values)	Earthworms (extreme values)	Moles
Population density ($g\ m^{-2}\ a^{-1}$)	20–300	up to 600	–
Diggings depth (m)	2–3	up to 7	up to 0.7
Transport to surface ($kg\ m^{-2}\ a^{-1}$)	1.9–2.8	up to 7.0	0.7 (up to 1.9)

The main reason for this method of construction is the exclusion of an absolute drying of the bentonite material that otherwise would cause damage to the geotextile above, which is additionally compacted by the superimposed load of the covering soil. Furthermore, a damaged geotextile in combination with a shrinking bentonite layer causing fissures and cracks increases the preferential flow downward. Because of the latter aspect it is generally recommended to use a combined construction consisting of both bentonite and geotextile.

Moreover, a combined construction might prevent bioturbation and reduce permeation of e.g. polar substances. Permeation is defined as migration of contaminant molecules through the geotextile based on diffusion from a solution with high concentration to a solution with less concentration. This process can occur downwards, upwards and laterally irrespective of the mass flow. The permeation decreases with increasing thickness of the geotextile and with increasing size of the molecules. Thus, pollutants chlorinated on a higher level such as compounds consisting of more than four chlorine atoms are less susceptible to permeation. Furthermore, contaminants with polar properties tend to migrate more intensely than non-polar substances.

The potential of the bioturbation to disturb and mix the constructed layers in the absence of artificial geotextiles should not be underestimated. As shown in Table 5.1, earthworms and moles are capable of deep digging and transporting considerable soil masses in the course of time. It is calculated that in soils up to 1.3 kg m^{-2} $year^{-1}$ are transported from a depth of 10 cm to the surface and approximately 0.5 kg m^{-2} $year^{-1}$ from a depth of 50 cm to the surface because of the mixing process of both earthworms and moles. Moreover, further animals such as mice and the muskrat (*Ondatra zibethicus*) may take part in bioturbation. Thus, the involvement of geotextiles such as geomembrane and fabric seems to be adequate to prevent mixing of contaminated and uncontaminated soil.

Impermeable clay layers are usually made of bentonite, a clayey material of different colours. The aim of using the material is to achieve extreme physico-chemical properties, namely hydraulic conductivity and cation exchange capacity (CEC). Generally, clay materials can achieve a low permeability of 10^{-7} m s^{-1}, if they are compacted (NJDEP 1998). Materials reaching a hydraulic conductivity of less than 10^{-9} m s^{-1} are optimised bentonite layers. The clay mineral kaolinite containing bentonite would already achieve this aim. However, illite-rich or smectite-rich bentonite is more beneficial, since the CEC reaches a high value simultaneously.

Table 5.2 Characteristics of clay minerals (components of bentonite) (Data from Black 2009; Bradl and Weil 2008, modified)

	Kaolinite	Illite	Smectite (montmorillonite)
Specific surface ($m^2\ g^{-1}$)	5–20	50–100	700–800
CEC ($mmol_c\ 100\ g^{-1}$)	3–15	10–40	80–150
Achievable hydraulic conductivity ($m\ s^{-1}$)	10^{-9}	10^{-10}	10^{-12}

Table 5.3 Geotextiles used for soil remediation purposes; data based on solicited offers from different vendors

Product	Properties	Thickness (mm)	Pore width (mm)	Material
Fabric	Permeable, low elasticity	2.4–8.0	0.08–0.53	PP, PEHD
Geomembrane	Permeable, high elasticity	0.2–1.5	<10	PP, PEHD
Synthetic liner	Impermeable	0.8–6.0	–	PEHD, PE, PVC
Geosynthetic clay liner	Impermeable	6–7 (dry)	–	PP, PEHD, between these materials pulverised sodium bentonite is filled in
Geonet	Permeable	2.9–4.9	>10	PP, PEHD

The CEC of kaolinite (3–15 $mmol_c\ 100\ g^{-1}$) and halloysite (5–14 $mmol_c\ 100\ g^{-1}$) does not indicate optimum adsorptive properties. Consequently, most bentonite materials consist of smectite based on the montmorillonite type or on vermiculite, which displays a CEC of >40 $mmol_c\ 100\ g^{-1}$ (NJDEP 1998). A low hydraulic conductivity leads to a reduced water percolation, while high CEC values are responsible for a high adsorption potential for cationic substances such as heavy metals. In Table 5.2 the physico-chemical characteristics of the three clay minerals which are components of bentonite are summarised.

Geotextiles (Table 5.3) which additionally minimise the water percolation to a great extent are called impermeable synthetic liners. Some authors use the term geomembrane as an alternative for impervious thin sheets of rubber or plastic (Nathanail and Bardos 2004). Their thickness ranges from 0.8 to 6.0 mm only and visible pores cannot be discovered by human eyes. Geosynthetic clay liners which contain pulverised sodium or calcium bentonite sandwiched between two geotextiles or bonded to a geomembrane might reduce the hydraulic conductivity strongly. They are constructed in wet conditions, although they exhibit enormous swelling properties caused by the clay mineral montmorillonite.

The geotextiles mentioned are responsible for very low water percolation. In contrast, fabric and geomembrane are permeable by water and vapour, but they can surely prevent bioturbation. These geotextiles vary considerably with regard to thickness and pore width. They consist of synthetic fibres made into a flexible porous fabric, allowing movement of fluids and accordingly used for separation and

Table 5.4 Minimum thickness of uncontaminated soil deposits to exclude human health risk (Data based on BBODSCHV 1999, modified)

Land-use type	Minimum thickness (cm)	Recommendations
Playgrounds	35	Geotextiles, permeable by water
Gardens (vegetable plots)	60	Sometimes geotextiles, permeable by water No difference between flower-gardens and vegetable or kitchen gardens No plantation deeper than 60 cm
Agricultural sites (cropland)	100	Possibility to plant crops with intensive root development
Parks	10 (–35)	Areas without digging activities, tillage, cultivation Complete plant cover (herbs, grasses)
Sports fields	5 (–10)	Mechanically compacted layers

reinforcement of layers, filtration as well as drainage. They are mostly formed in the case of soil contamination impacting the pathways soil-human and soil-plant and are not focused on the groundwater pathway. In addition, they are often used to prevent migration of fine particles to the drainage layers below.

The fabric exhibits a relatively low elasticity, hence it is not recommended in the presence of stony and sharp-edged materials underneath it. In contrast, geomembranes are more tolerant to sharp-edged material due to their higher elasticity. Nevertheless, the geotextiles should be installed on a smooth surface and the presence of sharp objects which may puncture the liner should be minimised.

Moreover, geonets (geogrids) are plastics formed into a very open configuration and used for reinforcement. They are aimed at the protection of other geotextiles from damage during the construction phase.

The materials common to all geotextiles are polypropylene (PP) and high-density polyethylene (PEHD). Generally, the construction of fabric and geomembrane requires careful handling. In particular, once in a while the use of wheel loaders and trucks is responsible for damage during the construction period. The danger of unnecessary damage can be decisively reduced, if a manually operated protective sandy layer of up to 15 cm thickness is formed before the following deposits are technically made.

Both fabric and geomembrane are intended to be permeable by water in the course of time. As time goes on the permeability will decrease, since fine particles block the openings (blocking) or even finer particles migrate into the material (clogging). This problem can be minimised by constructing a thin sandy layer above the geotextile, which will prevent the downward movement of silty and clayey grains.

The thicknesses of the top layers vary in relation to the planned use of the area of concern. For instance, the humic topsoil can be reduced to grow a lawn and the subsoil thickness depends on the plants to be chosen (see Sect. 2.3). Recommended thicknesses of uncontaminated soil layers are shown in Table 5.4. They are in accordance with the regulations published in the German Soil Protection Ordinance (BBODSCHV 1999) to a great extent. For instance, for playgrounds frequently

Table 5.5 Recommended soil handling depending on the soil moisture

Soil moisture (hPa)	Classification	Recommendation
<100	Very moist	Use of the machines on paved paths and construction roads, no excavating or depositing
100–150	Moist	Use of crawler vehicles on gravel paths and excavator support mats, soil handling using excavator support mats
>150–250	Moist to dry	Soil handling in consideration of vehicle weight and soil texture, crawler vehicles preferred
<250	Dry	Soil handling generally feasible with wheel vehicles

consisting of sand-pits 35 cm are estimated to be sufficient, in gardens used as vegetable plots approximately 60 cm are mentioned. In parks where no digging is carried out and where the vegetation cover is usually dense the thickness can be limited to 10–35 cm, and sports fields with mechanically compacted top layers require a thin uncontaminated layer amounting to only 5–10 cm.

Nevertheless, the thickness appears to be more problematical in the presence of vegetation that needs deep rooting zones. Cropland, for instance, cultivated with species such as wheat or maize demands at least 100 cm rooting depth and gardens in which shrubs and trees are planned should have a similar thickness as well. It is well-known that the roots of plants used for human consumption are able to take up nutrients as well as contaminants in layers below 60 cm. In urbanised areas a thick cover makes also sense because of infrastructure pipes which will be planned, constructed or repaired at a later date.

The covering process with uncontaminated humic topsoil and subsoil has to be carefully carried out to avoid unnecessary soil compaction. It should be noted that after the remediation the site is going to be re-used. Regarding future use, for instance as a residential area with gardens, the clean-up procedure is not expected to cause compaction which would impede usability later on. For this reason, the intended planning should include soil protection, as would generally be done in relation to construction measures. The excavation and deposit operation must pay attention to the weather conditions and the subsequent soil moisture. During rainy periods time-related alternative planning is required. In Table 5.5 the ability to handle soil depending on the soil moisture is introduced.

To minimise compaction of the covered and clean material the equipment used (see Sect. 4.2.2) should have the following characteristics:

- The compression to the contact surface of the vehicles (relation between weight of vehicle and soil surface) should be lower than 0.5
- Crawler excavators are preferred
- The wheel load of wheeled vehicles should not exceed 2.5 t and the tyre pressure should be adjusted so that it is as low as feasible.

The use of the clean topsoil deposited at the top of the rehabilitated sites at the end of the remediation procedure should obey strict rules including not mixing with

Table 5.6 Demand for mineral fertiliser (kg ha^{-1} a^{-1}) in landscaping related to different vegetation (Data from Beier et al. 2003, modified)

Vegetation	N	P	K	Mg
Perennial herbs, high demand	50–100	17–26	50–65	5–7
Perennial herbs, low demand	25–50	9–17	35–50	4–5
Ornamental shrubs	25–50	13–17	50–65	5–6
Shrubs, low demand	0–30	0–17	0–50	0–5
Roses	50–100	26–44	65–130	6–12
Representative lawn (English lawn)	150–200	24	73	13
Lawn for sunbathing	60–100	20	62	6
Sports turf	240–320	35	100	18

subsoil material as well as absence of technogenic substrates such as construction debris, slag, ashes, waste and mineral oil stemming from leaky construction vehicles.

In any case, future use has to be facilitated depending on the vegetation planned. Therefore, attention should be paid to the future plantation in relation to the macronutrients. The growth conditions are less favourable in the presence of nutrient deficiency. As long as the substrate mixture appears to be deficient in nutrients, fertilising should be considered based on the standards mentioned in Table 5.6.

The topsoil quality can be ameliorated by mixing of the mineral soil and some additives and by measures aimed at different quality characteristics. The addition of ameliorants such as organic manures and mineral fertilizers sometimes coupled with physical treatment like scarifying is common. As exhibited in Table 5.7, the additives and measures are associated with the generation of beneficial soil properties. Subsequently, they should be applied depending on the soil quality present. For instance, loamy-clayey material requires an increase in the air capacity, which can be conducted after addition of gravel, pumice, polystyrene, etc., whereas sandy material poor in nutrients might be treated using, for example, compost and legumes.

The use of topsoil derived from agriculturally used areas shows disadvantageous features, in particular the cost of transportation from areas far away from the site to be treated and, from the ecological point of view, the fact that the topsoil is a non-renewable resource. In contrast, the amendment of organic by-products which tend to be increasingly available in urban environments might be a cheap alternative. Apart from the classical organic fertilizers such as compost and bark mulch, paper mill sludge that indicates high organic matter and carbonate content is a good amendment applied in a quantity of 50–200 t ha^{-1} to the upper 15 cm of the soil. This by-product usually exhibits low concentrations of contaminants such as heavy metals and PAH (Boni et al. 2004).

The predominant purpose of the surface cover is low water percolation downward, thus avoiding leaching of contaminants into the groundwater. Accordingly, several measures should be taken into account to reduce the danger to the groundwater. The soil surface should not be flush, a slope gradient of 3–5% leads to accelerated runoff and subsequently minimised water infiltration. The runoff should be collected

Table 5.7 Additives and measures to rehabilitate mineral soil after depositing

	Aim	Remarks
Additive		
Compost	Increase of organic matter content, biological activity, accelerated water capacity	Well-biodegraded, poor in artefacts like plastic, hygienically acceptable, odourless, uncontaminated, C/N ratio <25
Sand, gravel, pumice, lava	Accelerated air capacity and hydraulic conductivity	No slag used
Polystyrene	Accelerated air capacity, structural improvement	
Lime	pH increase, structural improvement, enhanced biological activity	
Measures		
Soil pre-treatment	Accelerated air capacity, optimised mixing	Top down deposit, shallow tillage (15 cm depth), establishing of an accurate formation level
Mulching	Protection form erosion, deflation and compaction	Thickness 2–3 cm only, compost, straw or bark mulch application
Legume seed	Accelerated air capacity, increased organic matter content and nutrient capacity (in particular, nitrogen)	E.g. clover, lupines, alfalfa

in ditches located laterally. On the other hand, because of the reduced soil stability on steep slopes the slope gradient should be limited to 33% in the context of soil erosion (NJDEP 1998).

The plantation of vegetation with high transpiration might help to reduce percolation as well. In general, the avoidance of ponds, wet biotopes as well as rain storage reservoirs is necessary with reference to additional percolation.

Another aspect to be taken into account is the influence of groundwater in alluvial floodplains or at depressed sites. Due to the capillarity it is expected that groundwater can rise upwards and accordingly reach the uncontaminated layers constructed above. In this case groundwater first comes into contact with the contaminated material below and secondly with the unpolluted soil. For this reason, attention should be paid to the groundwater capillarity and therefore the total thickness should be altered according to the capillarity expected. The capillarity depends mainly on the texture as seen in Table 5.8.

In the context of soil cover the adaptation to the adjacent sites might play a further detrimental role. Because of the limited area, particularly within densely constructed residential and industrial areas, the handling opportunities are strongly restricted. If the construction of L-shaped stones is not feasible, the only way seems to be a combination of excavation and cover on the periphery of the contaminated land. The soil management that is very important for the containment solution and particularly for soil cover systems is explained in Sect. 4.2 in detail.

Table 5.8 Calculated thickness of soil cover for cropland purposes preventing pollutant transfer into plant roots (Data from AG Boden 2005, modified)

Texture	Effective root depth (cm)	Groundwater capillarity (cm)	Calculated thickness (cm)
Coarse sand	50	60	110
Medium sand	60	80	140
Fine sand	70	90	160
Loamy sand	80	120	200
Sandy silt	90	200	290
Clayey silt	110	170	280
Loamy silt	110	160	270
Clay	100	90	190
Peat	30	90	120

5.1.2 *Sealing*

The complete sealing of the surface can also be a technical solution for highly contaminated land. The asphalt or concrete layer must be properly designed to withstand the possible car traffic on the surface. Thus, the cover must have a total thickness of 10–20 cm, the subsoil must be well compacted and the cap surface must be graded and must incorporate a drainage system to avoid water ponding (NJDEP 1998). The sealing does not prohibit future redevelopment of the contaminated land because structures can be built on top of the sealing system (see Sect. 2.1).

From the soil protection point of view, a complete sealing of the contaminated area using concrete and asphalt might be a less preferred way of soil cover. There is no doubt that even overbuilding can be a solution as well (Fig. 5.7), if gas migration is not of concern and the performance is monitored with e.g. observation wells. The direct contact soil – human is interrupted and a groundwater contamination caused by percolating water can be excluded. However, the result of that treatment is the transformation from a medium enabling the soil functions to an artificial structure that cannot be called soil any more.

5.2 Side Barriers Installation and Encapsulation

5.2.1 *Side Barriers*

Soil cover or soil cap is a treatment approach that is reduced to the top of the soil only. The side barrier installation is focused on the lateral migration of contaminants but the construction of a cap and side barrier system is less efficient, if contaminants can migrate below the barriers. Therefore, in most of the case studies a bottom sealing is added based either on a natural impervious layer originally present or on an

Fig. 5.7 Contaminated land overbuilt and sealed to interrupt the direct contact soil-human; photograph taken during the construction process in Oulu, Finland

artificially created bottom sealing construction. Clearly, to be effective the barrier systems must be keyed into a natural layer with very low hydraulic conductivity or artificially created horizontal barriers must be constructed. The complete encirclement of the contaminated soil is termed encapsulation. The techniques have been adapted from water-tight encapsulation in the field of construction.

Apart from lateral migration of contaminants, the construction of a side barrier will become necessary, if the contaminated soil is located on a slope. Side barrier installation is also carried out in combination with pneumatic measures (see Sect. 7.2) or it is intended to isolate contaminated material, while microbiological treatment takes place. Moreover, side barrier systems are occasionally combined with the groundwater treatment (see Sect. 7.1). For instance, the walls can shield the contamination source against the groundwater flow and they are consequently upgradient barriers. They can be also downgradient barriers by trapping the groundwater flow. In cases where an insoluble contaminant phase floats on the groundwater surface (e.g. LNAPL) a hanging wall without bottom sealing can even be sufficient to trap the contaminants.

One must differentiate between various executions of the construction. The sheet-pile wall of steel which includes joints between the single piles reduces the water permeability but a hydraulic conductivity of less than 10^{-4} m s^{-1} cannot usually be achieved. The long-term blocking of the small joints, however, decreases the permeability in the long term. But some patented approaches to sheet piling have been developed where the joints between the different piles are sealed. This special lock system reduces the conductivity to more acceptable values (Bradl 2005) (Figs. 5.8 and 5.9). The walls are driven into the ground with a steam-driven or

Fig. 5.8 Sheet pile wall of steel to protect a stream against erosion of contaminated soil

Fig. 5.9 Lock system of a sheet-pile wall of steel (*top view*)

pneumatic pile driver. It is possible to install the piles in various types of soil from gravel to clay. The thickness of the sheet-pile walls measures only 1–2 cm but a maximum depth of approximately 20–30 m can be assumed. An Achilles heel of the sheet-pile walls of steel is the corrosion of the steel, in particular in soils indicating low pH values (e.g. pH 2–3). Thus, the pH requirements must be taken into account. On the other hand, by using hot-dip galvanising protection against corrosion can occur. The life expectancy of the walls is between 7 and 40 years (NJDEP 1998; Genske 2003). Furthermore, sheet pile walls can be constructed using aluminium and wood as well, but the material usually used is steel.

Another technique used is termed construction of a bored pile wall. The thickness varies from 0.6 to 0.8 m and the maximum depth measures approximately 25 m. The rounded piles (soilcrete columns) are constructed in a cased manner. In the borehole a high pressure jet is used and the material is transported upwards by a rotary casing. After retracting the equipment the resulting void is filled with the suspension. Usually, concrete containing gravel, sand and cement is used. The piles are introduced on an overlapping basis similar to the pilgrim's pace method to prevent permeable joints in future.

Moreover slurry walls with and without excavation are a common treatment. In the case of missing excavation a displacement technique is applied. The displacement treatment produces walls of a thickness smaller than 10–20 cm and can be operated to a depth of 20–25 m in skeleton-poor soils. After soil dredging the emptied voids are then filled with a cement-water-bentonite-mixture (slurry mix) that can additionally contain sand and gravel. Each section is driven in with an overlap to the previous one in order to provide a continuous barrier. This technique is faster than the sheet piling mentioned above.

As an alternative to the displacement technique the so-called jet grouting can be applied. Grouting is the pressure injection of a variety of special fluids into pores and voids of the soil to seal and strengthen it. First a line of holes is drilled to the planned depth and afterwards the fluid is injected. The injected grout should connect the adjacent holes, leading to the formation of a continuous and impervious barrier. This process results in the development of a so-called grout curtain, subsequently reducing the water permeability greatly. The technique is favoured in the case of porous and fractured material where other approaches are impractical. This technique is available for nearly all soil texture classes from gravel to silt but jet grouting in clayey material is less efficient. Injected grout curtains were even used in bedrock.

Most barrier systems are based on excavated trenches which are filled with the barrier material (slurry wall). To achieve depths of more than 2 m the trenches must be filled during excavation. For this reason, a slurry is simultaneously infiltrated that exerts sufficient pressure to stabilise the trench walls but this suspension must not drain away quickly. This material is necessary firstly to maintain the excavation and secondly it must harden in order to complete the barrier construction. But in some examples a second introduction of barrier material is applied, displacing the material originally infiltrated. While the second slurry is introduced from the bottom to the top, the first suspension will be pushed up and collected aboveground (see below).

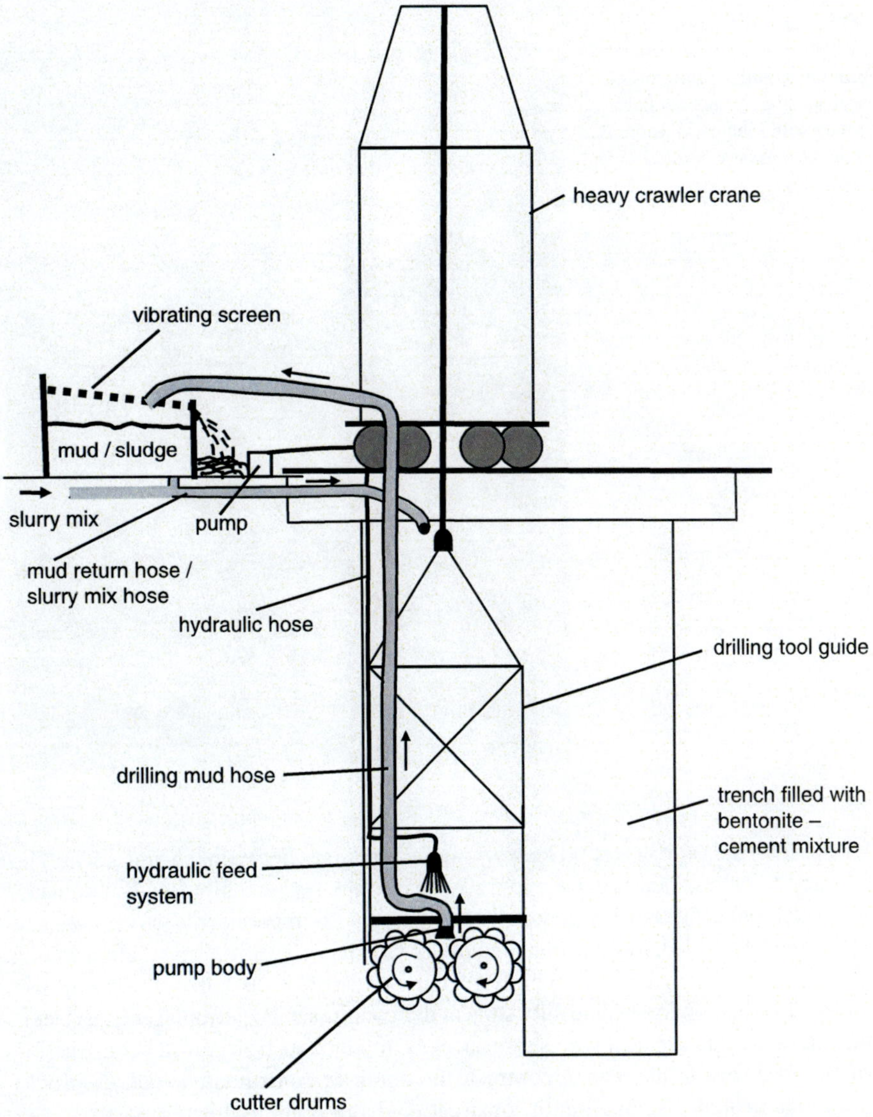

Fig. 5.10 Principle of a hydro-rotary machine (hydrofraise)

The trenches are not usually excavated by shovel excavators. The trenches are excavated to the desired depth by grab buckets, clamshells and hydrofraises. For the construction of slurry walls as shown in Figs. 5.10 and 5.11 a hydro-rotary machine (hydrofraise) must loosen the soil by using high pressure jets. Water that cut into the subsoil is used and changes the soil into a fluid suspension that is pumped off the ground and treated or piled aboveground. The displaced soil is replaced by the

Fig. 5.11 Hydrofraise installing a side barrier to minimize pollutant migration into an adjacent residential area (With kind permission of IGfAU company, Melle, Germany)

barrier material (slurry mix). The working width ranges between 0.5 and 3.5 m. The emptied trenches measuring up to 150 cm in thickness are filled with a cement-water-bentonite-mixture, occasionally combined with impermeable geotextiles and achieving a feasible hydraulic conductivity of less than 10^{-8} to 10^{-9} m s^{-1} after drying. The hydro-rotary machine is mainly focused on sandy soils without a strong soil compaction and on loamy and silty soils in dry conditions. An advantageous aspect is that the technique can deal with subsurface obstructions such as gravelly construction rubble to a certain degree. In contrast, the depth of consolidated rock (bedrock) cannot be treated. The maximum construction depth is approximately 80 m.

In general, a distinction is made between two execution approaches called one-phase and twin phase operation. In the case of the first one the trench is stabilised during excavation with the help of single slurry consisting of bentonite and water. The slurry remains in the trench and tends to solidify. In the case of the twin-phase technique the previously filled bentonite slurry is replaced by a second slurry suitable for final solidification because of the hardening additions as mentioned below. The exchange of the slurry is conducted using the contractor method as a basis, meaning that the filling process of the second phase occurs from the bottom to the top while simultaneously the first slurry is pushed upwards. Both slurries must indicate density differences of at least 0.5 g cm^{-3} (Bradl 2005).

The barrier is constructed using overlapping sections as shown in Fig. 5.12. Contaminant migration can be minimised with reference to a sequence of excavation 1-3-5, etc./2-4-6, etc., which is called pilgrim's pace method. After approximately two days the primary slurry has considerably solidified, enabling the construction of the secondary panels in an overlapping manner. Therefore, the contact between the panels is very closely in line with very low hydraulic permeability. Again, the targeted depth should ensure that impermeable natural layers are reached in order to prevent migration under the wall.

Irrespective of the technique used for construction of slurry walls there are distinct injection materials which are preferentially related to the soil texture (Fig. 5.13). It has been found that the following mixtures are recommendable:

- 30–50 kg Na-bentonite + 170–220 kg cement + 800–900 L water + aggregate (sand, gravel)
- 150–310 kg Ca-bentonite + 180 kg cement + 800–900 L water + aggregate (sand, gravel).

For example, these slurries reveal resistance against flocculation of solid material, since the solid phase and the liquid phase must not be separated during operation. If it is necessary to enhance the impermeability of the slurry, additionally geotextiles such as HDPE plastic liners can be inserted.

The technique of cryogenic barriers is used for groundwater control and to strengthen walls at excavation sites. Freeze pipes are normally used that fill e.g. liquid nitrogen into the ground. The barriers are constructed by artificially freezing the pore water in the soil to decrease the water permeability. Afterwards, the refrigeration system can be turned off if the construction process is finished (Nathanail and Bardos 2004). Previously, this technique was considered to be a temporary method to isolate mobile contaminants but in the meantime the method is also used for longer periods till alternative methods can be applied, because the hydraulic conductivity potentially feasible reaches 10^{-12} m s^{-1} (Bradl 2005). The main advantages of freezing the ground are no soil excavation, rapid removal by stopping the cooling as well as adaptability to depths of more than 100 m.

The use of the specialised technique can cause a number of problems which must be taken into consideration. High compaction exacerbates the drilling progress, particularly in compacted sandy soils, silty and clayey soils with low moisture content and skeleton-enriched soils. After drying of the suspension discontinuities and holes can arise, since the sand particles of the suspension tend towards a quicker sedimentation than the silty and clayey components. Furthermore, water can flow laterally in the surrounding soil, leading to a decreasing suspension level. Accordingly, additional suspension has to be infiltrated. In general, the procedure should occur quickly because the slurry shows thixotropic behaviour.

The emission of noise and vibration during the operation time means a disadvantageous impact as well. The most problematical point to be considered, however, relates to the long-term safety and the limited opportunity to monitor and repair the wall. In particular, acidified organic soils (organic acids) and soils exposed to intensive weathering appear to be difficult with regard to the construction of the usually alkaline

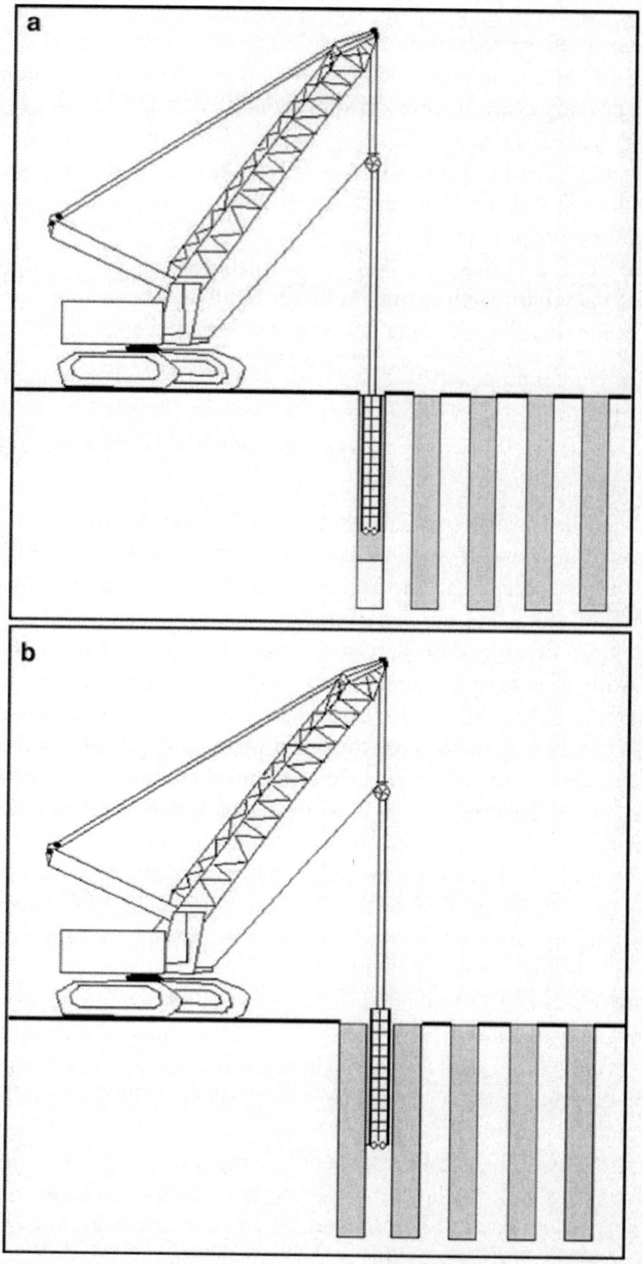

Fig. 5.12 Sequence of slurry wall construction based on the pilgrim's pace method and the contractor method. (**a**) Excavation of panels 1-3-5 etc. and simultaneous infiltration of the first slurry. (**b**) Excavation of panels 2-6-10 etc. and simultaneous infiltration of the first slurry.

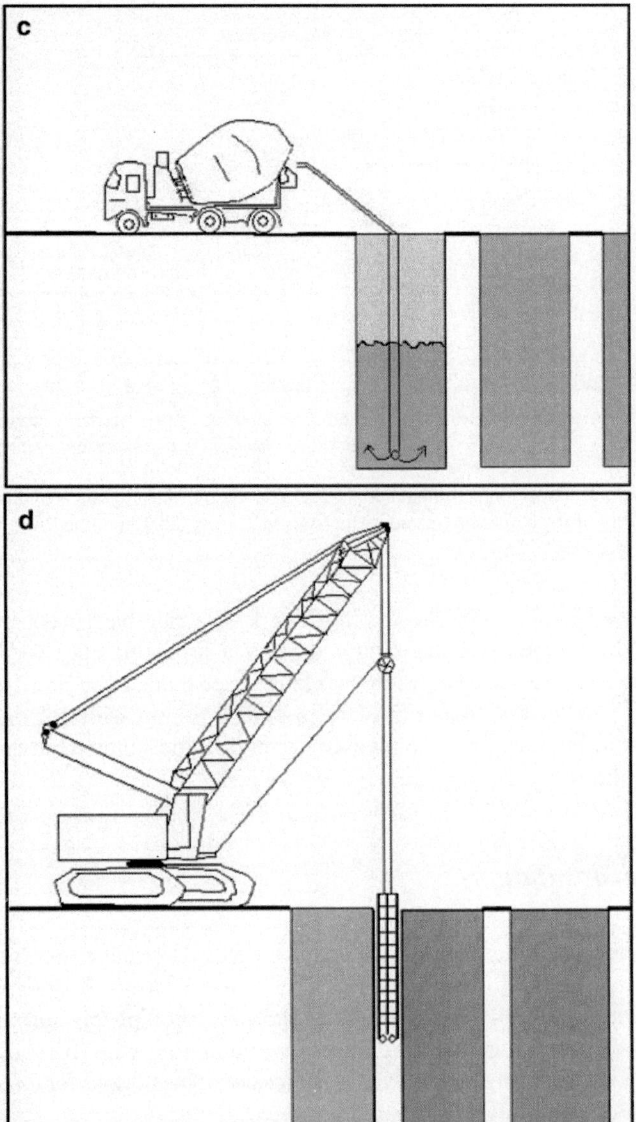

Fig. 5.12 (continued) (**c**) Infilling of the second slurry using the contractor method in the panels 1/2/3, 5/6/7, etc. (**d**) Excavation and slurry infiltration procedures in the remaining panels

side barriers. In general, the long-term resistance to aggressive substances should be taken into account.

Moreover, the theoretical deformability of the wall, leading to visible damages, should be mentioned, particularly in mining areas, where subsidence is frequently

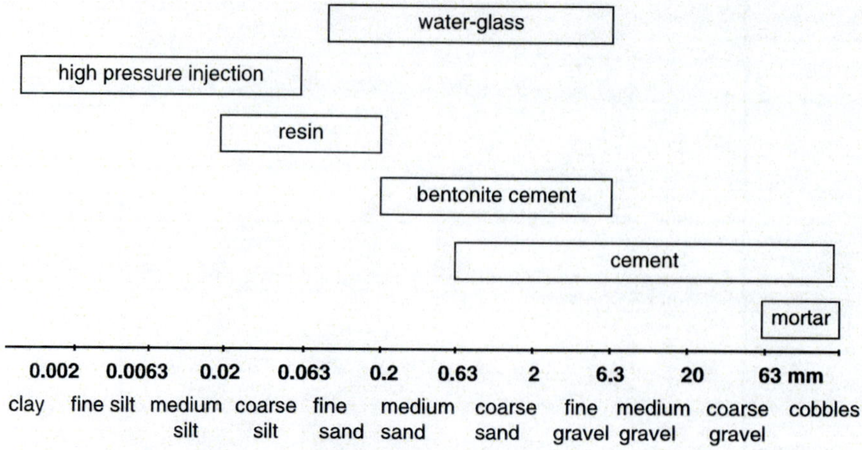

Fig. 5.13 Recommended injection materials for side barrier installation and bottom sealing regarding texture of the soil to be treated (Data from McCarthy 2007, modified)

observed (see Sect. 3.4). Within the framework of a side barrier construction the working width of the machines used can cause a problem, since for instance in urbanised and industrialised areas the working space tends to be limited. Hence, if the space is smaller than about 8–11 m (normal working width of the machines concerned), the big machines cannot excavate or drill the soil without danger to the adjacent facilities.

5.2.2 *Encapsulation*

As already mentioned, the term encapsulation is defined as a measure which includes soil cover, side barrier installation as well as bottom sealing. A synonym for this is the creation of a fossil subsurface, since the soil is completely surrounded with impermeable material and accordingly an interaction with the adjacent soil (exchange of nutrients, organisms, etc.) is not feasible. The reasons for the subsequent bottom sealing are as follows:

- The bottom of the contaminated soil is located underneath the groundwater table (direct contact between contaminated soil and groundwater)
- The distance between the bottom of the contaminated soil and the groundwater table is short, leading to an expected accelerated leaching of pollutants
- It was not possible for the side barrier installation to reach impermeable natural layers which prevent the lateral migration of pollutants under the walls (see Sect. 5.2.1).

There are uncertainties in using the bottom sealing technique, since it is problematical to verify that a really complete sealing has been achieved. In order to

Fig. 5.14 Horizontal sealing by tube injection; the slurry (stored in the container) is pumped into the white tubes which are arranged close together (With kind permission of Wessling company, Altenberge, Germany)

construct a subsequent bottom sealing several techniques are applied. Firstly, a pore or void injection takes place with tubes in which hardening cement solution is grouted. This process can be complicated, if the added material causes a lifting of the soil surface. Alternatively, the jet-grouting treatment is chosen. The jet located at the basis of the tube loosens the soil under pressure and afterwards the suspended material is transported upwards. Simultaneously, the hardening cement solution is infiltrated to the pores and cracks caused by the high pressure jet. In any case, the injection and extraction of the suspensions occurs from the surface.

The pressure usually applied amounts to 400–500 bar. The pressure technique is preferentially applied to sandy soils but coarse silty soils have also been successfully treated in this way. The distance between the installed tubes ranges from only 1.2 to 1.5 m, requiring a high number of tubes (Fig. 5.14). To configure the tube arrangement soil permeability, viscosity of the used suspension and the maximum feasible injection pressure must be taken into consideration. The hydraulic conductivity that can theoretically be achieved varies between 10^{-9} and 10^{-10} m s^{-1}. These target values, however, should generally be challenged.

In areas influenced by groundwater the presence of remaining voids and holes causes disadvantageous impacts, since groundwater can flow into the encapsulated zone. It is well-known that hydraulic measures are essential components of soil remediation. For instance, groundwater control enables excavation to take place and some treatments such as soil vapour extraction are only possible, if groundwater level and flow direction are controlled. But groundwater level control is also of importance in conjunction with physical barriers as part of a long-term remediation strategy.

Fig. 5.15 Sound insulating wall filled with encapsulated contaminated substrates close to a motorway in Rotterdam, The Netherlands

Extraction wells are involved to increase the effectiveness of the system by creating an inward groundwater gradient. The barrier system means a complete encirclement of the contaminated site, since the side barriers are keyed into the impervious layer below and a low permeability cap is constructed at the surface. To prevent a lateral contaminant migration a continuous inward hydraulic gradient should be achieved. For this reason, it is necessary to pump the groundwater continuously out of the encapsulated zone, so that an inward gradient occurs. The extracted water must be treated on site or in the local sewage treatment facility. In addition, long-term monitoring using observation wells upstream and downstream may be required (see Sect. 7.1).

One more problem is connected with the development of toxic gases in an encapsulated zone, if present. To avoid gas emission into buildings a permanent gas drainage system must be involved, since no-one can guarantee that a slow gas migration through the walls occurs, damaging the aboveground structures such as buildings.

Recently there has been an increasing tendency to use the encapsulation approach after excavation at the soil surface. This treatment appears to be important in relation to the soil management as introduced in Sect. 4.2. Piles which contain hazardous material based on the surface lining as discussed in Sect. 5.1.1 are constructed and vegetated. The migration of contaminants is completely excluded by means of surface, side and bottom sealing. For instance, in relation to the construction of sound-insulating walls the technique is appropriate, since the artificial wall must be installed in a secure manner in any event (Fig. 5.15).

5.3 Solidification

5.3.1 Cement-Based Solidification

While encapsulation is focused on the construction of a barrier system, solidification is aimed at enclosing the contaminants in a compacted solid mass such as concrete and converting them into an insoluble and immobile form. Sufficient material is added to form a new solid mass of high structural integrity. Usually large blocks are produced which show increased strength and decreased permeability. In contrast, stabilisation is a treatment where the contaminants are converted into a less soluble form without any change of the nature of the soil and of the main soil characteristics. Generally speaking, the result of solidification is more a monolithic block than soil, while stabilisation results in a continuous soil-like mixture.

Both techniques are frequently termed stabilisation/solidification (S/S) because they use partly the same additives such as calcium-based agents and fly ash and they are applicable to both soils and sludges (USEPA 1997). Exceptionally, if both S/S techniques are required, in the first instance stabilising agents, which have sufficient time to react before solidifying agents are used, are added. Otherwise, the solidifying agents inhibit stabilisation.

Solidification is a mixing procedure of contaminated soil with binding materials and water. The treatment usually occurs *ex situ*. After excavation the material is classified and oversized material is either removed or crushed and subsequently also used. Generally, rotary drums and pug mills are used to mix the contaminated soil properly with the agents applied. The mixing procedure only takes between one and 15 min. Up to 1,000 t day^{-1} can be mixed in large plants (Bradl 2005). An optional module deals with the off-gases treatment.

In situ applications are also of relevance. This involves injecting the solidifying mass by tubes or by an auger system to a depth of up to approximately 10 or 30 m respectively (NJDEP 1998; Bradl 2005). The injection is carried out from the top similar to the bottom sealing, as described in Sect. 5.2.2 in more detail. The main difference between both approaches is the fact that the bottom sealing requires a relatively thin impervious layer that is constructed at the desired depth, while solidification refers to the entire contaminated zone, which can reach a thickness of several metres. The delivering of the agents to the subsurface is often difficult because a uniform mixing is hard to guarantee and control. Therefore, the auger system dealing with augers of 1–3.7 m in diameter is more effective, since the chemicals are pumped in a slurry form into the soil after hollows have been drilled. The borings creating a circular column overlap each other to cover the entire contaminated site. Alternatively, the jet grouting technique is occasionally applied as described in Sect. 5.2.2 in the context of bottom sealing.

Solidification is effective for soil and sludge mainly contaminated with metals. Most of the binding agents are less effective in relation to organic pollutants. In general, anionic components such as cyanides must be removed before the solidification process starts. Anionic organic compounds serving as chelating agents as well as sulphates and chlorides should not be present. The solidifying agents are very alkaline materials acting as a pH buffer and subsequently ensuring the reduction of the leachability of cationic compounds. The negatively charged oxyanions, however, reveal an enhanced solubility, since anions cannot be absorbed in alkaline conditions. In particular, remediation by solidification is only recommended to a limited extent for soils containing elements which form oxyanions such as antimony, arsenic, beryllium, chromium, selenium and vanadium. It should be noted that the disappointing mobility reduction of anions is usually not the case, if secondary minerals like ettringite are formed (Chrysochoou and Dermatas 2004).

With reference to the formation of insoluble precipitants arsenic and mercury do not form low solubility precipitants at the high pH values which are associated with this technique. Toxic chromium (CrVI) behaves in the same way but water addition tends towards a reduction to the non-toxic species CrIII, which forms an insoluble hydroxide. Moreover, soil-containing oil and grease might also be problematical to operate. Thus, the treatment is mostly applied to soils contaminated with cationic metals.

Clayey soils and humus-rich soils are less interesting for this technique, since clay can adsorb the reactants, leading to a disturbance of the polymerisation of the binding agents, and the organic matter reduces the solidification chemistry due to the biodegradability of the substance.

In general, the volume of the treated material increases after addition of the agents. Advantageously, the permeation is lowered and the weathering effects are minimised. The main targets with reference to the solidification are a low hydraulic conductivity, which mostly varies from 10^{-5} to 10^{-8} m s^{-1}, and a strongly reduced leaching of the contaminants.

There are different solidifying agents used. The alkaline binders Portland cement and cement consisting of Ca silicates and Ca aluminates are crystallised and hardened. After water has been added hydration takes place, leading to hydrated calcium aluminosilicates such as $3CaO*2SiO_2*3H_2O+3Ca(OH)_2$ and $3CaO*Al_2O_3*6H_2O$. Moreover, the generated $Ca(OH)_2$ reacts with carbon dioxide as follows:

$$Ca(OH)_2 + CO_2 \rightarrow CaCO_3 + H_2O \quad (5.1)$$

This process describes the carbonation increasing the strength of the solidified product. In the presence of sand or coarse materials such as gravel and crushed stones concrete based on cement and water addition is produced.

Cement-based solidification reduces the mobility of metals significantly due to the formation of immobile hydroxides and carbonates, but also due to the adsorption and substitution of the metal into the mineral structure. In particular, a new hydrous calcium aluminium sulphate mineral called ettringite $(CaO)_6$ (Al_2O_3) $(SO_3)_3*3H_2O$ is created. It is a needle-like, octahedral crystal formed predominantly in hydrated Portland cement, where calcium aluminate and calcium sulphate are present. Problematically, it tends towards expansion, leading to the failure of the binding process. For this reason, attention should be paid to the velocity of ettringite formation, since a delayed formation causes unwanted expansion.

Ettringite stability is limited to the pH range from approximately 10.5–13. Below pH 9.0 it is assumed that a complete dissolution of the mineral takes place in spite of its high mechanical strength. Various isomorphous substitution of the aluminium cation in the octahedral structure is feasible, e.g. by chromium. The strongly reduced mobility of metals is related to the fixation of the metals in the ettringite lattice. Polar substances are adsorbed between the charged mineral layers and anions and cations are adsorbed to the interface layers electrostatically. Hence, ettringite and some other secondary minerals show a high fixation rate for both (organic) anions and heavy metals. They are also used for the fixation of cyanides, phenols, PAH, PCB, chloride and sulphate. In the presence of formed ettringite even the adsorption of oxyanions such as chromate, selenate, arsenate and vanadate were observed (Tewelde 2004; Chrysochoou and Dermatas 2004).

The cement-rich products show a relatively low hydraulic conductivity that can easily reach values of $<10^{-7}$ m s^{-1}. A disadvantage is that the volume of the treated material increases considerably and the solidified material is susceptible to aggressive acids. In general, physical stress can break down the solidified material, e.g. wet-dry cycle, freeze-thaw cycle and contact with reactive materials. Consequently, a long-term disintegration based on physical and chemical weathering must be taken into consideration, because a quick metal leaching possibly occurs in such conditions (La Grega et al. 2001).

The solidification based on cement improves in the presence of sodium silicate (Na_2SiO_4). Because of the high water absorption caused by sodium silicate the process can also be applied to polluted sludgy substrates. The volumetric expansion is lower, since the percentage of cement used is reduced.

Apart from the advantages mentioned in the context of cement (CaO), the solidification is better in the presence of e.g. fly ash, because it is well-known for its pozzolanic properties that enhance the solidification additionally. Fly ash can replace between 25 and 35% of the expensive Portland cement (Bradl 2005). Besides, there are further pozzolans alternatively used like volcanic ash, tuff, kiln dust and blast furnace sand. The contaminated soil is mixed with water, alkali additives like cement or CaO and the pozzolanic-based materials. In particular, polar substances are solidified to a greater extent, because the organic components are enclosed. On the other hand, the process of solidification can be disturbed by biologically degradable organic ingredients, the process takes more time and the pozzolanic additives themselves are possibly contaminated with heavy metals (fly ash). Unfortunately, the volume of the material treated with fly ash will increase more significantly.

Another method deals with polyethylene binders that are heated and then left to cool as an adequate means for the solidification approach. After the mixing the new solid material can be backfilled or disposed of. The thermoplastic binders are applied to a number of contaminants such as metals, PCB, PCDD/F and soluble salts (nitrate, chloride, sulphate).

Some solidification approaches are based on material heating. In the context of the application of thermal treatment the generation of a new minerals reservoir that is possible to fix detrimental cations and anions in the lattice is the basis for the high degree of effectiveness. In particular, the mineral ellestadite Ca_{10} $[(SiO_4)_3(SO_4)_3(Cl, OH, F)_2]$ reveals a high fixation potential.

Furthermore, different resins are used for solidification purposes:

- Epoxide resin: molecule polyaddition without dehydration
- Polyester resin: formation of alcohol and carboxylic acid by condensation with dehydration
- Urea resin: formation of formaldehyde and urea by condensation.

In general, the resin volume needed is relatively small, leading to a low volumetric expansion. The period of time needed to solidify the soil is relatively long and some agents like urea-formaldehyde are potentially degradable by microorganisms, resulting in problematic time-dependent effects.

5.3.2 Asphalt Batching

Distinct materials are used for the thermoplastic treatment. Oil-based bitumen and hard coal-based tar asphalt are the most common additives applied. In the so-called asphalt batching technology the contaminated material is embedded in molten bitumen or tar that afterwards cools down and solidifies. The process uses pug mills for soil mixing. After the loader has brought it into the hopper the tar emulsion is added. The mixing process occurs in an extrusion machine at temperatures ranging from 130 to 230 °C (cold-mix asphalt), in which the contaminated soil and the binders are properly mixed (USEPA 1997). Because this process does not involve the excessive heating of the material, volatilisation and air emissions are expected to be limited. Fine-grained soil requires a higher percentage of the emulsion to coat all particles. On the other hand, gravel is mixed inside the pugmill to stabilise the asphalt paving material for further use. The added liquid and viscous bitumen mixture tends towards a relatively rapid solidification lasting 48–72 h. The solidification process can be improved by addition of stabilising agents such as fly ash or cement.

The material is resistant to weathering processes and the contaminants will not leach out. The contaminants are changed to an immobile form that cannot enter the environment any more, since the contaminants are bound up in the oily emulsion. Nevertheless, the initial concentrations should not be too high with reference to the gaseous losses and long-term release. Contaminants such as TPH (recommended maximum concentration: 5,000–60,000 mg kg^{-1}), VCHC (30–1,800 mg kg^{-1}), PCB (2.0 mg kg^{-1}) and metals, e.g. arsenic (30 mg kg^{-1}), cadmium (30 mg kg^{-1}), chromium (500 mg kg^{-1}), mercury (10 mg kg^{-1}), lead (1,000 mg kg^{-1}), are treatable. Afterwards, the solidified material can be used for road and car park construction purposes (see Sect. 5.1.2). Asphalt batching is very common in the USA, where many facilities operate with a capacity of up to 500 m^3 $year^{-1}$ each (EBS 2011).

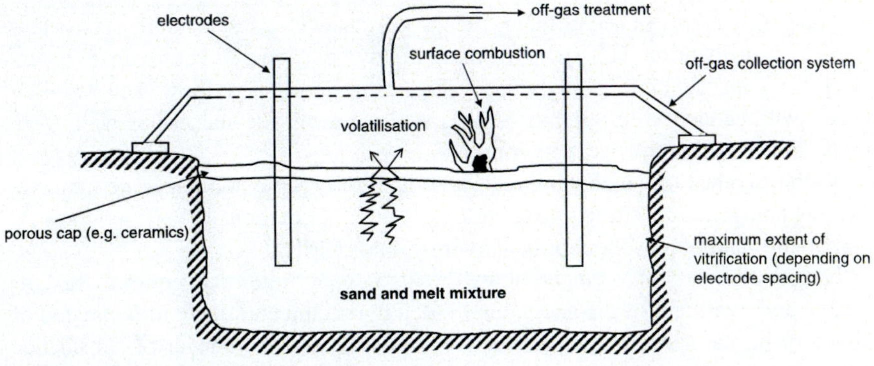

Fig. 5.16 Schematic of vitrification

5.3.3 *Vitrification*

A common technique usually applied *in situ* is termed vitrification (Fig. 5.16). Vitrification is also applied in a heated ceramic melter after soil excavation, since it is easier to control in this way. In the case of an *in situ* treatment electrodes are introduced into the soil in 1 m^2 and heated electrically by generators to a temperature ranging between 1,000 (1,600) and 1,800 °C (Mirsal 2004; Bradl 2005). Graphite surrounding the electrodes is used to connect the equipment. When it reaches these temperatures the soil melts downwards and is changed into a glassy structure. The hot liquid solidifies into an amorphous mass incorporating inorganic contaminants, so that they remain enclosed in the stabilised mass. Organic pollutants are pyrolysed or combusted during the melting procedure. After the soil has cooled it has been changed into an inert, stable and glassy product exhibiting low leaching properties.

Glass formation is only feasible if some component elements are available, which might not always be sure in the different soil types. For this reason, additives must be added. Subsequently, the process is done with sodium silicates, borosilicates and aluminosilicates which help to form the glass structure more efficiently. Soils containing sand (SiO_2) and limestone ($CaCO_3$) as well as dolomite stone $(CaMg(CO_3))_2$ usually require less additives. The glass consists of a three dimensional silicon-oxygen tetrahedron. The metals replace partly the silicon atoms in the interior of the tetrahedron or replace the linkages between the neighbouring tetrahedral structures. In this way, covalent bondings of metals to oxygen atoms of the tetrahedron and ionic bondings to the oxygen atoms outside the tetrahedral are created. Thus, the metal solubility in the silicate structure is reduced to a great extent. For instance, the solubility of arsenic, chromium, antimony and tin is 1–3%, cobalt, copper and nickel amounts to 3–5%, zinc to 15–25% and only lead to more than 25% respectively (Bradl 2005). The technique influences the leaching characteristics of metals, while the metal toxicity remains unchanged.

The *in situ* treatment needs homogeneous soil characteristics to assure a consistent impact of vitrification. Hot spots meaning highly concentrated contaminant layers are just as unacceptable as ingredients disturbing the heating front such as bulky refuse with sudden void volume. Scrap metal can limit the success as well. Very moist soils increase the operating time.

With reference to the contaminants a number of organic pollutants are treatable such as halogenated volatiles (e.g. VCHC), fuel hydrocarbons (TPH), monocyclic compounds and polycyclic aromatic hydrocarbons (PAH).

Negative aspects of vitrification are the high energy intensity required, the collection and treatment of off-gases, the limited maximum soil depth to be treated of about 10 m, the special equipment rarely available and the danger of subsidence caused by densification.

5.4 Stabilisation

The stabilisation approach is usually mentioned in the context of the S/S (solidification/stabilisation) techniques as discussed in Sect. 5.3.1. The main difference to solidification is the fact that the nature of the soil remains more or less intact, but the contaminants are converted into a least soluble and mobile form. In contrast to solidification the soil is not completely changed into an artificial form such as a concrete block or asphalt. Stabilisation minimises the danger associated with the pathways soil – plant and soil – groundwater, which are strongly linked to the solubilisation of contaminants.

Nevertheless, typical additives used for solidification are also applied in relation to stabilisation. Fly ash is one opportunity. The possible heavy metal contamination, however, should also be taken into account. Thus, the use possibilities appear to be limited instead of the beneficial effects related to biological and physical soil parameters.

In the first instance heavy metals are of importance with reference to stabilisation. Table 5.9 summarises the most important immobilising agents used for soil remediation. They are applied in different amounts depending on their effectiveness. Some agents are natural materials excavated and only anthropogenically worked (e.g. clay, lime, phosphate), some are produced chemically (e.g. synthetic zeolite, precipitants). There are examples stemming from industrial processes (e.g. fly ash) and some organic additives belonging to the organic waste components (e.g. compost, sewage sludge).

The metal mobility reduction is strongly influenced by the pH value of the soil. The use of alkaline additives such as lime (or fly ash) means an adequate increase in the pH combined with a considerably decreased mobility of cationic elements. It is caused by the metal complexation on hydroxylated surfaces, an increased cation exchange after deprotonation and the precipitation of metals as hydroxides and carbonates. Moreover, the improved biological activity takes part in an accelerated degradation of organic pollutants. However, the pH rise increases the mobility of

Table 5.9 Origin, application, advantages and disadvantages of agents used for soil stabilisation (Data from Feldwisch et al. 2004)

Means	Origin	Application (kg m^{-2})	Advantages	Disadvantages
Clayey substrates	Clay excavation (sludges, granulates)	1–27 (sludges), up to 350	High stability in soil High adsorption capacity (Cd, Cu, Ni, Pb, Zn)	Swelling/shrinkage dynamic High application rate Pollutant dilution
Fe/Mn oxides	Wastewater sludges, sludges from aluminium production, Fe oxides from industry	0.1–12 (Fe), 0.04–2 (Mn)	High specific adsorption potential (As, Cd, Pb, Zn) Cost-saving (wastes) High stability in aerobic soils	Contaminant input possible Liming necessary (Fe-hydroxides) Phosphate fixation
Al oxides	Zeolithe and montmorillonite excavation, synthetic zeolithes, fly ash	4–20	High stability in soil High adsorption capacity (Cd, Cu, Ni, Pb, Zn)	Contaminant input possible Cementation possible (synthetic zeolithes)
Phosphates	All types of mineral manures	0.5–25 (P)	High stability in soil Low leachate High adsorption potential (Cd, Pb, Zn)	Cd input possible Liming necessary Nutrient accumulation (eutrophication) Enhanced As, Se, Tl solubility
Organic matter	Sewage sludge, compost, bark mulches, etc.	1–50	High adsorption capacity (Cd, Cu, Ni, Pb, Zn) Cost-effective	Biodegradation Pollutant input possible pH value control (mobile complexes, DOC)
Lime	CaO, $Ca(OH)_2$	Indifferent	Cost-effective	Enhanced As solubility Purified nutrient supply No applicability to acid or organic soils Permanent procedure
Precipitants	$CaCO_3$, $Ca_3(PO_4)_2$, FeS	Indifferent	Formation of completely unavailable compounds (phosphate > sulphide > carbonate)	Cost-intensive

some toxic substances with anionic character such as arsenates and chromates. For liming different agents, namely CaO, $Ca(OH)_2$ and soda ash Na_2CO_3, are applied. If the particle size of the materials exceeds 0.05 mm the reaction time will increase, so that finer-textured limes are preferred. Moreover, sugar beet lime originating from the sugar purification with lime is sometimes taken. A repeated application appears to be necessary because the chemical reaction with acids normally generated, in particular, in humic topsoils and additionally accumulated by acid precipitation reduces the lime concentration in the run of time.

In a German field trial the effectiveness of lime application to cadmium-contaminated soils was tested (Table 5.10). The plant available cadmium concentration decreased with time and increased with soil depth. Thus, there was less danger of the plants (grassland) taking up high cadmium content from the uppermost layers (0–10 and 10–20 cm), where the roots are strongly developed. The reduction of the plant available portion was observed irrespective of the contamination level of the soil (König and Tonk 2008).

Another study carried out in the United Kingdom (Surrey and Staffordshire) attempted to find out the impact of lime, zeolite and iron oxides with an application rate of 1% (pot) and 0.2% (field) on cadmium mobility and plant uptake in greenhouse pot experiments and field experiments. A soil containing on average 57 mg kg^{-1} Cd and originating from a sewage treatment farm was used. As shown in Table 5.11, all amendments reduced the plant concentration of most of the vegetable plants in comparison with the control plot. With reference to lettuce even a significant reduction was observed in association with the application of lime and Fe oxide. Moreover, the pot experiment results showed more decisive differences. Broccoli, however, did not react satisfactorily. Additional investigations dealing with carrots, spinach and beans were focused on possible genotypic differences. Between species the ability to tolerate the toxicity of certain heavy metals is well-known. In the experiments highly significant differences between carrot species and, with reservations, between spinach species were detected, while beans did not indicate any differences. Accordingly, mineral adjustments to the soil in association with appropriate management of crops, especially the right choice of the species to be cultivated, may reduce cadmium uptake of more than 50% (Alloway et al. 2000).

Fertilisation with lime and organic matter is well-known in the agricultural sector and accordingly their application is a widely practised method. A wide range of organic alterations are used, since they are beneficial in different ways. They provide nutrients such as nitrogen, phosphorus and sulphur, enrich the organic matter content, establish microbial populations in line with higher microbial activity and enhance the water holding capacity as well as aeration. With reference to the impact on contaminants the metal adsorption is significantly accelerated by organic fertilisers due to the sorption to oxygen-containing functional groups such as -COOH and -OH. Nevertheless, the use of organic manures for the specialised purpose of reducing heavy metal mobility shows negative aspects as well. For instance, dissolved organic carbon (DOC) may form organo-metallic complexes, considerably accelerating the mobility of metals. Apart from composts as products from aerobic decomposition of organic material, biosolids, which are produced by wastewater treatment processes,

Table 5.10 Average plant available cadmium concentrations of different plots (loamy soil) fertilised with lime (53.8% CaO, 36.2% MgO) in relation to plots untreated with lime (= 100) (Data from König and Tonk 2008)

		No. 1		No. 2		No. 3	
Depth (cm)	Liming	1st–3rd year	4th–6th year	1st–3rd year	4th–6th year	1st–3rd year	4th–6th year
0–10	Without	100 (= 0.41 mg kg^{-1})	100 (= 0.59 mg kg^{-1})	100 (= 0.75 mg kg^{-1})	100 (= 1.16 mg kg^{-1})	100 (= 24.0 mg kg^{-1})	100 (= 29.3 mg kg^{-1})
	Moderate	29	23	41	26	42	26
	High	23	16	16	10	30	18
	Very high	17	14	14	9	22	13
11–20	Without	100 (= 0.32 mg kg^{-1})	100 (= 0.46 mg kg^{-1})	100 (= 0.43 mg kg^{-1})	100 (= 0.64 mg kg^{-1})	100 (= 21.7 mg kg^{-1})	100 (= 18.9 mg kg^{-1})
	Moderate	67	55	84	51	42	48
	High	57	36	46	22	27	24
	Very high	38	21	32	17	14	8
21–30	Without	100 (= 0.28 mg kg^{-1})	100 (= 0.23 mg kg^{-1})	100 (= 0.20 mg kg^{-1})	100 (= 0.37 mg kg^{-1})	100 (= 7.3 mg kg^{-1})	100 (= 5.0 mg kg^{-1})
	Moderate	76	104	69	50	61	66
	High	73	73	81	36	52	44
	Very high	56	50	65	27	30	18

Analysis plant available concentration: NH_4NO_3 extraction
Total cadmium concentrations (aqua regia): No. 1: 4.5–7.9, No. 2: 8.0–15.1, No. 3: 62–706 mg kg^{-1}
pH value before liming: No. 1: 5.9–6.2, No. 2: 6.2–6.6, No. 3: 5.6–5.9
Lime application: moderate: 6.3 (No. 3: 12.0), high: 14.0 (No. 3: 23.0), very high 24.5 (No. 3: 38.7) CaO t ha^{-1}

Table 5.11 Cd concentration of plant dry matter (mg kg^{-1}) for vegetables grown in soil of a sewage farm (pot experiment, field experiment) with mineral amendments added to the soil at 1% (pot) and 0.2% (field) (Data from Alloway et al. 2000)

Pot experiment				
Treatment/Plants	Cauliflower	Broccoli	Red cabbage	Spinach
Control	1.62	2.29	3.23	12.56
Lime	1.23	2.10	3.38	5.67
Zeolite	1.42	2.41	2.50	7.49
Fe oxides	1.47	2.26	2.77	5.87
Field experiment				
Treatment/Plants	Broccoli	Lettuce		
Control	9.78	28.82		
Lime	8.80	15.76		
Zeolite	7.48	19.98		
Fe oxides	9.24	18.05		

are also taken into account. Organic amendments should be analysed before applying because some show toxic levels of contaminants or salts depending on their origin.

With regard to organic fertilisers the high adsorption potential can be achieved by using clay minerals but the extremely high application rate of clay has to be assessed negatively, often leading to the exclusion of this variant.

The amendment of Fe, Mn and Al (hydro)oxides commonly present in natural soils appears to be beneficial, although the material that is usually deposited can be termed waste. The hydroxyl groups serve as a template for bridging a number of metals including arsenic. Some macronutrients (e.g. Ca, K, P, Mg), however, tend towards reduced availability and the high manganese concentration might cause phytotixicity (Mench et al. 2006), whereby the use in the agricultural sector appears to be problematical. In this context, red mud as a by-product of aluminium manufacturing (bauxite), which is alkaline and contains iron and aluminium oxides, is an adequate, but rarely applied agent to immobilise cadmium, copper, lead and zinc (SUMATECS 2008).

The crystalline aluminosilicates zeolites which are formed in nature or industrially synthesized also play a remarkable role as agents for stabilisation purposes. The zeolite framework has tetrahedral structures that are linked at the corners. On the one hand, the structure enables the immigration of metal cations through channels and cavities in the structure and, on the other hand, a high cation exchange at the tetrahedral surface occurs (Oste et al. 2002). Thus, a strong binding capacity of zeolites with metals like cadmium, copper, lead and zinc can be concluded. With reference to the agricultural application nutrient deficiencies and physical disaggregation of the soil can happen, while the biological activity does not indicate detrimental effects due to the alkalising properties of zeolites. Pozzolanic synthetic zeolites can cause cementation and hardening of the soil surface after application.

A very effective approach deals with phosphates that are applied to the soil by spraying or mixed with the soil in a mixer. The first method can be done using a normal agricultural manure distributor. The most common phosphates are chloroapatite ($Ca_5(PO_4)_3Cl$), hydroxylapatite ($Ca_5(PO_4)_3OH$) and fluorapatite ($Ca_5(PO_4)_3F$). The mobile metals are initially adsorbed on the apatite surface and then a cation exchange between calcium of the apatite structure and the metal occurs according to (example hydroxylapatite and Pb^{2+}):

$$Ca_{10}(PO_4)_6OH_2 + xPb^{2+} \rightarrow Ca_{10-x}Pb_x(PO_4)_6OH_2 + xCa^{2+} \tag{5.2}$$

This chemical reaction has been studied for a number of heavy metals such as Cd, Pb and Zn, while oxyanions like AsO_4^{3-} and SO_4^{2-} are simultaneously replaced with the structural phosphate PO_4^{3-} (Bradl 2005). The metal solubility tends to decrease enormously due to the evolution of the new precipitants and hence corresponds with the main aim of the phosphate application. On the other hand, the apatite addition can cause cadmium accumulation because apatite containing raw material deposits (phosphate rocks) are frequently enriched with the toxic cadmium (Knox et al. 2006; McGrath and Tunney 2010). Alternatively, synthetic apatite and phosphoric acids are provided. Furthermore, repeated application of phosphates can cause acidification and subsequently inhibition of microbial activity which has to be simultaneously corrected by the addition of alkaline material such as lime (SUMATECS 2008).

There is no doubt that the most effective agents for restricting the metal mobility are precipitants which have already been successfully used in waterworks. Dissolved metals are precipitated as hydroxides, carbonates, silicates and phosphates, which are low solubility compounds. Hydroxides react best at pH values between 7.5 and 11 (Bradl 2005). In the case of a lower pH value alternative precipitants are chosen for the selected metal precipitation. For instance, sulphides (e.g. Na_2S, CaS) are used but their precipitation rate is lower than that of hydroxides and a resolubilisation cannot be excluded in oxidative soil conditions. Besides, the pH value should not be too low, since gaseous H_2S can be generated. Indeed, hydrogen sulphide can be used for the precipitation. The handling of the gaseous additive, however, is rather difficult. In relation to the mercury stabilisation organosulphur compounds such as dithiocarbamates and thioamides are used (Grasso 1993).

Most of the stabilisation agents are suitable for soil contamination in the upper soil depth until 60 cm, since they are applied and afterwards ploughed-in. Only the means used in moist conditions such as the oxides containing sludges and the water-soluble precipitants are suitable for slightly deeper soil layers. Therefore, the immobilisation might be mainly interesting in relation to agriculturally used areas and gardens because of the pathway soil – plant. Moreover, choosing stabilisation can be a good alternative, if the contamination is reduced to the upper portion of the soil and the groundwater table lies very deep.

References

AG Boden (2005). *Soil survey manual*. Stuttgart: Schweizerbart'sche Verlagsbuchhandlung (in German).

Alloway, B., Warren, G., Lepp, N., Singh, B., Penny, C., Bocherau, F., Gregory, L., & Dourado, A. (2000). *Minimizing the risk of heavy metal exposure from contaminated soil by reducing bio-availability using adsorptive minerals and selected garden vegetables*. Proceedings Vol. 3. Paper presented at 1st international conference on soils of urban, industrial, traffic and mining areas, Essen, Germany.

BBODSCHV. (1999). Federal Soil Protection and Contaminated Sites Ordinance. http://www.umweltbundesamt.de/altlast/web1/berichte/pdf/bbodschv-engl.pdf. Accessed 1 Apr 2012.

Beier, H.-E., Niesel, A., & Pätzold, H. (2003). *Manual of practice for gardening, landscaping and sports ground construction*. Stuttgart: Ulmer (in German).

Black, P. M. (2009). Geologic inventory of North Island aggregates recourses. University of Auckland, New Zealand. http://www.aqa.org.nz/documents/Ni%20GEOL%20INVENTORY%20.pdf. Accessed 30 Mar 2012.

Boni, M. R., D'Aprile, L., & De Casa, G. (2004). Environmental quality pf primary paper sludge. *The Journal of Hazardous Materials, 108*, 125–128.

Bradl, H. P. (2005). *Heavy metals in the environment*. Amsterdam: Elsevier.

Bradl, N. C., & Weil, R. R. (2008). *The nature and properties of soils*. Pearson International Edition, New Jersey.

Chrysochoou, M., & Dermatas, D. (2004). *Application of ettringite on heavy-metal immobilization: A literature review*. Conference protection and restauration of the environment VII, Mykonos, Greece.

Delschen, T. (2000). *Soil covering as remediation technique for heavy metal contaminated gardens*. Proceedings Vol. 3. Paper presented at the 1st international conference on soils of urban, industrial, traffic and mining areas, Essen, Germany.

EBS – Environmental Bio-systems. (2011). *Cold mix asphalt treatment of sewer sludges and contaminated soils and river sediments*. http://www.ebsinfo.com. Accessed 1 Apr 2011.

Feldwisch, N., Müller, I., & Marschner, B. (2004). Immobilization of contaminants – A new approach of hazard prevention for soil contamination related to the pathway soil – Plant. *Bodenschutz, 4*, 124–131 (in German).

Genske, D. D. (2003). *Urban land – Degradation, investigation, remediation*. Berlin: Springer.

Grasso, D. (1993). *Hazardous waste site remediation – Source control*. Boca Raton: Lewis Publisher.

Hartmann, R. (2003). The meaning of bioturbation with reference to the soil cover of contaminated sites. *Bodenschutz, 1*, 21–24 (in German).

Knox, A. S., Kaplan, D. I., & Paller, M. H. (2006). Phosphate sources and their suitability for remediation of contaminated soils. *Science of the Total Environment, 357*, 271–279.

König, V., & Tonk, W. (2008). Immobilization of cadmium by liming of contaminated grassland. *Bodenschutz, 4*, 108–112 (in German).

La Grega, M. D., Buckingham, P. L., & Evnas, J. C. (2001). *Hazardous waste management*. New York: McGraw-Hill.

McCarthy, D. F. (2007). *Essentials of soil mechanics and foundations*. New Jersey: Pearson.

McGrath, D., & Tunney, H. (2010). Accumulation of cadmium, fluorine, magnesium and zinc in soil after application of phosphate fertilizer for 31 years in a grazing trial. *Journal of Plant Nutrition and Soil Science, 173*, 548–553.

Mench, M., Renella, G., Gelsomino, A., Landi, L., & Nannipieri, P. (2006). Biochemical parameters and bacterial species richness in soils contaminated by sludge-borne metals and remediated with inorganic soil amendments. *Environmental Pollution, 144*, 24–31.

Mirsal, I. A. (2004). *Soil pollution – Origin, monitoring & remediation*. Berlin: Springer.

Nathanail, C. P., & Bardos, R. P. (2004). *Reclamation of contaminated land*. Chichester: Wiley.

NJDEP. (1998). *Revised guidance document for the remediation of contaminated soils*. Department of Environmental Protection State of New Jersey, USA.

Oste, L. A., Lexmond, T. M., & Van Riemsdijk, W. H. (2002). Metal immobilization in soils using synthetic zeolites. *Journal of Environmental Quality, 31*, 813–821.

SUMATECS. (2008). *Sustainable management of trace element contaminated soils –Development of a decision tool system and its evaluation for practical application*. Final Research Report, Vienna.

Tewelde, M. (2004). *Secondary mineral formation and alinite production by MSW fly ash, Mansfeld area copper slag and limestone meal*. Dissertation, University of Halle-Wittenberg, Germany. (in German)

USEPA (United States Environmental Protection Agency). (1997). *Engineering Bulletin: Technology alternatives for the remediation of soils contaminated with As, Cd, Cr, Hg, and Pb* (EPA/540/S-97/500). Cincinatti: Office of Research and Development.

Chapter 6
Soil Decontamination

Abstract All relevant decontamination approaches are introduced in this chapter. In particular, the widely used technical devices for soil washing, bioremediation and thermal treatment are explained in detail by using flow charts and a number of pictures. The approaches are also discussed in relation to the soil properties and contaminant parameters which are important for the selection of these strategies. Furthermore, the soil washing principle considers extraction methods involving solubilising agents. Regarding bioremediation, biochemical details about the kinetic of the degradation processes for pollutants like BTEX aromates, TPH and PAH are also included. The problem of material reuse after thermal treatment is a further point of consideration. In relation to the techniques, necessary soil preparation steps such as screening, magnetic separation and light fraction removal are introduced. Apart from the common techniques mentioned, the idea of phytoremediation is discussed, taking all relevant approaches such as phytoextraction and phytodegradation into account. Besides measures oriented towards decontamination, the phytostabilisation method aimed at a physical stabilisation of the damaged soil is mentioned. Moreover, the electrokinetic treatment of contaminated soils is also a part of this chapter, in which technical devices and the dependence on soil properties and contaminant characteristics are discussed. In general, advantages and disadvantages in conjunction with the different approaches are contrasted. Accordingly, the description of the decontamination measures makes it possible to obtain a comprehensive overview of the possibilities to remediate polluted soils. Information about treatment centres completes the chapter.

Keywords Bioremediation • Contaminant treatability • Electroremediation • Phytoremediation • Soil washing • Thermal treatment

H. Meuser, *Soil Remediation and Rehabilitation: Treatment of Contaminated and Disturbed Land*, Environmental Pollution 23, DOI 10.1007/978-94-007-5751-6_6,

6.1 Soil Preparation

Before application of the decontamination measures soil washing, bioremediation (*ex situ*) and thermal treatment dry preparation work steps need to be included due to disturbing artefacts impeding the smooth running of the procedures and possibly damaging the facilities. Usually a mixture of soil and coarse materials, particularly sharp-edged technogenic substrates like construction debris and masonry residues, have been excavated and transported to the decontamination equipment. These materials can cause damage to the facility and reduce the effectiveness of the clean-up procedure. Moreover, a lot of artefacts present in the excavated soil such as iron and steel residues, plastics, wood, etc. are responsible for hindering the clean-up process and limited opportunities to re-use the treated material at a later date.

In the conditioning facilities the material flow is operated by conveyer belts which are encased to prevent dust development and dispersion to the adjacent areas. Additionally, some facilities are automatically irrigated depending on the weather conditions in order to prevent the deposition of contaminated dust.

State-of-the-art facilities directly attempt to separate the contaminated material originating from different contaminated land. Figure 6.1 shows the entrance to a soil preparation area. A truck has arrived and the driver is asking about the box or container to be used for the contaminated material. The different boxes contain distinct kinds of contamination. Some are roofed because of anticipated contaminant leaching and some are closed because of pending legal investigations.

The flow chart of conditioning facilities consists of a number of work steps explained and illustrated as follows. In the beginning, bar screens separate construction debris such as blocks of concrete and foundation residues from the mixed material (Fig. 6.2).

Fig. 6.1 Entrance area of a soil preparation plant exhibiting different boxes for soil storage before the preparation process begins and a sealed bottom to prevent leaching of contaminants

Fig. 6.2 Grizzly device to separate construction debris, masonry and bulky refuse

The large pieces are crushed into smaller pieces by jaw breaker, hammer crusher and impact mills. Jaw crushers using electric motors are heavy machines that can crush the material into granularity when requested. The machine enables both a high reduction ratio and a high productivity. Usually, the maximum feeding size varies between 125 and 1,000 mm and a production capacity of up to 600 t h^{-1}can be achieved. The equipment is flexible because the discharging openings have a wide adjustment range. They range from 10–40 mm to 150–300 mm. The movable jaw of the machine corresponds to the desired trajectory, so that the material in the crushing chamber is processed to the targeted size. Finally, the finished material is discharged from a discharging hole.

Alternatively, hammer crushers driven by motors are used to crush the material of different sizes into small particles of more or less equal sizes. Various large, medium-sized and small- sized stones and concrete are treatable and the maximum feeding size amounts to 2,000 mm. The production capacity is classified to be high ranging between 10 and 350 t h^{-1}. Normally, the granularity indicating cubic shape principally does not exceed 120–500 mm according to a high reduction ratio. The working principle is as follows: rotors rotate at high speed and simultaneously the hammers on the rotor rotate also at high velocity. The material is firstly crushed by hammers, and then it is thrown to the opposite plate for secondary crushing and finally crushed again by the hammers after rebounding. This process can be repeated until the desired size is achieved before discharging of the crushed material through the outlet.

In addition, ball mills are used to continue the processing. The ball mill is a horizontal rotating device with steel balls in the chamber. The centrifugal forces bring the balls to a certain impact level and grind the material. Usually, coarse grinding is processed, but in some cases the material enters a second chamber through separating plates where grinding occurs again. Sizes of less than 0.4 mm can be achieved and the production capacity amounts to approximately 8–60 t h^{-1}. This preparation is rarely and exceptionally applied to soil washing and thermal treatment facilities.

Next a light fraction separator termed a wind sifting device is used to remove materials with a relatively low specific gravity such as the waste constituents plastics, textiles and leather as well as technogenic substrates which have been industrially produced like coke, blast furnace pumice and some kinds of ashes. The light fraction separator is based on a low-pressure suction apparatus that can remove light materials, which are collected and recycled without any difficulties. The separation is based on the size of the particles, their density and their shape impacted by an air-flow. Lighter materials are entrained by the air, while heavy materials sink downward against the air-flow. The separation technique is applied to heterogeneous materials, but the separation does not achieve 100%. Two kinds of wind sifting devices are widespread, the so-called zig-zag separator (counter-flow separator) and the cross-flow separator. Both consist of separator housing where the separation takes place by means of air-flow. The air volume is generated by a fan apparatus. The material is moved with the help of conveyor belts. In the case of the zig-zag wind sifter consisting of a multi-deck channel the material is fed from the top and drops down according to the gravity through the upward flowing air. Light and very fine particles are entrained upwards and afterwards removed at the top, while heavy and large particles are discharged at the bottom. The movement of the light fraction occurs from one deck to the next, improving the efficiency of the treatment, because the dwell time is relatively long, while the air velocity is kept constant. Accordingly, optimum results can be obtained. A capacity of 25 t h^{-1} is feasible similar to the cross-flow separator (Fig. 6.3). In this technology the material is fed into the housing using a conveyor belt, where air is blown from below using blast nozzles. Heavy materials are unaffected by the air-flow, but the light fraction is captured by the air-flow and subsequently transported into a fluff plenum. The air return to the blast nozzles is operated by the fan apparatus.

In spite of the use of these separation units some artefacts remain in the material, for instance residues of bulky refuse and large pieces of household waste such as old shoes, tyres and rubber blankets. For this reason, manual sorting may sometimes be necessary. In Fig. 6.4 a small building can be seen in which workers are trying to remove disturbing materials manually.

Materials containing iron such as construction steel and ferrous slag are removed by one or up to four magnetic overhead belts (Fig. 6.5). Ferrous material is compounded with cobalt magnet. The processing capacity varies from 120 to 200 t h^{-1}.

Fig. 6.3 Cross-flow separator to remove the light fraction

Fig. 6.4 In-door picture of the small housing for removal of waste components; workers are busy collecting large waste materials transported on a conveyor belt

The dry conditioning procedures can be combined with particle size separation by using screens. Apart from static screens, there are portable vibratory screeners with screen decks rapidly moving up and down (up to 400 times per minute), allowing the angle of the screen deckform to be changed, for example from 45 to 35°.

Fig. 6.5 Magnetic overhead belt separator to detach ferrous metals

In both the static and vibrating versions one screen can be fitted with a number of additional screens with distinct mesh widths. The differentiation of the material can also be conducted by sedimentation and flotation, as described in Sect. 6.2.1 in more detail. Accordingly, each pile can be treated separately with reference to the following clean-up process.

6.2 Soil Washing

6.2.1 Technical Devices Used

6.2.1.1 Stationary Washing Plants

Soil washing is a clean-up technology which is well established in some European countries such as Belgium, Germany, The Netherlands, Switzerland and United

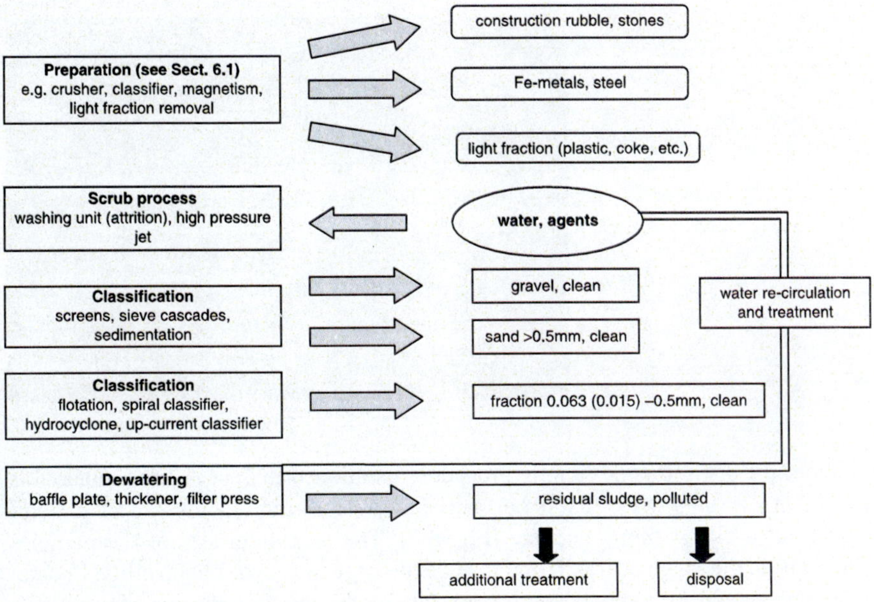

Fig. 6.6 Flow chart of a soil washing facility

Kingdom as well as in North America and Japan (Pearl et al. 2006). In the last decade until 2010 the number of soil washing vendors did not show increasing tendency, but the market will continue to be important. Some spectacular case studies are running, e.g. the soil remediation process in the Olympic Park in London, United Kingdom, where the Olympic Games were held in 2012 and where soil washing was applied as an important remediation technique (Anonymus 2008).

Generally, soil washing is a volume reduction or waste minimisation treatment process where those soil particles which contain the contaminants are separated from the bulk soil fractions or contaminants are removed from soil by aqueous chemicals. Fundamentally, soil washing with water and soil extraction using water and solubilising agents must be differentiated. In order to be effective soil washing and soil extraction must either transfer the contaminants to the wash fluid or concentrate the contaminants in a defined fraction using a size separation technique. Thus, the separation of soil particles according to the size differences is always involved. At least the differentiation into a clean and a dirty fraction is clearly recognisable.

In Fig. 6.6 the scheme of a soil washing and soil extraction process is shown. The design of the soil washing process configurations of the various types of equipment varies widely and is based upon the ease with which the contaminated fractions can be separated. Most soil washing companies use modular equipment configurations which can be adapted to the site-specific conditions. Most of the washing plants, whether the stationary or mobile type, are automated processing systems with online measurement (DEC 2011).

Fig. 6.7 Attrition equipment in a soil washing plant

After the dry soil conditioning procedure mentioned in Sect. 6.1 the material is treated in a scrub process that is normally based on attrition facility which is reminiscent of a big washing-machine (Fig. 6.7). The maximum size of the particles should be in the range of 10–50 mm to avoid damage to the washing facility. Coarser materials should be removed by e.g. grizzly, if this process did not already take place during the soil conditioning. Where contamination occurs predominantly in the coarse fraction, it may be necessary to crush the material to release the contaminants from the bulk.

The high-intensity attrition scrubbing uses (high-pressure) water and centrifugal acceleration or vibration to detach the contaminants and contaminated particles from larger particles such as sand and gravel. The washing process results in a hydro-mechanical separation and in a disintegration of soil agglomerates under the influence of high shear and friction forces (Werther et al. 2001). Several attrition units, e.g. three or four, can be combined to improve the cleaning process. The soil is fed into the first chamber, transferred as slurry from unit to unit and ultimately discharged from the last chamber. The volume of the chambers made of steel, stainless steel or polyurethane coated with steel ranges from 0.5 to 3 m^3. The coarser fractions (predominantly gravel and coarse sand) can be treated based on a simple rotating washing drum with vertical inside agitators (e.g. sword washers), but the attrition module associated with high energy input is favoured in the presence of soils containing finer particles. Between the treatment of the gravelly texture and the sandy one hydrocyclones (see below) can be integrated. Hydrocyclones should be integrated before the material undergoes the next washing step.

Alternatively, some vendors use water sprays or the high pressure steel jet tube technique (usually 250–350 bar, maximum 500 bar) instead of rotating washers to clean the material. In general, contaminants occurring as coatings require high intensity treatment such as attrition scrubbing or high pressure water jets.

The scrub process is conducted by water and, in the case of the extraction type, different agents. The water can be re-circulated after passing a treatment module to remove suspended, dissolved and emulsified contaminants. For instance, sand filters for suspended fines, activated carbon absorbers or exchange resins are applied

Fig. 6.8 Sieve cascade in a soil washing plant

(see Sect. 7.1.4). Approximately 10% of fresh tap water has to be added and 90% of the water used comes from the recirculation process. The encapsulated stationary facilities are partly equipped with an air exhaust system to allow volatile contaminants to be treated. Afterwards, the exhaust air has to be treated with activated carbon (see Sect. 7.2.3).

Afterwards, the material first has to be washed with water once again to remove any remaining and adhering fine soil and dust particles and secondly classified with the help of vibrating screens or a series of screens termed a sieve cascade (Fig. 6.8). Results from the sieving procedure are the completely clean fraction gravel. Sand particles larger than 0.5 (0.25) mm are easily separated by using settling chambers due to their relatively high settling velocity. The clean fractions are dewatered on a screen.

In contrast, the classifying of finer fractions is a much more complex task. Between 0.5 and 2 mm particle size the washing process is assessed as being very effective and between 0.063 and 0.5 mm as effective to a limited extent. Nowadays, technically re-equipped facilities can even clean fractions ranging from 0.015 to 0.063 mm (in an approximately similar way to the texture class coarse silt).

Alternatively, the following techniques are feasible for the separation process of fractions smaller than 0.5 mm:

- Sedimentation
- Use of hydrocyclones
- Flotation.

Sedimentation means the separation of soil particles according to density differences, because coarser particles sink to the bottom of the water basin. The sedimentation is usually oriented to particle sizes ranging from 0.2 (= 200 μm) to 10 mm. In general, it cannot be avoided that particulate material <0.5 μm remains in the water. In some cases the contaminant density may play an important role, if contaminated particles can be separated from uncontaminated ones. For instance, lead pellets coming from shooting ranges are extracted in this way.

The separation of different fractions occurs in a conic vessel called a hydrocyclone, in which a tangential flow is generated. Heavier particles sink to the bottom of the vessel, while lighter particles are transported to the upper outflow. The hydrocyclones are mainly associated with particle sizes from 20 to 200 μm (Dennis et al. 1994; Wilichowski 2001).

Flotation occurs prior to the particle settling with the aid of some flocculating agents. In a water basin hydrophobic particles rise upward adsorbed to the rising bubbles, which are caused by detergents. The floated particles are removed at the surface of the suspension in a froth which is stabilised by frothing agents. The flotation is limited by the particle sizes. Particles exceeding a certain size cannot be transported upwards and small particles tend to disturb the process by unselective adsorbing to the reagents. The preferred particle sizes vary in a small range between 15 and 100 μm. Thus, the flotation step must follow the separation by screens and hydrocyclones. The sludge at the bottom of the basin is highly contaminated and is enriched with the flocculants used. It must be dewatered to obtain a solid product, which must be treated separately (Werther et al. 2001; Neeße 2001).

Furthermore, in the presence of released magnetic particles such as slag residues and iron oxides which were not possible to separate during the soil conditioning a separation based on a magnetic separator unit can be added.

Up-current classifiers and spiral classifiers are also used for separation purposes. They are preferred to the fraction <0.5 mm but they can also replace the separation units mentioned above (screens) with reference to the sandy fraction. The first one is applied to remove the light-density materials such as wood, organic matter and tar particles from the hydrocyclone underflow. The module ensures cut sizes from 0.1 to 1.0 mm. Nozzle plates lead to a good distribution of the upstream water. Varying water feed and altered nozzles enable a sharp separation of the fractions. Apart from the light fraction, which tends to float, material exceeding the specific gravity of water up to 1.9 g m^{-3} is theoretically capable of up-current classification.

Spiral classifiers are sorting devices according to the different densities. They are used for grain sizes from 0.04 to 4 mm with a limited capacity of up to 4 t h^{-1}. If a higher throughput is required, several spirals are combined. The material is fed into the spiral classifier from the top and moves downward to the bottom. Particles with

Fig. 6.9 Filter press in a soil washing plant

a lower density such as wood, coal and organic matter are concentrated in the outer areas of the flow, while particles with a higher specific density are moved towards the spiral axis. At the bottom two or three splitters may separate the fractions – in the case of three splitters as heavy, medium-density and light-density fractions.

The clean fractions are recycled on the contaminated site, used for other sites or disposed of as material less polluted than the original one and consequently the disposal costs are lower. Instead of the processing costs the total costs are reduced due to the lower amount of material requiring expensive disposal. In summary, *ex situ* soil washing alters the soil to a great extent. The clean fraction manifests a disaggregated structure, affected soil organisms, strongly reduced clay, silt and organic matter content and changed pH values in the case of the extraction method. Subsequently, the clean material can mostly be re-used in the building sector.

The moisture content of the residual sludge should be as low as possible, since the disposal in landfills is based on the weight of the material received. The highly polluted sludge containing the clayey and silty fraction as well as the organic matter must be treated after drying, which is done by belt filter presses (Fig. 6.9). The presses consist of textile or metallic tissue and their results are termed filter cake. The solid matter can also be separated by a baffle plate thickener before starting the filter press operation. Air drying requires a very large area, which is usually not available because of space limitations in urbanised areas. The residual sludge has to be treated microbiologically or thermally, since it carries higher levels of contamination than the original soil. Alternatively, it can be used for solidification as a substitute raw material in the cement industry (see Sect. 5.3.1) or be disposed of.

The stationary soil washing plants are designed to treat different quantities. A maximum of 100 t h^{-1} is feasible when operating 24 h day^{-1}. Most units have a capacity of 25 t h^{-1} (Dennis et al. 1994) and 30–70 t h^{-1}, respectively (Pearl

Fig. 6.10 Mobile soil washing plant

et al. 2006). Moreover, small mobile plants designed in a modular system and with a throughput of 10–20 t h^{-1} (Neeße 2001) have been constructed for on-site operation.

6.2.1.2 Mobile Washing Systems

In some countries mobile aqueous soil washing systems in up to 13 containers are offered, reaching a maximum quantity of 20 t h^{-1} but electricity, water and compressed air are utilities required on-site. Assuming wastewater recovery is feasible, the quantity of water consumed is estimated at 3–7 m^3 t^{-1}, whereas the fresh water demand is reduced to 0.1–0.3 m^3 t^{-1} (Neeße 2001). The mobile equipment consists of all units usually found in soil washing facilities such as feed unit, washing drum, screening machines and hydrocyclones for sizing, spiral separator for density separation, wastewater treatment unit and screen belt press for drying of the remaining sludge (Fig. 6.10).

6.2.1.3 *In situ* Soil Washing

The soil washing approach can also occur without excavation. Figure 6.11 illustrates an *in situ* soil washing process in a sandy soil by means of the high pressure

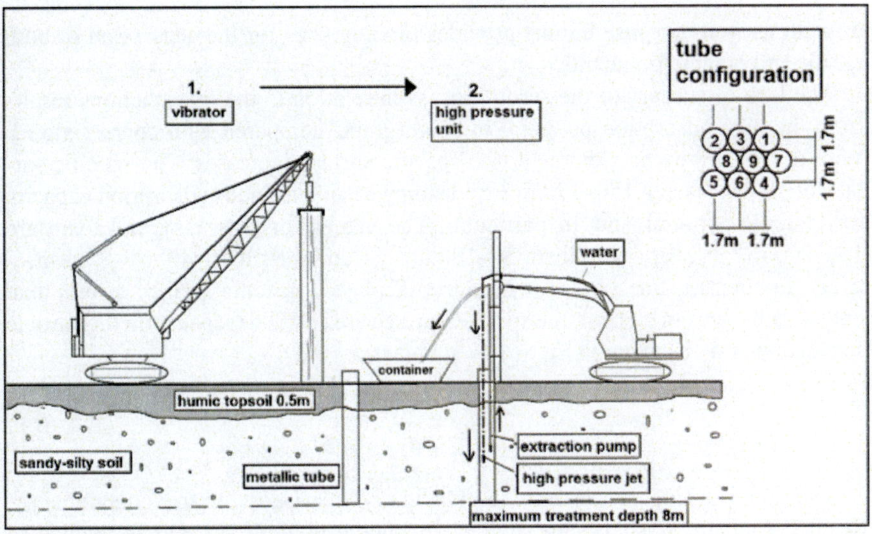

Fig. 6.11 *In situ* soil washing process

technique. In the first instance, steel tubes measuring about 1.7 m in diameter and about 10 m in height are driven by a vibrator based on the tube arrangement as shown in the Figure. Afterwards, a high pressure jet is used to loosen the soil and to form a soil-water-suspension that is pumped out of the ground and separated over ground (not shown in Fig. 6.11). Afterwards, the clean gravelly to sandy fractions are backfilled into the emptied tubes as soon as possible.

6.2.2 Required Soil Properties and Treatable Contaminants

Before introducing a full-scale commercial unit it makes sense to arrange a bench-scale unit or pilot-scale study. Because of different types of soil to be treated and the uncertainty about how efficient removal from the soil will be, the risk of going to a full-scale unit and processing thousands of tonnes is unwarranted. The study results provide assurance of the system performance planned. Besides, pilot-scale testing gives a sophisticated insight into difficulties relating to the material handling. The tests should involve particle size separation tests conducted by e.g. grain size distribution analysis, a settling velocity test as well as a float and sink test to determine specific gravity. Furthermore, contaminant dissolution can be evaluated by solubilisation tests using chemical extraction methods (Pearl et al. 2006).

In any case fractions smaller than 0.015 mm cannot be treated in this way. For this reason, a percentage of more than about 30–40% silt plus clay might result in an exclusion of the soil washing procedure (Dennis et al. 1994; Nathanail and Bardos 2004; Pearl et al. 2006). High humus content makes the removal extremely difficult as well, because humus provides binding sites for the adsorption of both metals and organic pollutants.

The lack of success of the scrub process in relation to the fine fractions results from the high adsorption potential of small grains compared with coarser grains. With decreasing particle diameter the specific surface increases. The specific surface of clay measures 150–250 $m^{-2}\ g^{-1}$ leading to an enormous adsorption capacity for cationic contaminants in particular. The values for silty clay and loam are 120–200 $m^{-2}\ g^{-1}$, for sandy loam 50–100 $m^{-2}\ g^{-1}$ and for silt 10–40 $m^{-2}\ g^{-1}$ respectively. In contrast, fine sand, for instance, displays a specific surface of less than only 0.05 $m^{-2}\ g^{-1}$. In general, the specific surface area Sv is dependent on the particle size diameter d_p, as given in Eq. 6.1 (Wilichowski 2001):

$$Sv = \frac{6}{d_p} \tag{6.1}$$

The soil washing processes will potentially treat most of the contaminants. Results from soil washing case studies are shown in Table 6.1. When applied to sandy soils soil washing achieves 90–99% contaminant removal. These are predominantly the organic pollutants. Heavy metals are capable of achieving a decontamination rate of 40–90% but the washing process is mainly targeted at a reduced leachability of the metals rather than their absolute removal (Pearl et al. 2006). However, it is of importance to compare the results from soil washing with the quality standards required. Hence, the presentation of the results should be preferentially based on the concentration in mg kg^{-1}. Soil washing is only successful, if the level of contamination in the treated material is below the specific quality standards.

Apart from the texture the efficiency of the washing process is limited due to other factors. As already mentioned, high organic matter content appears to be detrimental for the washing process due to the adsorption capacity, especially for organic pollutants. Furthermore, metals that are tightly connected to the material concerned such as tar asphalt and ashes are not treatable. Metals incorporated into oxides like goethite and hamatite, which are found in gleyic soils and reductosols, are difficult to wash as well. The effectiveness of soil washing is also low, if contaminants such as ashes of high temperature coal power stations and sand derived from foundries are included in a crystalline matrix resulting from high temperature operations. Moreover, gaseous contaminants cannot be eliminated by soil washing.

Contaminants adsorbed to the surface of fine soil particles such as clay and humus are very amenable to physical separation processes, because they are only associated with these fractions. Discrete particles such as metal grains and tar balls are also suitable for separation processes. In contrast, contaminants coating the outside of particles due to the precipitation of the contaminant must be removed by abrasion, which can be done successfully by attrition scrubbing or by chemical solution. Liquid coatings like fluid oils and tars are, in principle, treatable in the same manner but sticky fluids may block the technical equipment such as screens and hydrocyclones, causing disadvantages. Contaminants which penetrate into

Table 6.1 Results from soil washing case studies in the Netherlands, Germany, United Kingdom and USA (Data from Dennis et al. 1994; Pearl et al. 2006; CL:AIRE 2007)

Country	Contaminant	Decontamination rate (%)	Residual concentration (mg kg^{-1})	Technique
USA	Oil and grease	50–83	250–600	Screw conveyors
USA	Oil and grease	90–99	<5–2,400	Drum screen washer
USA	Cr	81	159	Mobile equipment
	Cu	77	259	
	Ni	80	83	
The Netherlands	PAH	98	2	Gravity separation (ashes)
The Netherlands	Pb	98	50–60	Gravity separation (shooting range pigeons)
The Netherlands	Cd	95	0.4–0.8	Gravity separation
The Netherlands	Cd	74–81	1.3	Gravity separation (heavy metal slag)
	Pb	57–90	60	
	Zn	69–93	200	
The Netherlands	Crude oil	97	2,300	High pressure jet
The Netherlands	Cyanides	93–99	<15	Mix tanks and soil fraction equipment
	Heavy metals	approx. 70	<200	
The Netherlands	Cyanides	>95	5	Froth flotation tanks
	Heavy metals	>90	>150	
The Netherlands	CHC	>99	20	Scrubber, upflow classifier
	Cyanides	95	5–15	
	Heavy metals	75	75–125	
	CHC	98	<1	
Germany	Phenols (total)	86–94	7–22.5	Vibrating screw conveyor
	PAH	86–90	91.4–97.5	
	PCB	84–88	0.5–1.3	

(continued)

Table 6.1 (continued)

Country	Contaminant	Decontamination rate (%)	Residual concentration (mg kg^{-1})	Technique
Germany	CHC	>75	<0.01	High pressure jet
	PAH	95	15.5	
	Phenols	>99.8	<0.01	
Germany	TPH	47	80	Attrition scrubbing
	PAH	76	7	
	Cd	70	1	
	Cr	96	3	
	Cu	82	34	
	Ni	78	7	
	Pb	84	129	
	Zn	79	277	
United Kingdom	TPH	93–95	209	High pressure jet, attrition scrubbing
United Kingdom	PAH	74–98	48	High pressure jet, attrition scrubbing
	TPH	78	205	

Table 6.2 Assessment of the contaminant mobility

Contaminant	Water solubility	Degree of desorption	Volatilisation
Volatile chlorinated hydrocarbons (VCHC)	+	+	+
Monocyclic aromatic hydrocarbons (BTEX)	+	+	+
Phenol compounds	+ (2)	+	– (1)
Mineral oil hydrocarbons (TPH)	O	+	O
Polycyclic Aromatic Hydrocarbons (PAH)	–	O	–
Polychlorinated Biphenyls (PCB)	–	O	–
Free cyanides	+	+	+ (3)
Complex cyanides	–	+	–
Heavy metals	O (4)	O (4)	–
Dust particles	–	+	–

+ = high
O = moderate
– = low
(1) = apart from some compounds
(2) = under alkaline reaction
(3) = under acid reaction
(4) = dependent on pH-value

porous soil material are problematical and require crushing in order to make those accessible (Pearl et al. 2006; CL:AIRE 2007).

In general, the possibility to mobilise contaminants depends on the water solubility and the desorption potential. An overview of the assessment of the mobility is summarised in Table 6.2. Moderate to high water solubility or a moderate to high degree of desorption are acceptable for soil washing technology. It should be noted that the volatilisation of some organic pollutants may lead to a decrease in the contamination. The result, however, is mainly a consequence of the gaseous losses and subsequently not related to the intrinsic soil washing process.

The nutrient content of the material to be treated should also be taken into account because the washing process can result in enhanced nitrate and sulphate leaching, which might cause additional problems to the water treatment.

6.2.3 Solubilising Agents

Some agents can be chosen to enhance the solubilisation of contaminants (Table 6.3). Since these agents have to be removed and recovered, the addition of the means should be reflected in more detail. In general, bases and complexing agents are rarely applied because of the difficulties of treating the wastewater. The presence of additives in the wash fluid makes it more difficult to treat by conventional treatment processes such as

Table 6.3 Agents used for the soil extraction method

Means	Impact	Parameter
Mineral (e.g. hydrochloric, sulfuric, nitric acids) and organic acids	pH decrease	Heavy metals
Bases (e.g. sodium hydroxide, sodium carbonate)	pH increase	Petroleum hydrocarbon derivates
	Metal precipitation	Heavy metals
Complexing agents (e.g. EDTA, DTPA, ammonium acetate)	Formation of mobile organo-metallic complexes	Heavy metals
Detergents	Dispersion, enhanced solubility of hydrophobic substances, flotation additive (impact up to 1.9 g cm^{-3})	Organic pollutants (CHC, BTEX, TPH)
Organic solvents (e.g. acetone, ethanol, isopropyl alcohol)	Enhanced solubility (use without water after drying prior to extraction)	Non-volatile hydrophilic and hydrophobic organic pollutants
Water temperature rise	Increased solubility, decreased viscosity	Organic pollutants (CHC, BTEX, TPH)

settling, chemical precipitation and activated carbon adsorption. Furthermore, the opportunity to dispose of the residual sludge might be less favourable. An overview of surfactants used in soil washing technology is given by Chu 2003.

For instance, by selecting the right surfactant, non-volatile organic pollutants can be removed to an extent varying between 40 and 90% (Dennis et al. 1994). With reference to the effectiveness of extraction means associated with heavy metals ethylenediaminotetraacetic acid (EDTA) was tested in an experiment dealing with soils of abandoned mines in Lavrion, Greece. The soil exhibited high metal contamination varying between 12 and 159 mg kg^{-1} for Cd and between 3,200 and 35,600 mg kg^{-1} for Pb. The adsorption potential of the soil was supposed to be high due to the relatively high organic matter content (2.0–3.9%), the high CEC (25–45 mol_c 100 g^{-1}) and the slightly alkaline pH value (7.7–8.3). EDTA preferentially forms soluble chelate complexes but the effectiveness of EDTA depends upon different properties such as molarity of the solution and pH value of the soil to be treated, to which attention should be paid. For instance, the 0.1 M EDTA solution showed a high effectiveness in removing Cd (72–100%) and Pb (70–100%), while the 0.01 M EDTA solution indicated less effective results. Furthermore, soil material with pH values <6 were more suitable because in the range of pH 3–6 with $H_2[EDTA]^{2-}$ 1:1 stoichiometric metal complexes were formed. In the presence of high soil pH values and the subsequent high buffer capacity $H_2[EDTA]^{2-}$ is not dominant (Karvounis and Kelepertsis 2000).

The nature of the organic contaminant can generally hinder the effectiveness. A contaminant with a high aqueous distribution coefficient (K_{OC}) (e.g. PCB) is more difficult to remove from soil particles than a contaminant with a lower K_{OC} (e.g. phenols).

Increasing water solubility corresponds to decreasing partitioning into soil organic matter or a lower K_{OC} value. Surfactants are needed to improve removal efficiency. But in the case of used additives larger volumes of wash fluid are required.

Organic pollutants are influenced by detergents to a great extent, since the hydrophobic side will be oriented to the organic pollutant, while their hydrophilic side is oriented to the surrounding water. Hence, detergents can transfer the organic pollutants into the water. Detergents produce foam, which hinders separation and settling characteristics and subsequently decreases the material throughput. On the other hand, they are used for the flotation process during the soil classification because they form bubbles on which organic pollutants are adsorbed.

With reference to the organic pollutants oxidation without any accelerated solubilisation appears to be an alternative and adequate method. The soil is conveyed into mixing drums where water and some specific additives, particularly hydrogen peroxide and caustic soda, are added. An oxidation of the pollutants, in which temperatures up to 60°C may occur, begins. The chemical reaction removes the organic pollutants to a great extent (see Sect. 7.1.6).

The mobility of heavy metals is strongly enhanced in the presence of mineral and organic acids, because with decreasing pH value the solubilisation of the metals generally increases. The impact of the pH value, however, is related to the elements, e.g. Cd mobility increases significantly at $pH < 6$, while Pb reacts $< pH\ 4.5$.

Organic solvents have been demonstrated to be effective in treating sediments, sludges and soils containing organic pollutants such as Polychlorinated Biphenyls (PCB) and TPH. Moreover, they are used for the separation of organic pollutants in paint wastes, coal tar wastes, wood-treating waste, pesticide waste and petroleum refinery oily sludges (EPA 1990). Solvents like ethanol and acetone are also used exceptionally. Complex contamination may often require a combination of surfactants.

A water temperature rise can also improve the contaminant mobility, since reduced viscosity in combination with increased solubility is beneficial with reference to the extraction method. Moreover, gentle heating raises the volatilisation of certain organic pollutants.

Complex mixtures of organic contaminants and metals make it difficult to find a suitable wash fluid that will remove all the various types of contaminants in one washing step. This might require sequential washing steps with distinct additive modifications.

6.3 Bioremediation

6.3.1 Principles of Bioremediation

The principal basis of the biodegradation is the bioavailability of the pollutant, which is determined by the mass transfer from the pollutant source to the microorganisms, the spatial distribution of pollutants and microbes, the rate of the microbial substrate consumption and the growth and decay of the microbial cells. The degradation only occurs when the pollutants are dissolved in soil solution, from which they enter the microbial cells. For this reason, organic pollutants indicating low water solubility (e.g. PAH, PCB) are less biodegradable than compounds manifesting higher solubility (e.g. TPH, phenols). Thus, extremely insoluble and hydrophobic compounds are more or less untreatable, as observed in connection with the dioxin 2,3,7,8-TCDD, the Seveso poison. The bioavailability depends on the chemistry of the compound, e.g. aqueous solubility and sorptivity to hydrophobic surfaces. Pollutants with low aqueous solubility tend towards crystallisation once the maximum solubility is exceeded and additionally to high sorption onto hydrophobic soil constituents.

A minimum concentration of the soluble pollutant should not fall below 0.1–2 mg L^{-1} depending on the parameter concerned. Below this concentration the bacterial growth might be impossible. The general solubility of hardly soluble contaminants can be enhanced by means of agents, as mentioned in Sect. 6.2.3. Microbial naturally generated biosurfactants and bioemulsifiers can accelerate the solubility to a certain extent. Moreover, artificial surfactants like detergents that consist of molecules with a hydrophilic and a hydrophobic side considerably facilitates the desorption of organic pollutants from the contaminated soil. Apart from the solubilisation, facilitated transfer of hydrophobic substances may also result from the lowering of the surface tensions in the water. However, it should be taken into consideration that an enhanced apparent solubility of the pollutant does not necessarily mean an increase in the availability of the contaminants for the microbial cell. Accordingly, in spite of an improved solubility, speeding up of the biodegradation and complete elimination of the toxic substance might not be guaranteed (Tiehm et al. 1997). Moreover, one more obstacle related to an *in situ* approach is the unknown biodegradation rate of the surfactant itself.

If acceptable maximum concentrations of organic contaminants are exceeded, a biological degradation is reduced or completely stopped. Critical maximum values are assumed at 50,000–100,000 mg kg^{-1} for TPH, 5,000 mg kg^{-1} for Σ BTEX, 10,000 mg kg^{-1} for phenols, 20,000 mg kg^{-1} for Σ PAH_{EPA}, 1,000 mg kg^{-1} for CHC and 100 mg kg^{-1} for Σ PCB respectively. The lack of bioavailability of most of the toxic substances can be considered as advantageous, since it is in accordance with the fact that only seldom are the organisms attacked by an overdose of the pollutant concentration. The toxicity appears to be defined by the octanol-water partitioning coefficient K_{ow}. It has been found that organic pollutants indicating a K_{ow} >4.0 do not negatively influence the microbial disintegration, whereas K_{ow} values <2.0 inhibited the biodegradation in a number of case studies. Although the BTEX monoaromates degrade well, in principle, the biodegradation can become problematical due to the possible toxicity to microorganisms, since the K_{ow} values are in the order of magnitude 2.0 (benzene), 2.4 (toluene) and 3.1 (xylene) respectively. The K_{ow} of chlorinated pollutants tends to decrease during the biodegradation, leading to an increased toxicity of the metabolites (e.g. pentachloroethylene (PCE) < trichloroethylene (TCE) < dichloroethylene (cis-DCE) < vinyl chloride (VC)) (Khan et al. 2004; Grotenhuis and Rijnaarts 2011).

One problem may arise because in many cases the organic pollutant cannot enter the microbial cell due to its large size. Subsequently, the contaminant is physically not capable of being transported into the microbial cell. Bacteria capable of hydrolysing large molecules into a diffusible fraction that can be assimilated into the cells are absolutely necessary. These bacteria will, for example, convert complex structures into soluble compounds such as complex carbohydrates into sugar units, proteins into amino acids and fats into fatty acids (Bandyopadhyay et al. 1994).

Another problem is related to the pollutant migration. The mass transfer rates depend on the pollutant solubility. The gradient of the dissolved pollutant concentration over a defined distance is affected by the maximum solubility of the pollutant as well as the diffusion coefficients. The latter vary from 5×10^{-10}–10^{-9} $m^2\ s^{-1}$ (low molecular weight compounds) to 9×10^{-12}–2×10^{-21} $m^2\ s^{-1}$ (high molecular weight compounds). Consequently, the average daily diffusion distance in soils is expected to be as low as only 1 μm day^{-1}. For this reason, the pollutant quantity reaching the microbial cells is strongly limited because of the circumstances mentioned. A reduction of the distance between the pollutants and the living bacteria which occurs in the case of a homogenised contaminant distribution or in the case of a considerably enhanced density of the degrading organisms might accelerate the mass transfer due to the steepening of the diffusion gradient (Wick et al. 2001).

Furthermore, the soil properties such as amount of sorptive surfaces and amount of micropores with low accessibility for microorganisms play a significant role. If the molecules of the pollutants are much smaller than the microorganisms, the bioavailability will be reduced, since the pollutants diffuse into the soil pore system where the bacteria and fungi have no access because of their size (Mahro et al. 2001). While the pollutants are capable of migrating into very small pores, the microbes are able to live in pores with a diameter of >2 μm. Therefore, the spatial distribution of the microbes reveals unequal patterns, thus exacerbating the bioavailability.

There are different strategies which bacteria and fungi pursue. In general, the degradation process takes place on the basis of enzymatic activity. The enzyme activity takes place either adsorbed into cells or on an extracellular basis outside the confines of the cells. The enzymes may generate substances such as free radicals which attack the pollutant.

The decomposition of organic pollutants can be strongly limited or entirely stopped. During the degradation process some compounds can be created which are resistant to continuous degradation. A typical example is vinyl chloride, which originates from the chemical degradation process of tetrachloroethylene. Such compounds act as dead end products which interrupt the complete degradation of the organic contaminants. It is possible that the toxic intermediates accumulate and cause the biodegradation process to come to an end abruptly.

During the degradation process integration of degraded substrates (e.g. fragmented benzene and phenol cores) into the organic matter cycle can occur, leading to the formation of so-called bound residues. Bound residues are generally less bioavailable with increasing age and subsequent humification, bioaccessibility and bioavailability decrease significantly. These macromolecular non-extractable pollutant residues are difficult to determine analytically, so that the decrease in contamination may indicate falsification of results.

Pollutants can be degraded to chemical structures which are very similar to the natural organic matter constituents such as carbohydrates, amino and fatty acids and which are not considered as bound residues. Moreover, during the degradation phase a certain part of the pollutant is incorporated into the microbial biomass. This temporary fixation of the fragments is also not considered to be formation of bound residues, since the biomass will be transformed to the humic substances afterwards. Technogenic carbon, as results from combustion processes, does not tend to cause bound residues due to a lack of interaction with the pollutants. However, this carbon fraction affects the bioavailability of the organic pollutants.

Bound residues are defined as compounds which persist in the matrix in the form of the parent substance or its metabolites. They are results from interaction between complex pollutants like PAH and the natural organic matter. The residue formation equals a naturally occurring humification process, in which xenobiotic carbon is involved. This xenobiotic carbon is sequestered together with the natural organic matrix and is thus analytically not separable. While the toxicity of the former pollutant declines, the xenobiotic carbon might be accumulated in the environment.

There are different ways to explain the incorporation of bound residues into the organic matter. The organic matter consists of hydrophobic aromatic and aliphatic nuclei linked by covalent bonds and carrying reactive functional hydrophilic groups. The organic pollutants produce metabolites just after the initial degradation processes, which are chemically different from the original pollutant. On the one hand, the hydrophobic metabolite structures are adsorbed to the hydrophobic nuclei. Furthermore, the reactive functional groups of the organic matter act as binding sites for the chemically altered pollutants. Ultimately, the metabolites are covalently bound by chemical linkages within the organic matter (Kästner and Richnow 2001).

Besides the approach that covalently bound metabolites become an integral part of the organic matter, a physical explanation gained attention. Accordingly, a slow partitioning of organic pollutants and their metabolites in micropores of the organic matter which form reactive inner surfaces with a hydrophobic character takes place. There, the hydrophobic pollutants are encapsulated and retarded unless the structures are modified (Eschenbach et al. 2001).

Using ^{14}C-labelled substances as a basis the percentage of formed bound residues was investigated. The non-extractable residues of several PAHs after incubation periods of up to 176 days ranged from 5 to 74% (Kästner et al. 1999). Another publication mentioned a percentage of 15–68% based upon an incubation time of up to 291 days (Eschenbach et al. 2001). Studies dealing with TNT, the metabolites of which are covalently bound to the organic matter, resulted in bound residue percentages between 84 and 99% after an incubation time of up to 170 days (Eschenbach et al. 2001). Nevertheless, the main elimination pathway of the PAH was usually the mineralisation but the formation of bound residues started at once and was associated with the mineralisation. Subsequently, in time periods with high microbiological activity both the mineralisation (CO_2 production) and the bound residue formation showed enhanced values. Nevertheless, the formation of bound residues did not increase after addition of organic substrates such as compost, bark chips and forest litter, although there was a tendency for the microbial activity (CO_2 production) to increase.

The formation of bound residues varies between the different pollutants. Chlorinated compound such as PCB and chlorophenols do not tend to form residues because of the chemically less reactive substituents. In contrast, pollutants consisting only of functional groups such as carboxyl, hydroxyl and nitro groups react with the natural organic matter more easily and it is assumed that they create bound residues preferentially.

One problem associated with the formation of bound residues might be the question whether this functions as a temporary sink or whether a remobilisation must be expected. In particular, if the physical entrapment is the most important reason for the bound residue creation, an availability of the toxic fragments cannot be excluded in future. Different tests were performed to find out the long-term remobilisation of bound residues. In summary, it was not possible to determine a significant remobilisation under physical stress (freezing/thawing, wetting/drying) in soils contaminated with PAH and TNT (Eschenbach et al. 2001). The approach based on the covalent binding to the organic matter would mean that the release will occur continuously but slowly, since the turnover rate in soils is usually reduced to 2–5% per year.

In field conditions the degradation of contaminants does not often occur satisfactorily. This can be caused by a lack of appropriate microorganisms being able to attack the pollutants of concern under the environmental conditions. For garden topsoils approx. 9.8×10^6 bacteria g^{-1} soil and 1.2×10^5 fungi g^{-1} soil were reported. With increasing depth the number decreased and 10^3–10^5 bacteria g^{-1} soil and 10^3–10^4 fungi g^{-1} soil lived in the subsoil (Fletcher 1994). Other sources state 10^8 g^{-1} bacteria and 10^5 g^{-1} fungi on average (Paul and Clark 1996) as well as 10^9–10^{10} g^{-1} soil bacteria and 10^6–10^8 g^{-1} soil fungi. Apart from the mentioned microorganisms, other microbes such as actinomycetes (10^7–10^8 g^{-1} soil) and algae (10^4–10^5 g^{-1} soil) may contribute to the degradation rate. The biomass of the microorganisms varies between 40 and 500 g m^{-2} soil (bacteria and actinomycetes), 100 and 1,500 g m^{-2} soil (fungi) and 1 and 50 g m^{-2} soil (algae) (Bradl and Weil 2008). It should be noted that in less fertile soils the bacteria and fungi density is much lower, in particular in the subsoil. Here, high microbe concentrations are limited to the upper 10 cm. In general, the metabolic activities are related to the number and biomass of organisms. The microbe distribution shows a fluctuating picture. Strong concentrations of microbial activity simultaneous to the biological activity are present in the vicinity of living roots (rhizospehre), in earthworm tubes and where degradable detritus is located. In the rhizosphere within 50 μm from the root surface the microbial population can be up to 50–100 times greater than in unvegetated topsoil. With regard to the whole soil matrix, rhizosperic soils contained a microbial population that was 1–2 orders of magnitude higher than in soils without vegetation (Karthikeyan and Kulakow 2003). Consequently, in the areas indicating high bacterial concentration (hot spots) the biodegradation potential might be significantly higher.

Microorganisms are classified into aerobic, anaerobic and facultative ones. The first group needs only oxygen as electron acceptor, while the second group uses alternative electron acceptors such as nitrate, manganese, iron and sulphate. Aerobic microbes are not able to metabolise organic pollutants in the absence of oxygen, but facultative microorganisms utilise oxygen if present. Otherwise, they switch to the other electron acceptors mentioned.

The species used for bioremediation are usually derived from the site that has to be treated. They are adjusted to the environment where they are reintroduced in a great number. Population densities of 10^7–10^8 cfu (colony forming unit) g^{-1} are considered to be optimum and the minimum density should exceed 10^6 cfu g^{-1} in any case. The indigenous genera *Arthrobacter, Azotobacter, Burkholderia, Comamonas, Chromobacter, Clostridium, Mycobacterium, Nocardia, Pseudomonas, Ralstonia, Rhodococcus* and *Sphingomonas* are applied in differing densities and compositions. They are added in the form of dry compounds, suspensions and foaming agents.

There are specific affinities between different genera and the pollutants. Aliphatic hydrocarbons (e.g. TPH) are preferentially attacked by the genera *Achromobacter, Acinetobacter, Aeromonas, Alcaligenes, Arthrobacter, Bacillus, Brevibacterium, Co rynebacterium, Flavobacterium, Mycobacterium, Nocardia, Pseudomonas, Vibrio* and by actinomycetes. For aerobic phenol treatment *Pseudomonas putida, Trichosporon cutaneum, Nocardia* sp., *Klebsiella pneumoniae, Serratia liquefaciens, Flavobacterium chromobacter, Chlamydomonas ulvaensis, Phoridium fuveolarum, Scenedemus basiliensis* and the genera *Vibrio* and *Spirillum* were mentioned (Vipulanandan et al. 1994).

The degradation of the PAH naphthalene occurs with the help of some fungi, cyanobacteria and algae but most of the degraded naphthalene results from bacteria such as *Acinetobacter calcoaceticus, Alcaligenes denitrificans, Bacillus cereus, Corynebacterium renale, Mycobacterium* sp., *Moraxella* sp., *Rhodococcus* sp. and *Streptomyces* sp. Furthermore, *Pseudomonas* species (namely *P. cepacis, P. fluorescens, P. putida,* and *P. vesicularisare*) are capable of degrading naphthalene. For the degradation of the PAH phenanthrene the genera *Acinetobacter, Aeromonas, Alcaligenes, Arthrobacter, Beijerinckia, Falvobacterium, Micrococcus, Mycobacterium, Nocardia, Pseudomonas, Rhodococcus, Streptomyces* and *Vibrio* were described (Feitkenhauer et al. 2001).

In another study it has been observed that highly soluble compounds with a low molecular weight like naphthalene are degraded preferentially by *Pseudomonas*, medium-soluble compounds like phenanthrene by *Sphingomonas* and *Mycobacterium* and compounds with a high molecular weight like benzo(a)pyrene mainly by *Mycobacterium*. It is obviously possible to discover relationships between the PAH congeners and the bacterial species (Kanaly and Harayama 2000).

Mycobacterium belongs to a number of species which are predominantly associated with the organic matter and which have a strong tendency to adhere to hydrophobic substances. In general, the specific affinity between bacteria and pollutant allows an efficient decontamination even at relatively low pollutant concentration. The highly specific bacteria are responsible for steeper concentration gradients and subsequent higher substrate transfer rates (Wick et al. 2001).

After identification and selection the biomass of the useful bacteria population should be increased. In a fermentation tank the bacteria are inoculated and afterwards incubated to reach higher turbidity. The treated culture is used as seed for a larger vessel, in which the process is repeated until the volume of the broth is acceptable (Fletcher1994).

Sometimes for bioremediation purposes additional bacteria are used. In particular, such selection is essential for less degradable contaminants such as PAH and chlorinated hydrocarbons. The contaminated soil has to be inoculated by bacteria that are able to survive in the given environment. The best solution is bacteria isolation from the same geographical area and comparable soil types. Again, the selected microbes inoculate a nutrient culture that is incubated until the broth is turbid. The turbidity shows the increased population from several thousands to millions. It is possible to repeat the process until the population density is sufficiently high for the application.

The process of selection and population increase is termed bioaugmentation. It is frequently accompanied by the addition of nutrients, oxygen or water in order to stimulate and improve the living conditions for the microorganisms. In this case, biostimulation and bioaugmentation complement one another.

Results from case studies with bioaugmentation are different. Many studies have shown the low efficiency of bioaugmentation with regard to common contaminants such as TPH, PAH and chlorinated compounds (Grotenhuis and Rijnaarts 2011). In general, bioaugmentation can only be continued, if the pollutants are bioavailable to the microorganisms. It has been found that inoculated communities are more successful in relation to the degradation of freshly contaminated soils, while contaminated sites looking back over a long-term contamination history are less affected by the microbes. Obviously, the old contamination is no longer available to the added (and the indigenous) bacteria. On the other hand, after accidents, for example, when there is not enough time for the autochthonus bacteria to adapt but the contamination is relatively fresh, bioaugmentation might make sense in order to degrade the pollutants quickly (Müller and Mahro 2001).

The selection of appropriate strains of bacteria is critical because they have to be non-pathogenic for humans and animals. For instance, the species *Pseudomonas aeruginosa,* which is capable of degrading TPH, is supposed to be pathogenic for man and animals and consequently no useful microbe for soil remediation purposes. In general, known pathogens should not be selected since they are adjusted to higher temperatures (about 37°C), which are usually not present in terrestrial ecosystems. Accordingly, these pathogens are not able to be adapted to the environmental conditions in soils. Pathogens such as *Salmonella* sp., *Shigella* sp., *Streptococcus* sp. and *Staphylococcus aureus,* in particular, are non-reliable microbes. Furthermore, some *Clostridium* bacteria, namely *C. botulinum, C. tetani, C. perfringens*, and *Bacillus anthracis,* which manifest a high density in soils, are less suitable for bioremediation due to their pathogenic character.

Apart from the selective enrichment approach, genetic manipulation is alternatively targeted at accelerated degradative capability. The manipulation means an alteration in the genetic coding, resulting in the formulation of more efficiency

Table 6.4 Assessment of the principal biodegradation of contaminants

Well-biodegradable	Moderately biodegradable	Non-biodegradable
Mineral oil hydrocarbons (TPH)	Long-chain aliphatic hydrocarbons	Heavy metals
BTEX-aromates and phenols	Polycyclic Aromatic Hydrocarbons (PAH)	Radionuclides
Volatile chlorinated hydrocarbons (CHC)	Polychlorinated Biphenyls (PCB)	
Nitroaromates	Polychlorinated Dioxins/ Furans (PCDD/F)	
Free cyanides	Complex cyanides	
	Complex nitroaromates	

metabolic opportunities. With reference to biodegradation of contaminated sites, this approach might play a minor role in Europe, whereas in the USA more research projects are running.

6.3.2 *Kinetic of the Pollutant Degradation*

Table 6.4 provides information about the contaminants capable of biological treatment. In summary, mineral oil hydrocarbons (TPH), mono-aromatic compounds (BTEX, phenols), volatile chlorinated hydrocarbons (VCHC), nitroaromates and free cyanides are presumed to biodegrade well because in the majority of the case studies a remediation time of 1–2 years has been sufficient. In principle, the decomposition of TPH, BTEX and phenols always occurs in aerobic conditions, whereas the degradation of the substituted compounds requires more complex conditions related to the role of the electron acceptors. In the first instance, the success of the elimination of TPH, BTEX and phenols depends on the oxygen supply (Grotenhuis and Rijnaarts 2011).

In contrast, parameters such as long-chain aliphatic hydrocarbons, polycyclic aromatic hydrocarbons (PAH), polychlorinated biphenyls (PCB), polychlorinated dibenzodioxins and dibenzofurans (PCDD/F), complex cyanides and complex nitroaromates are, in principle, biodegradable. However, the period of time needed might lead to the exclusion of the treatment in a great number of cases. In general, an incomplete biodegradation might occur in the presence of a mixture of compounds, since the more degradable pollutants are quickly eliminated, while the refractory part of the contaminants remains at the end of the treatment period.

Generally speaking, metals and radionuclides (that are of reducing concentration in the long term due to the decay) are not adapted to microbiological decontamination. Indirectly, in the case of chromium bioremediation is of significant importance. Soils containing high levels of organic matter and consequently enhanced biological activity tend towards accelerated reduction processes, also resulting in the Cr(VI) reduction to Cr(III). This is carried out by bacteria strains resistant to raised Cr concentrations such as *Pseudomonas fluorescens, Pseudomonas aeruginosa* and *Enterobacter cloacae*. In its trivalent form Cr is a component of a

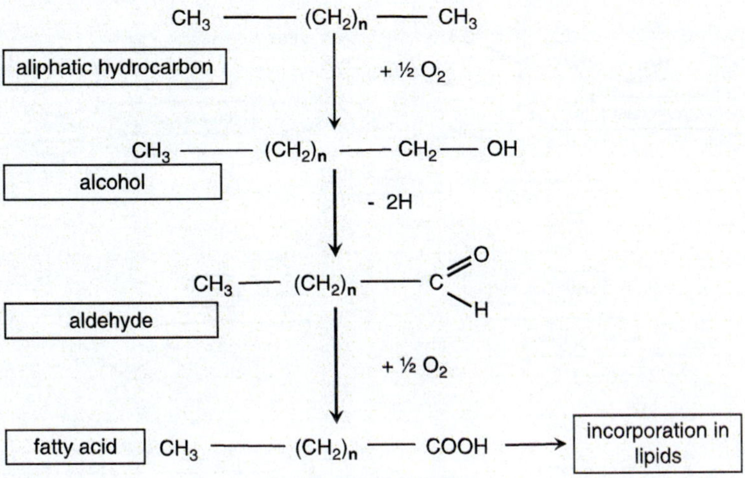

Fig. 6.12 Aerobic decomposition of aliphatic hydrocarbons

balanced human and animal diet, while the hexavalent Cr, which is normally from anthropogenic sources and does usually not occur in natural minerals, is toxic for human beings and animals in high doses. Both forms are supposed to be stable in soils. In anaerobic conditions chromium is reduced to the more harmless trivalent form. In contrast, in oxidative conditions the opposite process can occur slowly (Zayed and Terry 2003).

The aerobic degradation of aliphatic hydrocarbons which are the main constituents of mineral oils (TPH) is displayed in Fig. 6.12. Firstly, the compound is transformed into alcohol in aerobic conditions, followed by the metabolisation to aldehyde and finally to fatty acids in aerobic conditions which microorganisms are able to incorporate into lipids produced naturally in the body. Aliphatic compounds comprising double bond structures are degraded by hydration, which is responsible for the reduction of the double bond structure into a one bond structure. Within the aliphatic chemicals compounds with a one bond structure indicate a quicker degradation rate than the two bond ones. Similarly, the degradation increases with decreasing chain length (C_5–C_9>C_{10}–C_{19}>C_{20}–C_{40}). Very short-chain compounds (C_1–C_4), however, might result at an extremely low biodegradation rate due to microtoxicity.

The degradation of aromatic compounds like benzene and phenols appears to be more complex, as exhibited in Fig. 6.13. In addition, the biodegradation occurs in the presence of oxygen. In the case of benzene the process starts with the enzymatic reaction dioxygenase, leading to the adsorption of hydroxyl groups in combination with elimination of protons. In the end pyrocatechol results from this enzymatic process. The degradation process of phenols follows the same sequence. The hydroxylation, however, is limited to one OH group only (monooxygenase). Subsequently, the actual destruction of the ring structure must be carried out. Alternatively, for ring disruption purposes ortho cleavage and meta cleavage can be

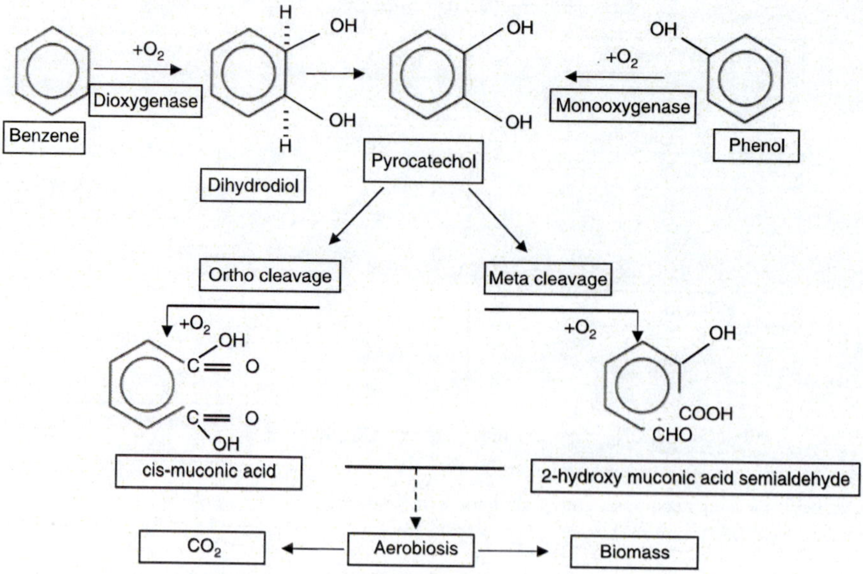

Fig. 6.13 Aerobic decomposition of benzene and phenol

carried out. In any case, the OH groups may configurate COOH groups that are transformed into acids which microorganisms can incorporate. End products of the following aerobiosis are carbon dioxide and microbial biomass.

In principle, organic compounds consisting of more than one ring structure show the same pathways in relation to the biodegradation. The reconstructed pathways of the three-ring PAH anthracene are displayed in Fig. 6.14. Ring after ring is disrupted analogously to the sequence explained above. The main difference refers to the duration of biodegradation, since the more rings are present, the longer the remediation time must be calculated. It has often been observed that high-molecular weight PAH like benzo(a)pyrene were omitted from biodegradation within the calculated degradation time period in contrast to low-molecular PAH like naphthalene. This result is related to the extremely low water solubility of benzo(a)pyrene (4 $\mu g\ L^{-1}$) compared with naphthalene, which indicates 30 $mg\ L^{-1}$ (Mahro et al. 2001). The remaining less degraded benzo(a)pyrene may reduce the risk to the environment due the low bioavailability but on the other hand this parameter is well-known for its carcinogenicity and consequently particular danger to humans and the environment.

A study dealing with artificially contaminated soil (tar oils) revealed the distinct degradation rate of PAH congeners and phenols (Table 6.5). While phenols and PAH with two or three rings were well-degraded, complex PAH like benzo(a)pyrene exhibited a significantly lower degradation effect within the same period of time (Vipulanandan et al. 1994).

In addition, there is one more aspect which has an impact on the principal degradation of PAH. It has been found that ageing of the compounds reduces the bioavailability of, for instance, phenanthrene, anthracene, fluoranthene and pyrene.

Fig. 6.14 Aerobic decomposition of anthrazene (*PAH*) (simplified)

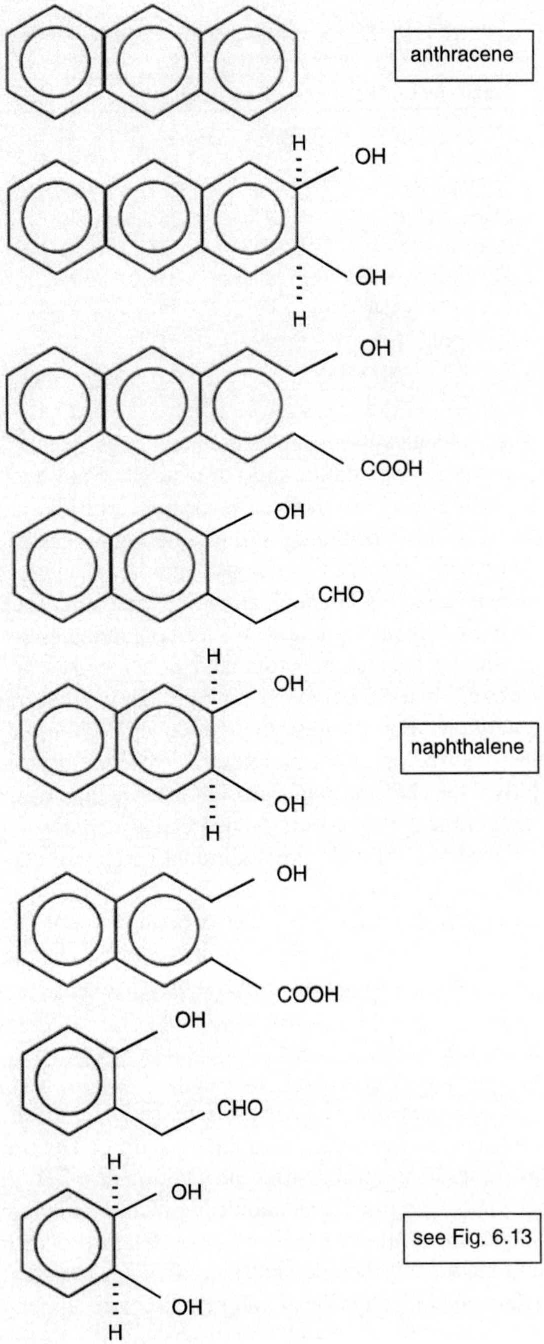

Table 6.5 Concentrations of individual coal tar oil constituents during biodegradation of an artificially contaminated soil (Data from Lajoie and Strom 1994)

Parameter (mg kg^{-1})	Time (weeks)					
	0	1	2	4	8	15
Phenol	37	<0.5	<0.6	<0.5	<0.3	<0.3
Naphthalene	202	2	<0.3	<0.3	<0.2	<0.2
Phenanthrene	774	254	39	15	7	1
Anthracene	247	138	32	15	7	1
Fluoranthene	406	254	198	35	12	5
Benzo(b)fluoranthene	84	86	55	91	82	71
Benzo(a)pyrene	30	31	16	11	13	15

The molecules were sequestered and subsequently hidden or inaccessible. Thus, a continuous degradation did not take place any more. With an increasing tendency towards sequestration the disappearance of the compounds was strongly reduced. However, after beginning soil weathering processes (wetting and drying cycle) aged compounds showed continuous degradation again (Alexander 2001). In conclusion, sites manifesting contamination derived from processes several decades ago might be more difficult to treat than sites contaminated recently.

Substituted and, in particular, halogenated compounds are generally biologically treatable but the complexity of the compounds requires more complicated basic conditions. For instance, the reductive de-chlorination leading to the separation of the chlorine halogens preferentially occurs in an anaerobic environment, particularly if the chlorine atoms amount to more than two. Because of the Cl removal from the pollutant the process is also called dehalorespiration. The R-Cl compound is reduced to R-H+Cl^-. An additional carbon source should be present. The chlorinated compound itself serves as electron acceptor. This means that it will be reduced during the conversion of other organic material (Nathanail and Bardos 2004). The process of de-chlorination may also occur in the presence of carbon dioxide or sulphate serving as electron acceptors (Middeldorp and Langenhoff 2002). While organic compounds consisting of one to two chlorine atoms can theoretically be degraded aerobically, compounds amounting to 3 or 4 Cl atoms are in any case degraded more effectively in anaerobic conditions (Table 6.6).

In the presence of a co-substrate the decomposition of volatile CHC can also be carried out by methanotrophic bacteria. The methanotrophic monooxygenase increases considerably after the addition of a CH_4-air-mixture.

The anaerobic degradation of tetrachloroethylene is represented in Fig. 6.15. It should be mentioned that the degradation procedures can be promoted biotically and abiotically. The de-chlorination is conducted step by step until vinyl chloride indicating one chlorine atom only is formed. Finally, ethylene results from the biodegradation process, which will subsequently be treated aerobically.

According to the sequence mentioned, the treatment of chlorinated contaminants requires both firstly an anaerobic iteration and secondly an aerobic one. The change from anaerobic to aerobic decomposition is necessary for a number of pollutants.

Table 6.6 Basic conditions for the degradation of chlorinated compounds

Number of chlorine atoms	1–2	3–4
Anaerobic degradation	(+)	++
Aerobic degradation	+++	+
Degradation with additional carbon source	(+)	+
Degradation without additional carbon source	+	–

+++ optimal
++ very good
+ good
(+) limited
– impossible

For instance, the pesticide pentachlorophenol (PCP) is anaerobically degraded step by step (sequence: tetrachlorophenol → trichlorophenol → dichlorophenol → monochlorophenol), followed by the aerobic decomposition of the residual phenol compound.

Not only halogenated pollutants are biodegradable but also other substituted chemicals are capable of biological decomposition. For example, trinitrotoluene (TNT) can generally be treated in a regeneration pit. The nitrite reduction also occurs step by step in anaerobic conditions, leading to the formation of amino-NH_2, which reacts in the presence of organic matter and water to volatile NH_3 (deamination):

$$R-NH_2 + H_2O \rightarrow NH_3 + R-OH \tag{6.2}$$

By comparison, compounds substituted with hydroxyl, carboxyl and amino groups are more easily and more rapidly degradable than aromates substituted with fluoride, chlorine and nitrite. Bromine containing contaminants are considered to be toxic substances which are particularly inadequate for degradation.

With reference to the most important organic pollutants as already mentioned attention should be paid to some criteria, if bioremediation is taken into account. The chain length and the number of branches present are of interest with reference to the TPH compounds. Related to the aromatic compounds the number of benzene rings is a decisive factor. In the context of chlorinated hydrocarbons the degree of the chlorination as well as the volatilisation potential may play a major role. Similarly, in the case of PCB and PCDD/F the degree of chlorination but also the number of adjacent C atoms which have not been substituted by Cl appears to be of importance. The criteria are listed in Table 6.7.

6.3.3 Technical Devices Used

6.3.3.1 Regeneration Pit (Biopile)

Of all the different strategies bioremediation is the most frequently used clean-up approach worldwide. In many countries, in particular in Europe (e.g. Denmark,

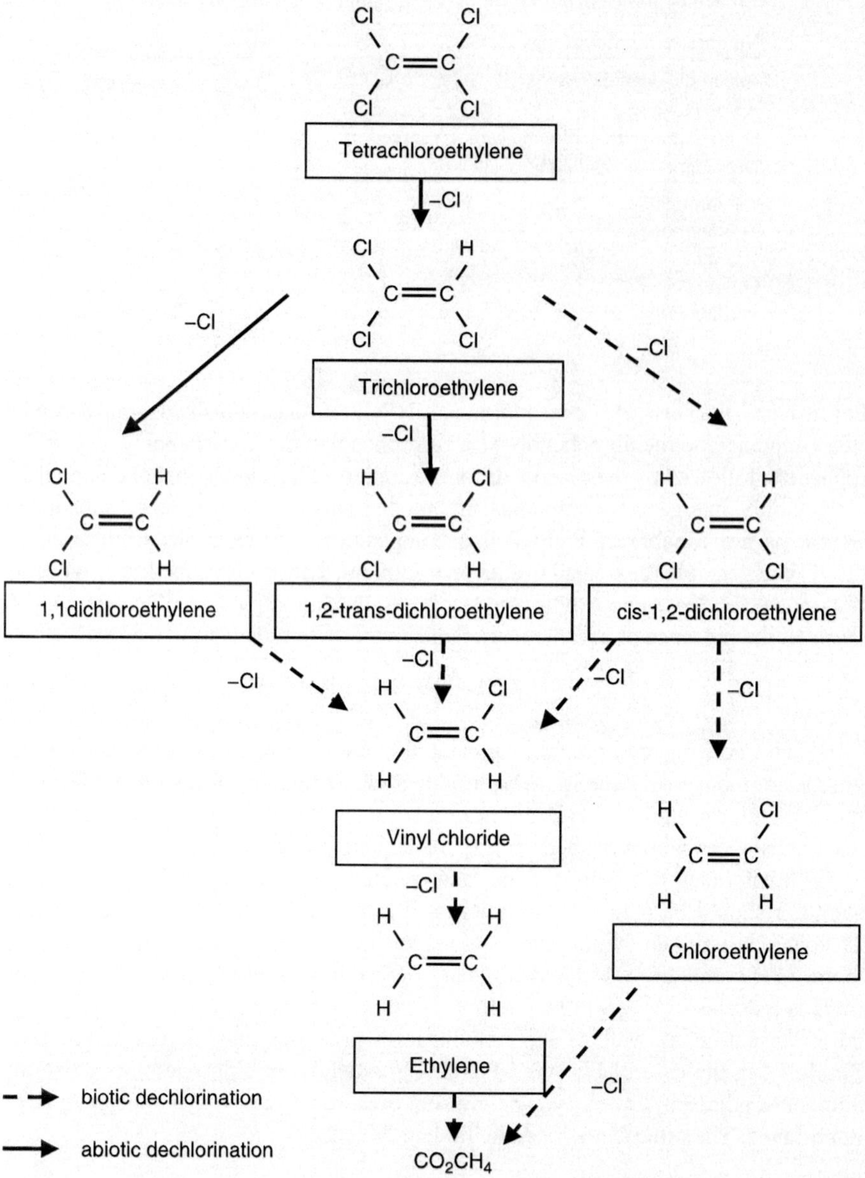

Fig. 6.15 Anaerobic degradation of tetrachloroethylene (*TCE*)

France, Germany, Sweden, Switzerland, The Netherlands and the United Kingdom), the USA and Australia, it has been adopted to decontaminate particularly sites of petrol stations, fuel depots, refineries, accident sites alongside motorways, etc. There seems to be a tendency towards using bioremediation as the most important decontamination strategy nowadays and this is also likely to be the case in future.

Table 6.7 Important remediation criteria for organic pollutants

	Optimum	Degradation excluded
Mineral oil (TPH):		
Chain length	C_{10}–C_{20}	>C_{30}
Number of branches	<3	>4
BTEX, phenols, PAH:		
Number of benzene rings	<4	>5
CHC:		
Degree of chlorination	<50%	–
Volatilisation (hPa)	<500	>1,000
PCB, PCDD/F:		
Degree of chlorination	<4	>5
Adjacent, not Cl-substituted C atoms	>2	<2

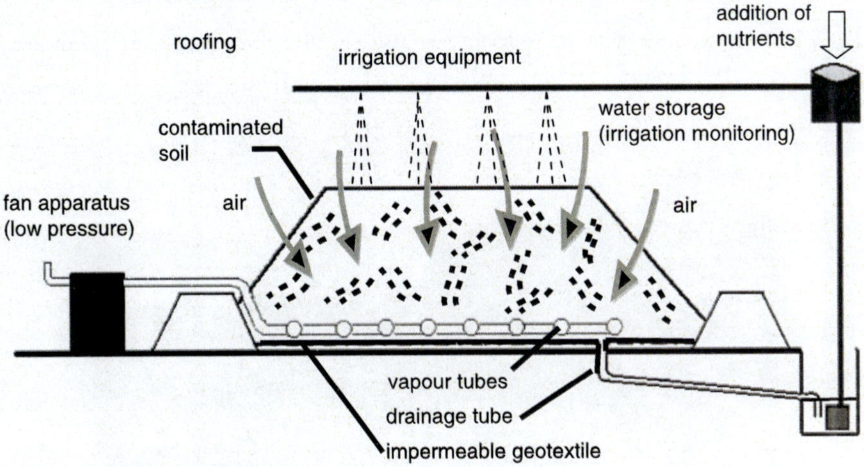

Fig. 6.16 Design of a regeneration pit for bioremediation purposes

In association with bioremediation the regeneration pit treatment (biopiles) is the most widely used version.

Piles in which the biological decontamination may occur (Figs. 6.16 and 6.17) are constructed. The height ranges from 0.6 to 5 m and they have a length of up to 100 m and a trapeze shape. To prevent erosion the pile located in the open-air is covered with a geotextile which is permeable by air or it is vegetated. In addition, a (dark coloured) geotextile facilitates the warming of the soil treated. During the establishment and homogenisation of the pile the contaminant concentration can already be reduced due to volatilisation (Fan and Tafuri 1994). Some volatile components tend to evaporate rather than biodegrade. In general, in proximity to the bioremediation plants terrible odour development caused by volatilisation should be taken into consideration in association with the planning process of the facility.

Fig. 6.17 Roofed regeneration pit (With kind permission of DHC company, Freiberg, Germany)

Fig. 6.18 Different substrates deposited before the biological treatment occurs in a remediation plant (With kind permission of DHC company, Freiberg, Germany)

Conditioning procedures are necessary with reference to the biopile application as well. In particular, because of the distinct origins of the material to be treated some separation procedures (see Sect. 6.1) should be carried out (Fig. 6.18). The homogenisation of the contaminated material is essential to make sure that the microorganisms are able to reach the entire contaminated soil. Thus, screened soils should be homogenised by mixer systems before the regeneration pile is deposited.

The living conditions for the microorganisms can be manipulated in different ways. In order to achieve aerobic conditions fresh air is continuously drawn in by a low pressure fan apparatus. Afterwards, it is necessary to treat the air using an activated carbon adsorber or by biofiltration before releasing it into the atmosphere (see Sect. 7.2.3).

It is important that the oxygen concentration never falls below 1 vol% (Hupe et al. 2001). To support microbial growth a content ranging from 2 to 4 vol% is preferable (Fan and Tafuri 1994). The technique is necessary in piles with a height exceeding 1.5 m, since the diffuse oxygen supply derived from the atmosphere is supposed to be insufficient. In a biopile measuring a height of 3.20 m the oxygen concentration decreased continuously during the TPH remediation process and the critical limit of 1 vol% was reached in depths below 2 m (Koning et al. 2001). An automatic biopile aeration device with intermittent aeration pulses reacting as soon as the oxygen concentration is below 2 vol% would prevent anaerobic conditions. It can usually be expected that the oxygen content fluctuates continuously, falling occasionally below the critical limit in combination with the development of anaerobic gases such as hydrogen sulphide.

Alternatively to the fan apparatus, the oxidising conditions of the regeneration pits can be achieved in the presence of earthmoving equipment. This windrow-based system consisting of an active turning and tilling of the material comes from the composting industry. Furthermore, the aerobic conditions can be improved, if structuralising organic materials with an accelerating air capacity such as wood chips, compost, bark mulch, sawdust, straw and peat are added in layers during the construction of the pile. Shredded timber and wooden railway sleepers which are used occasionally must not be processed with fungicides.

In the case of substituted compounds like TNT there is generally the opportunity to change the oxygen conditions (see Sect. 7.1). While initially a nitrite reduction occurs in anaerobic conditions caused by the increasing oxygen consumption due to the amendment of organic substrates such as compost to the biopile, the degradation of the remaining toluene is conducted in aerobic conditions after, for example, aeration by intensive mixing of the material (Winterberg 2001).

The water household is monitored by roofing (e.g. tents, hall) or by covers made of geotextiles permeable by air but impermeable by water. Since the piles tend to dry out during the remediation process, as observed, for example, in a pile treating TPH (Koning et al. 2001), irrigation equipment is installed. In a similar way, the pH value is changeable after addition of lime. The percolating water is collected by drainage tubes that are constructed above an impermeable geotextile and concrete base and partly re-circulated. Finally, an air and a water treatment module must be included in the bioremediation process (see Sects. 7.1.4 and 7.2.3). Furthermore, the nutrient status can be altered by the addition of some nutrients in case the supply of the latter is insufficient. Either the nutrient solution is sprayed on the pile surface or the nutrients are applied to the different layers during the pile construction.

Apart from the positive impacts on the air household of the piles, organic material supplies the piles with nutrients, particularly nitrogen and phosphorus, and microbes. In biopiles the addition of cultured microbes is mostly not required (Fan and Tafuri 1994). Nevertheless, the percentage of structuralising organic substrates should range between 10 and 15 vol%, but also a higher percentage is sometimes assessed to be beneficial. For instance, for the TNT degradation a concentration between 15 and 25% (based on dry matter) is recommended (Winterberg 2001).

It should be taken into account that the substrate addition causes exhaustion of oxygen, leading to an increasing lack of oxygen and possibly anaerobic conditions.

The presence of well-biodegradable organic compounds which should raise the co-metabolic degradation serves as an additional carbon source preferentially consumed by microorganisms. In particular, well degraded compost usually applied in the regeneration piles to improve the soil structure appears to be beneficial with regard to the decomposition of the organic compounds. Apart from the organic matter, some pollutants are only degradable in the presence of other pollutants which serve as co-substrates. For instance, 4-chlorophenol was not degraded by *Ralstonia eutropha* when it was the only substrate. In the presence of 2,4,6-trichlorophenol, however, the degradation occurred. Obviously, trichlorophenol induces enzymes that are useful for the degradation of 4-chlorophenol (Müller and Mahro 2001).

There is no doubt that the arrangement of a regeneration pit requires a lot of basic knowledge and experience. For this reason, depending on the kind and quantity of the soil contamination, pre-tests should be integrated to find the optimum physical and chemical conditions for a rapid and effective decontamination process. The tests are aimed at answering all questions about microbiological, physical and chemical soil properties and all relevant pollutants as well as about subsequent transferability into practice. The preliminary tests should be designed in a way that field conditions are simulated as closely as possible. However, the tests are problematical because they cannot simulate all given characteristics precisely such as the spatial heterogeneity and distribution of the pollutants and the physico-chemical parameters like particle size distribution, viscosity of the pollutants, microbial activity, etc. for every location exactly. Hence, it should be noted that every kind of pre-test modelling is considered to be more or less unrealistic.

With reference to the regeneration pits in optimised conditions the pollutant reduction was highly efficient. For instance, results from treatment with a TNT-contaminated soil were a reduction from 1,160 mg kg^{-1} to only 1.3 mg kg^{-1} within 6 months (Winterberg 2001). For TPH a pollutant reduction from 2,870 to 616 (40 days), 457 (125 days) and 357 (221 days) mg kg^{-1} using a biopile has been reported (Heely et al. 1994). Case studies in a 3.20 m high aerated biopile resulted in a TPH decrease from 8,200 mg kg^{-1} to 4,200 mg kg^{-1} (200 days) and to 3,900 mg kg^{-1} (300 days) at the depth 1–2 m and from 6,600 mg kg^{-1} to 2,300 mg kg^{-1} (200 days) and to 1,600 mg kg^{-1} (300 days) at the depth 0–1 m respectively (values estimated approximately) (Koning et al. 2001).

6.3.3.2 Bioreactor Technique

Except for the regeneration pit treatment alternative techniques dealing with the biological potential for degradation purposes are feasible. Even if mostly applied in a research context, although available as a full-scale technology, the bioreactor technique, alternatively termed bioslurry system, may successfully enable the degradation of organic compounds, since all determining factors can be manipulated to a great extent. With reference to mineral oil decontamination, results have been

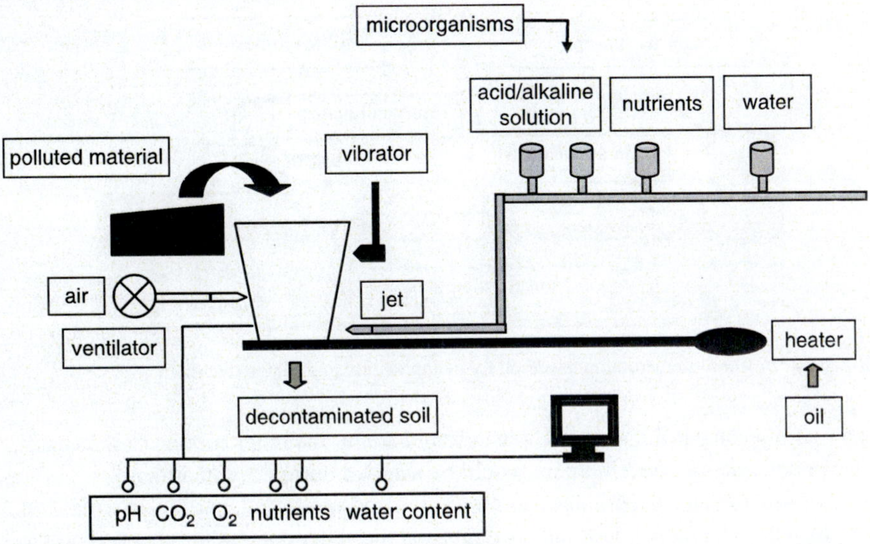

Fig. 6.19 Setup of a bioreactor technique

reported indicating a very fast and effective reduction of the pollutant concentration. An initial concentration of approximately 8,500 mg kg^{-1} TPH was reduced to about 2,200 mg kg^{-1} within 15 days and about 400 mg kg^{-1} within 50 days in a small bioreactor filled with a contaminated soil/compost mixture. Based on a carbon balance, up to 35% of the initial carbon content (TPH, compost, microbial biomass) was neither degraded (CO_2 losses) nor incorporated into the biomass. It is likely that this balance gap is associated with the formation of bound residues (Lotter et al. 2001) (see Sect. 6.3.1). Apart from petroleum hydrocarbons, VCHC, pesticides and PCB were treated with this technique as well (Khan et al. 2004).

As shown in Fig. 6.19, the polluted material is treated in a vessel with a limited capacity for a short period of time. For most approaches the material is mixed with water to form slurry. Non-uniform particle size and high clay content can reduce the effectiveness due to inhibiting microbial contact. Air injection conducted by a ventilator in addition to the impact of a vibrator makes the oxidising conditions safe and an oil-fired heater usually set at 15–35°C increases the reactive temperature of the material. Furthermore, microorganisms containing suspensions, acid or alkaline solutions (optimised pH value: 4.5–8.8), nutrients and tap water can be added. The microorganisms are indigenous genera or specially added cultures (NJDEP 1998). The water content depends on the texture. Sandy soils require water contents between 50 and 70% of field capacity, silty and clayey soils are treated with a water content of 70–90% of field capacity (slurry treatment). For the wastewater purification modules have to be added.

There are two types of vessel technique – the fixed-bed type and the rotary drum type. The latter one is more problematical, since the soil can be changed into a round pellet structure less susceptible to the degradation progress and, furthermore,

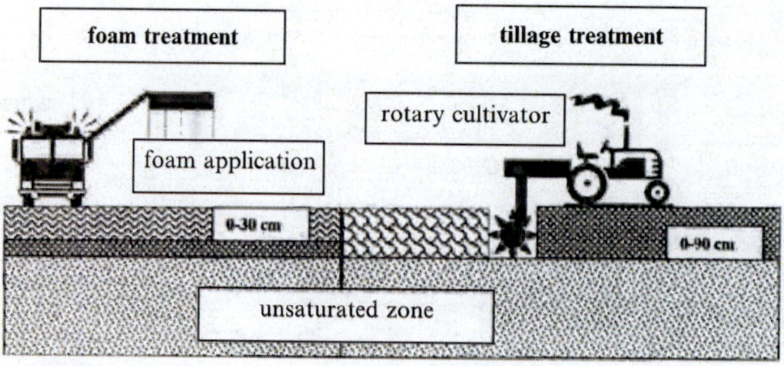

Fig. 6.20 *In situ* bioremediation based on foam application and tillage treatment

the rotating can cause damage known as abrasion to the inner surface of the vessel. The pellet structure development would be avoided during dry treatment.

Every parameter is permanently monitored and registered, particularly pH, CO_2 and oxygen. The most important nutrients and the water content are checked online. For this reason different sampling devices are installed. The volume flow of gases such as carbon dioxide is detected by an inductive flowmeter. The permanent measurement of the parameters leads to a rapid response automatically. In summary, the usefulness of dynamic reactors might be limited according to the cost and the relatively small volume that can be treated because of time constrictions. For this reason, this technique is applied to a few highly contaminated soils or to the after-treatment of soils previously treated by soil washing (see Sect. 6.2.1).

6.3.3.3 *In situ* Treatment

In the case of surface-near contamination *in situ* bioremediation is used as an alternative (Fig. 6.20). The irrigation treatment using a specified foam suspension containing microorganisms, nutrients and water must be considered as a method which does not reach great depths (usually the uppermost 30 cm), while the tillage treatment can reach a depth of approximately 90 cm using rotary cultivators and the same agents. The object of this idea is to avoid volatilisation and consequently the release of the pollutants to the atmosphere. In contrast, the oxygen supplied should improve the aerobic degradation process. In the case of deep-reaching contamination oxygen must be pumped through pipes into the subsoil. Both approaches – foam application as well as tillage treatment – are limited to the unsaturated zone and are frequently appropriate at accident sites alongside motorways and railway lines. By avoiding the tillage procedure the *in situ* approach can even be applied underneath existing buildings.

In situ bioremediation is frequently applied to crude oil spills at beach sites caused by ship disasters. For instance, after the terrible accident of the Exxon Valdez in Alaska the oily beaches were degraded by indigenous bacteria after amendment of oil-soluble fertilizers containing nitrogen and phosphorus. Compared to areas where fertilizers were not sprayed, the treated soils already showed a clear decrease in oil contamination within the first 70 days. After 110 days about half of the oil was even removed (Bradl and Weil 2008).

Attention should be paid to enhanced degradation of the humus in association with nutrient leaching of e.g. nitrate and sulphate. Investigations with artificially contaminated soil material (incubation tests), however, showed that the nitrogen mineralisation is strongly inhibited in soils highly contaminated with e.g. petroleum hydrocarbons. The nitrogen net mineralisation of material containing 1% or more lubricating or fuel oil was totally reduced and of material containing 0.1% oil already significantly affected. Consequently, ammonification and especially nitrification were comparatively low (Kiene et al. 2001). On the basis of this study after accidents the nitrate leaching at highly contaminated sites is likely to be negligible during an *in situ* bioremediation process.

As an alternative to the biological degradation of organic pollutants it is possible to carry out mineralisation using ozonation as also applied to the groundwater (see Sect. 7.1.6). Ozone is capable of direct attacks on carbon-carbon double bonds and of an unspecific reaction of hydroxyl radicals generated during the ozone decomposition. In the unsaturated zone ozone is relatively stable, indicating a long-term chemical reaction with the pollutants. Longer ozonation, however, can cause disadvantages such as oxidative destruction of the organic matter and reduction of edaphon and biological activity. Besides, the radical reaction may produce new compounds previously unknown (Tiehm and Stieber 2001).

Strong oxidation, which is, for instance, associated with the ozonation, can cause enhanced metal mobility, since the simultaneous oxidation of the organic matter may destroy the binding capacity (Grotenhuis and Rijnaarts 2011). Furthermore, degradation of the organic matter causes acidification, which is responsible for accelerated metal mobility as well.

6.3.3.4 Landfarming

The *in situ* soil bioremediation covers the so-called landfarming as well. This treatment is focused on the development of an enhanced volatilisation of some organic pollutants. The contaminated material, for instance oil sludge, is spread onto former farmland or brownfields. The piled material usually has a height of 20–50 cm only, exceptionally up to 1.5 m, and an impermeable plastic sheet is laid under it to prevent bioturbation and damages caused by rodents as well as to minimise leaching processes. Without the construction of a sealing base the potential mobilisation and downward migration of pollutants will be risky (Nathanail et al. 2002).

The oxidising conditions are processed by earthmoving equipment, for instance by ploughing. The soil should be well mixed to increase the contact between the pollutants and the microorganisms. Nutrients and moisture are continuously controlled and amended. Because a C/N ratio of 9:1 is aimed at, nitrogen, in particular, must be added. It has been found that the best moisture content for landfarming is 18% and the degradation rate will significantly decrease, if the water content is lower than 12% or higher than 33% (Khan et al. 2004). Normally, no special microbes are added. The construction of sidewise bunds reduces run-off as well as erosion.

Disadvantage is that a large area is required and the effectiveness in wintertime is strongly limited due to the unfavourable soil temperatures. The optimum air temperature ranges from 25 to 40°C. High molecular weight constituents such as heating and lubricating oils, diesel fuel and kerosene frequently require too long time periods for degrading. In general, a TPH concentration exceeding 50,000 mg kg^{-1} reduces the effectiveness enormously.

Besides, there is no control of volatile emissions during the operation. It is assumed that, for instance, lighter petroleum hydrocarbons tend to be removed preferentially by volatilisation instead of aerobic degradation.

6.3.4 Soil Properties Required

In order to carry out *ex situ* bioremediation the persons involved should have exact knowledge of the soil properties. The parameters described as follows are also of importance in the case of an *in situ* approach. The texture classes silt plus clay should be low, since permanent aerobic conditions might be badly to afford in a mainly cohesive material. Soils consisting of more than 90% silt plus clay, for example, are impossible to treat biologically. Moreover, the seepage tends to accumulate fine particles, which might block the percolating water below and cause stagnating water.

The water content in the regeneration pit should vary between 40 and 80% of the maximum water holding capacity (depending on the texture) to avoid too dry and too wet soil conditions, which may negatively influence the aerobic decomposition of the organic pollutants. For treatment of oil-contaminated soil a water content of 35–65% of the maximum water holding capacity is recommended (Hupe et al. 2001). Biopiles with earthmoving devices should work with water contents tending towards the lower value to prevent pellet and agglomerates. In the case of water content of 90% the aerobic degradation is always finished and a detrimental side effect of anaerobic conditions is the development of H_2S, causing a horrible odour in proximity to the regeneration pits. In contrast, more than 10% of the field capacity should be enabled in order to initiate bacterial activity (Fan and Tafuri 1994).

The soil temperature within the pile should range from 20 to 42°C, since the biological activity is considered to be most beneficial. In general, increasing temperature between 10 and 40°C improves the degradation of a number of pollutants, including non-aqueous phase liquids (NAPL), because their viscosity tends to decrease in conjunction with accelerated solubility and subsequently availability to the bacteria (Hupe et al. 2001; Koning et al. 2001). For instance, *Pseudomonas putida*, a species often used for the degradation of TPH, grows optimally at 30°C (Fletcher 1994). Within the biopile enormous differences of up to 12°C can exist (Koning et al. 2001). Most of the bacteria-degrading pollutants are mesophilic but

Fig. 6.21 Regeneration pits for the bioremediation of an oil-contaminated soil in cold climatic conditions in Oulu, Finland – the photograph was taken in November when the sun sets at 3 p.m. and the air temperature is −35°C

the thermophilic (mainly adapted to temperatures from 48 to 72°C) and psychrophilic (mainly adapted to temperatures from 10 to 24°C) microorganisms are also generally active. Because of the cold climates in the northern hemisphere only restricted seasonal periods exhibit surely appropriate weather conditions (Fig. 6.21). In the presence of the addition of organic amendments such as compost the temperature might normally increase up to 65°C comparable with the composting process.

In the case of an indoor treatment using ground heating on-site bioremediation can be applied very favourably. From the energy saving point of view the use of waste heat from other processes (e.g. composting) or from block-heating power plants nearby can be useful. The temperature rise reveals a number of advantageous properties that may improve the degradation process. Solubility and diffusion coefficient of the organic pollutants increase and the solubility is additionally enhanced because of decreasing viscosity of some pollutants such as tar oil. In particular, long chain aliphatic compounds are less attacked in mesophilic conditions but with increasing temperature the compounds are more present in liquid form and subsequently better degradable. However, parameters like phenols and naphthalene, which are well degradable under mesophilic conditions by *Pseudomonas,* showed efficient degradation (predominantly by *Bacillus thermoleovorans*) at elevated temperature as well. Furthermore, at accelerated temperatures exceeding 60°C pathogenic microorganisms are reduced. The metabolic processes of the microorganisms that are mediated by enzymes can take place at higher temperatures as well. For instance, the genera such as *Aquifex, Bacillus, Clostridium, Fervidobacterium, Thermoanaerobacter, Thermotoga* and *Thermus* are efficient at temperatures between 60 and 80°C (Feitkenhauer et al. 2001).

Because of the favourable living conditions of both bacteria and fungi the pH value should vary between 5.5 and 7.0. Fungi tend to prefer slightly to moderately acid pH values and bacteria neutral to slightly alkaline pH values. Since soils of contaminated land often contain alkaline technogenic substrates such as concrete and slag (Meuser 2010), higher pH values up to 12 might cause problems to the biological treatment. On the other hand, acidification based on e.g. pyrite oxidation usually existent in mining wastes, are responsible for too low pH values, which impair the biological treatment that is mainly conducted by the species *Thiobacillus*. In acidic soils with high organic matter content liming with calcium carbonate or dolomitic limestone is necessary (Fan and Tafuri 1994; Hupe et al. 2001).

Attention should be paid to the nutrient capacity, in particular the macroelements carbon, nitrogen and phosphorus. A C/N ratio of 20:1 or less should be adjusted and the C/P ratio reaches an optimum at 50:1 for oil-contaminated soils (Hupe et al. 2001). Consequently, the optimum relationship between the elements C:N:P amounts to <20:1:0.4. Other studies recommend a ratio of 30:1:0.1 (Fan and Tafuri 1994). Both nitrogen and phosphorus can be amended by mineral fertilizers usually applied in agriculture. In most of the soils there is usually a sufficient amount of the nutrient elements including sulphur, so that they do not become a limiting factor. Thus, many studies reporting about the addition of nutrients to the contaminated soil did not exhibit an improvement of the biodegradation (Norris 1994).

The organic matter content should be lower than 5%, since otherwise the adsorption potential of organic pollutants to the organic matter is too strong. It is difficult to achieve this limit because the addition of organic substrates which are beneficial for better aeration, supply of microorganisms, nutrient supply and pH buffer may easily cause higher values. For remediation practice a compost content between 10 and 33% dry weight has often been recommended but a remediation effectiveness might also be gained if the content is lower than 10% (Hupe et al. 2001). Irrespective of the detrimental effects of organic matter content with regard to the biodegradation potential, an increased quantity of compost added means a higher space requirement for the treatment and limited reutilisation options after the treatment.

In addition, the biodegradation process will be disturbed if the organic matter content is high because microbes possibly tend to prefer the decomposition of well biodegradable humus to the decomposition of complex organic pollutants. Dissolved organic matter (DOC), however, can function as a solubilising agent.

Technogenic carbon derived from substrates like ashes, coke, coal and asphalt should not be constituents of the soil, since these artefacts may adsorb organic pollutants to a great extent, so that the process of biodegradation will be minimised (Meuser 2010).

Because of microtoxicity attention should also be paid to the concentration of heavy metals and salts (e.g. chloride). For this reason, a mixed contamination with organic pollutants and metals or salts might considerably reduce the success of the bioremediation. The maximum salt concentration should not exceed 0.5%.

In the context of biodegradation an influence on the metals frequently concomitantly present in the soil matrix can also be observed. For instance, heavy metals or other toxic substances such as cyanides may reduce or inhibit the biological process, since they are supposed to be toxic to bacteria and fungi in great quantities. On the other hand, as already mentioned above, the degradation process has an impact on the mobility of metals, e.g. by destruction of robust bindings like chelates, acidification or methylation of mercury (Hg), which is linked to the degradation of aromatic compounds and may increase the Hg mobility.

6.4 Phytoremediation

6.4.1 Principles of Phytoremediation

The term phytoremediation includes both phytodecontamination measures and phytostabilisation processes. The former is differentiated into:

- Phytoextraction, which is defined as uptake of contaminants by plant roots and subsequent plant harvest
- Phytodegradation leading to the microbial metabolism in the plant rhizosphere and subsequent degradation of organic pollutants
- Phytovolatilisation leading to gaseous losses of the contaminants after metabolism in the plant tissue.

Phytostabilisation involves mainly physical processes which stabilise the contaminated soil and interrupt the pathways endangering the environment and human health:

- Physical stabilisation by intensive root development preventing erosion and deflation of contaminated material that can contaminate adjacent areas
- Evapotranspiration, causing the reduction of contaminant leaching and groundwater hazard.

The terms (soil) stabilisation (see Sect. 5.4) and phytostabilisation must be differentiated. The former focuses only on soil handling, with plants playing a minor role. Phytostabilisation involves vegetation as the main element of the soil treatment. In a way, both approaches, however, can create smooth transitions. For instance, residues of degraded plants, particularly roots, contribute to the humification in the soil. In this sense, plant residues bind contaminants into the soil organic matter with the help of microbial enzymes in conjunction with the development of bound residues (see Sect. 6.3.1) or they can adsorb inorganic contaminants to a greater extent.

The technique to remove contaminants from polluted waters is termed rhizofiltration. The rhizofiltration technique, which is frequently applied with regard to wastewater treatment, is highly effective, the costs are low and it can be easily handled.

The removal occurs in different ways as adsorption or precipitation to the roots or as incorporation into the root tissue until they are saturated with the contaminants. Some aquatic plants such as *Eichhornia crassipes* (water hyacinth),

Hydrocotyle umbellata (pennywort), *Lemna gibba* and *Lemna minor* (duckweed) are capable of removing metals from water courses in spite of their small size and slow growth. The efficiency, however, is restricted. For this reason, some hydroponically cultivated plants which are transplanted into the polluted water body show a high potential to remove metals from contaminated water. For instance, in this way it was possible to remediate water polluted with metals (e.g. As, Cd, Cr, Cu, Ni and Se using water hyacinth, Cu and Hg using duckweed) (SUMATECS 2008). A substantial load reduction in municipal wastewaters has also been observed by using other herbaceous and wooden plants such as willows.

Generally speaking, a number of advantageous and disadvantageous aspects should be taken into consideration with regard to phytoremediation. There is no doubt that the measure can be assessed as being soil-protective, since anthropogenic disturbances of the soils are avoided. Hence, this approach has a high level of acceptance in the general public. During the remediation process the brownfields are turned into aesthetically beautiful greenfields. Thus, particularly in urban areas with large brownfields, which cannot be remediated due to the costs, phytoremediation can be a realistic approach. It can be an alternative or an addition to the natural attenuation (see Sect. 8.2).

With reference to the economic assessment the treatment is solar driven and consequently energy-extensive and more or less emission-free. Taking the climatic relevance into account, phytoremediation appears to be a new option for biofuel production, if non-food crops are chosen. Moreover, the metal recycling from the harvested plants provides a further economic option. For these reasons, the disposal of the yielded biomass should not be problematical in the context of biofuel utilisation and metal recycling.

On the other hand, there are a lot of problems associated with the application of phytoremediation. Firstly, the treatment is very slow in comparison with alternative decontamination measures and depends on the seasons. Accordingly, it is normally accepted as long as economic restrictions for brownfield redevelopment are present. An alternative use of the area of concern is definitely excluded and existing buildings and structures must be demolished. Thus, it might be difficult to persuade investors and property owners to wait years and ultimately to accept an insufficient contaminant reduction (see Sect. 2.1.2). It must be usually expected that at the end of the phytoremediation process further conventional civil engineering techniques are applied.

The limited opportunities of the application give rise to negative aspects because only a few contaminants can be treated and, furthermore, the method is limited to a few species due to phytotoxicity problems. Moreover, the applicability is limited to surface-near soil contamination.

6.4.2 Phytoextraction

The widely pursued phytoremediation strategy focuses on phytoextraction. In botany history it has been found that some plants have adapted to accelerated heavy

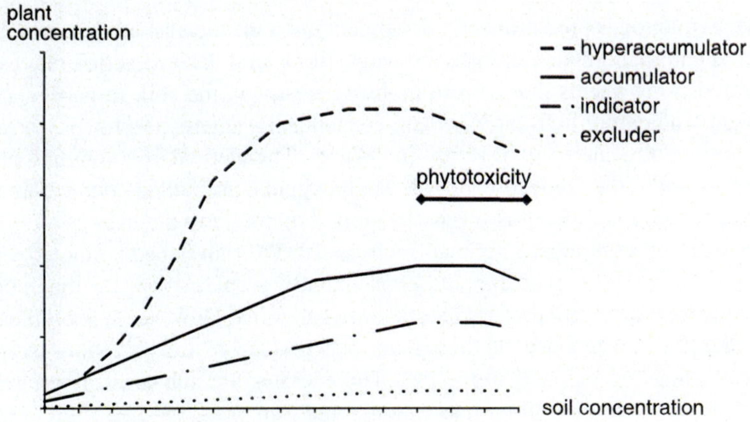

Fig. 6.22 Different plant strategies for growing in contaminated soils

metal concentration in soils with outcropping ore bodies. They are able to incorporate higher degrees of metals in comparison with other plant species. This plant characteristic is used to remediate polluted soils. In general, plant species can be divided into four groups which are distinctly responsive to the uptake of contaminants (Fig. 6.22):

- Excluders meaning plant species that cannot take up contaminants and transfer contaminants into aboveground plant tissue
- Indicators, which take up contaminants to an extent that indicates a linear relationship to the soil contamination level
- Accumulators, which take up higher amounts of contaminants, predominantly metals, as the soil contamination increases
- Hyperaccumulators which have the same effect but the increase of contaminants in the plants occurs exponentially.

The latter two groups are of immense interest in relation to phytoextraction. Organic acids derived from root exudates improve the ability to take up soluble metals. On the one hand, they decrease the pH value around the roots, causing a faster and more effective metal transfer into the roots, since the adsorption to clay minerals and the organic matter is significantly reduced with decreasing pH. On the other hand, the root exudates serve as complexing agents which are responsible for increased metal mobility due to the high solubility of organo-metallic complexes, particularly at a pH value of 4–6 (Meuser 2010).

Consequently, the increase in organic acid exudation experienced much genetic engineering interest. In relation to phytoremediation much attention was paid to the determination of enzymes which can be manipulated in order to change the organic acid biosynthesis and to enhance the exudation in plants. This approach is one opportunity to find and identify genes which are basically used for transgenic plants indicating optimum remediation capacity.

Phytoextraction is predominantly dependent upon the metal solubility, which can be altered and manipulated in different ways. In general, the extraction rate can be accelerated using agents that help to mobilise metals in the soil. In particular, the crop plants indicating high biomass and consequently interesting for phytoextraction would require chelating or acidifying agents. The method of combining phytoextraction with the amendment of such agents is called chelate-assisted phytoextraction. However, this is mostly applied to non-crop plants.

Chelating or complexing agents such as EDTA (ethylenediaminotetraacetic acid), which are less degradable and consequently problematical for the environment in the long term, are favoured to increase solubility. However, it is problematical to predict the influence of the agents exactly due to the variations between chemicals amended and soil properties. The chelates are subject to interferences with other cations in the soil such as calcium and iron.

Nevertheless, EDTA was successfully used in different projects as reported briefly by SUMATEC 2008. The accelerated uptake of metals refers not only to Pb, which is usually less mobile in soils, but also to elements such as Cd, Co, Cu, Ni and Zn in multi-contaminated soils. It should be noted that the chelators used are probably plant- and element-specific. The addition of agents does not make sense at a very high metal concentration of potentially phytotoxic elements.

Investigations into EDTA application to poplars grown in metal-contaminated sludge showed that the solubilisation of some metals was a higher one but the leachate led to drainage water pollution. Furthermore, there was only a low effect on leaves, bark and wood metal concentrations. While EDTA increased the root uptake, the transfer to the aboveground parts of the plants remained unaffected. Obviously, EDTA can go beyond the diffusion limit to the roots but the barrier from roots to shoots was not influenced positively. In conclusion, the EDTA application, which is normally in common use, was not recommended (Liphadzi and Kirkham 2006).

As an alternative to EDTA some biodegradable agents have been tested, for example ethylenediaminedissuccinate (EDDS) and methylglycinediacetate (MGDA), but the relatively rapid degradation of the amendment caused a limited effectiveness. Natural organic acids were reported to be less effective with reference to the solubility and plant uptake, in particular for the main element of concern lead (Pb), because the organic acids seemed to be consumed by the microorganisms or to be mineralised (SUMATECS 2008).

The metals are transported through the plant by the xylem after being taken up by the root system. The transport occurs either in the transpirational stream or actively by proteins. Compared with the uptake rate the translocation from the roots to the shoots appears to be significantly lower. It is caused by the plant property that the translocation of non-essential metals is basically lower than the translocation of nutrients required.

Plants producing high biomass in conjunction with an accumulation process may transfer a high quantity of metals from the soil to the plants harvested. In principle, farmers aim to achieve a high biomass production from a number of agricultural plants such as wheat, barley, oats, ryegrass and peas and consequently these are of interest but they are usually not adequate for a sufficient metal uptake. In contrast, the agricultural crops rice, maize, sunflowers and Indian mustard exhibited accumulator qualities associated with some elements.

Plants combining high biomass production and increased metal uptake such as *Brassica juncea* (Indian mustard) may remove higher quantities of metals such as Cd and Zn. Because of the extreme biomass production such plants are sometimes more effective compared with typical hyperaccumulators revealing restricted biomass development. Hence, the potential of hyperaccumulators (e.g. *Thlaspi caerulescens, Alyssum murale, Alyssum lesbiacum, Alyssum tenium, Silene vulgaris*) to remediate the soil is limited because of their slow growth rate.

Nevertheless, the extraction only occurs beneficially, if optimum growth conditions are guaranteed and additional, e.g. atmospheric, input of contaminants does not take place. The latter might pose problems, particularly in urbanised and industrialised areas. Moreover, it should be possible to cultivate plants in the area to be treated with typical agricultural equipment. Accordingly, a soil with low skeleton content, a sufficiently deep rooting zone, an acceptable topography and no disturbing artefacts such as foundation residues and old trees is required. Necessary measures, for instance ploughing and fertilizing, should be possible by using agricultural machinery. In particular, fertilizing might be a useful tool to improve the phytoextraction success. For example, after application of ammonium nutrition acidification in the rhizosphere accelerates due to the proton excretion by the root cells. Consequently, this process enhances the metal uptake by the roots significantly and stimulates the metal extraction capacity.

Phytoextraction might be a suitable approach for large contaminated areas, which are mainly contaminated at shallow depths and with a low to moderate level of contamination. From the botanical point of view the plants selected should be derived from the geographical area where the contaminated site is located, should develop an intensive root system, should be able to translocate metals to the shoots, should be tolerant to metals in relatively high amounts and should show high biomass growth. Environmentally, the plants should be unattractive to animals, particularly mammals and birds, and consequently they should not transfer the toxic elements within the food chain.

The accumulation of toxic metals means a strong reduction of the material that must be removed or landfilled. The ratio of excavated and deposited soil to disposed contaminated plant material is 200:1 (Black 1995). Normally, the harvested and highly contaminated biomass is landfilled or incinerated. Depending on the future use of the material, composting and re-use in areas of low sensitivity might also be a solution. Composting may reduce the volume to be re-used but the metal enrichment means strong limitations for the future utilisation. Commercial use of the biomass as an energy source combined with the following extraction of metals from the ashes (phytomining) could generate profitable business in the next few decades (see Sect. 2.3).

Regarding organic pollutants phytoextraction might play only a minor role, because it is hardly possible for polar substances to cross biomembranes and consequently for them to be taken up. The bioavailability that is the basis for plant uptake is reduced due to the adsorption preferentially to organic matter. If ever, organic compounds either indicating a K_{ow} value between 0.5 and 3.0 or present as a neutral

Table 6.8 Families and species successfully used for phytoextraction (without wooden plants) (Data from SUMATECS 2008)

Parameter	Species
Cd	*Brassicaceae* (e.g. *Thlaspi caerulescens, Arabis gemmifera, Brassica napus, Brassica juncea*), *Sedum alfredii, Solanum nigrum*, rice, maize
Cr	*Brassica juncea*
Cu	*Lamiaceae* (e.g. *Elsholtzia splendens*), *Leguminosae* (e.g. *Trifolium repens*), *Helianthus annuus, Brassica juncea*
Ni	*Bassicaceae* (e.g. *Alyssum murale, Thlaspi caerulescens, Brassica juncea*)
Pb	*Brassica juncea, Helianthus annuus*
Se	*Leguminosae* (e.g. *Astragalus racemosus*), *Brassica juncea*
Zn	*Brassicaceae*, (e.g. *Thlaspi calaminare, Thlaspi caerulescens, Arabis gemmifera, Brassica juncea, Arabidopsis halleri*), *Helianthus annuus*, maize

molecule are taken up to a minor extent. Apart from the advection (mass flow), diffusion depending on the concentration gradient, which increases with a fast metabolism in the root cells, may contribute to the uptake of organic pollutants. Hence, in the case of a quick metabolism inside cells there might be an increasing tendency towards uptake (Trapp and Karlson 2001).

6.4.2.1 Herbs, Grasses and Ferns

With reference to non-cultivated plants approximately 400 hyperaccumulators are known. They accumulate more than 1% Ni and Zn, more than 0.1% Co, Cr, Cu and Pb, more than 0.01% Cd and Se and more than 0.001% Hg of the dry weight shoot biomass. Most of the species accumulate Ni, whereas for the other elements approximately 30 species were identified. Suitable families and species are listed in Table 6.8. The highest rates are mentioned with reference to the botanical family *Brassicaceae*. The best-studied species belonging to that family are called *Thlaspi caerulescens* and *Arabidopsis halleri*, which are particularly good Cd, Ni and Zn hyperaccumulators without visible symptoms of toxicity (Baker and Brooks 1989; Milner and Kochian 2008).

In the meantime, even for lead a strongly accumulating plant species called *Fagopyrum esculentum* (buckwheat), achieving up to 4.2 mg g^{-1} in the shoots, has been found (Tamura et al. 2005). Furthermore, different accumulators apart from the spermatophytes which concentrate metals aboveground were discovered. For instance, some ferns like *Pityrogramma calomelanos* and *P. vittata* are well-known for the accumulation of arsenic, reaching up to 23 mg g^{-1} in the shoots. Another example is the fern *Athyrium vokoscense,* which is able to accumulate copper and iron (SUMATECS 2008).

Thlaspi caerulescens can enrich the shoot concentration for Cd up to 1,800 mg kg^{-1} and for Zn up to 39,600 mg kg^{-1}. Apart from the relatively

Table 6.9 Biomass production, BCF and metal removal in plants used for phytoextraction (Data from SUMATECS 2008)

	Aerial biomass (ton ha^{-1})	BCF (–)	Metal removal (g ha^{-1} year^{-1})
Cadmium			
Thlaspi caerulescens	0.9–2.9	1.5–208	35–4,204
Zea mays (maize)	9.4–15.6	0.08–0.3	3.7–16.1
Salix sp. (willow)	6.1–17.8	0.18–20	26–222
Zinc			
Thlaspi caerulescens	0.9–2.7	0.45–43	2,470–29,208
Salix sp. (willow)	7.6–17.8	0.05–0.83	821–5,034

mobile elements Cd and Zn, extreme accumulation rates were mentioned for Cu (*Aoellanthus biformifolius*: up to 13,700 mg kg^{-1}), Ni (*Phyllanthus serpentinus*: up to 38,100 mg kg^{-1}), Co (*Haumaniastrum roberti*: up to 10,200 mg kg^{-1}) and Se (*Astragalus racemosus*: up to 14,900 mg kg^{-1}) respectively (Siegel 2002).

The ratio of the metal concentration in shoots (mg kg^{-1}) to the metal concentration in soil (mg kg^{-1}) is called the bioconcentration factor BCF. A BCF value above 1 is at least necessary to evaluate the plant as an accumulator. Depending upon soil conditions (e.g. texture, organic matter content, pH value, etc.), contaminant concentration and binding form in the soil as well as climatic influences, the BCF of one plant species varies in a wide range (Table 6.9). For *Thlaspi caerulescens* and cadmium values ranging from 1.5 to 208 have been found and consequently the metal removal per year showed enormous differences between the field trials. At lower orders of magnitude differences are visible in relation to other species listed and the element Zn (SUMATECS 2008).

Different autochthonous plant species were tested for phytoextraction purposes (Table 6.10) under Mediterranean climatic conditions in the Southeast of Spain at mining sites and marsh sites which were both contaminated with Zn. The plant concentrations at the mining sites, which revealed moderate Zn contamination (soil values: 170–336 mg kg^{-1}), ranged from 15.1 to 755 mg kg^{-1} DW. For the contaminated marsh sites (soil values: 703–1,108 mg kg^{-1}) the plant concentrations amounted to 18.8–251.9 mg kg^{-1} DW. The BCF values varied between only 0.03 and 2.76 and fell mostly below the desirable value of 1 (Garcia et al. 2002).

Regarding chromium, which in hexavalent form means a particular danger to human health, attempts were made in an anaerobic environment (wetlands) to eliminate the metal by plant uptake after reduction. Species such as *Phragmites karka* and *Scripus lacustris* should take up the element and translocate it to the shoots but most of it accumulated in the roots.

Nevertheless, accumulators relating to Cr which normally grow on serpentine soils are well-known, e.g. *Leptospermum scoparium, Sutera fodina, Dicoma niccolifera.* The latter are able to accumulate up to 48,000 mg kg^{-1} DW Cr in the shoots. In highly contaminated soils grasses, vegetables and legumes may accumulate between only 0.04 and 9.6 mg kg^{-1} DW Cr. Otherwise, toxicity to plants would become visible, if the soil concentration exceeded approximately 200 mg kg^{-1} or

Table 6.10 Zn concentration in soils (mg kg^{-1}) and plant species (aboveground part) (mg kg^{-1} DW) as well as BCF results at Spanish mining and contaminated marsh sites (Data from Garcia et al. 2002)

	Zn (soil)	Zn (plant)	BCF
Marsh sites			
Phragmites australis	703	18.8	0.03
Sarcocornia fruticosa	975	85.2	0.09
Salicornia ramosissima	1,108	251.9	0.23
Mining sites			
Helichrysum decumbens	278	84.8	0.30
Lygeum spartum	185	15.1	0.08
Piptatherum miliaceum	336	16.7	0.05
Zygophyllum fabago	273	755.0	2.76
Hyparrhenia hirta	170	203.6	1.20

the leaf concentration exceeded 10 mg kg^{-1} DW. However, the toxicity depends on the element species because Cr(VI) is more toxic to plants than Cr (III). Consequently, after reduction a stronger accumulation rate can be supposed (Zayed and Terry 2003).

6.4.2.2 Wooden Plants

Some tree species show a high and fast growth rate and biomass production even though they are in an inferior position to hyperaccumulators with regard to the metal uptake. Species like willow (*Salix* sp.), however, can accumulate metals such as Cd and Zn, in particular after previously using hydroponics. The main part of the incorporated metals remains in the root system. The root biomass which cannot be yielded is at least five times smaller than the shoot biomass production under field conditions but the phytoextraction effect can be enhanced by using short-rotation coppice (every 2–3 years) and subsequent continuous biomass removal. Additionally, at the end of the remediation time the root boles can be excavated before starting the redevelopment of the site.

With reference to the genus *Salix* there are enormous differences between the individual species regarding their ability to extract metals from the soil. For this reason, a further specification of the plants used appears to be necessary. For instance, high effectiveness was found for *Salix alba, Salix viminalis, Salix caprea* and *Salix cinerea,* which are species normally invading polluted areas such as polluted dredged sediments (Vandecasteele et al. 2005).

The effectiveness of wooden plants such as willows, poplars and alders in relation to their phytoextraction potential has been predominantly investigated on sediments consisting of highly contaminated dredged sludge originating from different water courses like harbours, canals and lakes (Meuser 2010). In particular, willows are preferred for the phytoextraction approach in contaminated sludge fields, because they have a lot of advantageous characteristics. Growth is extremely productive at the juvenile stage, the root system is extensive, high evapotranspiration rates which reach up to 85% of the annual precipitation can be gained leading to a fast site dehydration, there is tolerance to saturated soils and oxygen shortage, the propagation occurs vegetatively, a re-establishment from coppiced stumps is possible and, last but not least, the accumulation rates for some metals are high. To avoid phytotoxicity soils with low and medium levels of contamination are the main focus of willow utilisation (Vandecasteele et al. 2005). The possible deep rooting may lead to the extraction of contaminants from deeper soil horizons.

While most of the metals such as Cu, Cr, Ni and Pb are not translocated to stems and leaves, Cd and Zn are accumulated in the aboveground part of the willows to a great extent. With regard to Cd and Zn average concentration in leaves, stem and roots were even positively correlated to the soil concentration (Vandecasteele et al. 2005). As found in cadmium investigations at dredged disposal sites in Belgium species like *Salix viminalis* can accumulate in leaves and stem 7.4–9.5 mg kg^{-1} DW (Meers et al. 2003) and 1.1–18.1 mg kg^{-1} DW (Vandecasteele et al. 2002), depending on the degree of soil contamination. This ranged from 1.0 to 5.0 mg kg^{-1} and 0.5 to 20 mg kg^{-1}.

Continuously dredged lake sediment with partly elevated metal concentration (total concentration Cd: 1.9–2.7, Cr: 120–143, Cu: 36–44, Ni: 28–38, Pb: 122–133, Zn 287–356 mg kg^{-1} in 0–30 cm) was pumped into different sludge fields surrounded by dams for recycling purposes in the future after long-term drying. Due to the proximity to the city of Hanover, Germany, and discharges of the leather industry cadmium, chromium and zinc, in particular, had accumulated in the sedimented sludge. Depending on the age of the fields distinct vegetation (*Typha latifolia, Phragmites australis, Salix caprea* and *cinerea, Betula pendula*) established in accordance with natural succession.

It was possible for *Salix* to accumulate high rates of Cd (mean values 14.5 g ha^{-1} a^{-1} DW) and Zn (2,960 g ha^{-1} a^{-1} DW), whereas the predominant concentration in *Betula* was Cr (196 g ha^{-1} a^{-1} DW) and Zn (2,445 g ha^{-1} a^{-1} DW). However, the plant uptake would need long periods of time (a minimum of 49 years) to reach metal values enabling material recycling based upon German directives. The main reason for the limited plant uptake appeared to be the real pH values of soil solution. Taking the results from suction cups into account, which ranged between 6.2 and 8.1, enhanced metal mobility was not to be expected (Meuser and Makowsky 2009).

The potential of phytoextraction poses the danger of rapid metal removal by leaching instead of plant accumulation, particularly in soils consisting of dredged sediments. In Warrington, United Kingdom, the potential was monitored over a period of 32 months using several field trials located at contaminated sludge fields. During oxidation of metal sulphides and subsequent generation of sulphuric acids the pH value usually drops significantly, whereby the metal mobility tends to increase. Compared to unplanted plots the total concentration of Cd, Cu, Pb and Zn decreased considerably as a result of leachate but to a less degree as a result of plant uptake. Though the total metal concentrations were extremely high (As 420 mg kg^{-1}, Cd 20 mg kg^{-1}, Cr 980 mg kg^{-1}, Cu 740 mg kg^{-1}, Pb 1,445 mg $kg^{-1}$and Zn

Table 6.11 Number of years to reduce Cd concentration by 5 mg kg^{-1} by willow planting related to different soil depths (Data from Dickinson 2006)

Depth (cm)	Plant tissue concentration (mg kg^{-1})			
	10	25	50	100
0–20	67	27	14	7
0–40	133	53	28	14

Assumptions: yields of 15 t ha^{-1}, bulk density of the soil 1.0 g cm^{-3}, consistent Cd uptake throughout period

4,285 mg kg^{-1}), the percentage of surviving shrubs during the 32 month-long period was found to be relatively high (e.g. at the end of the monitoring period alders 59%, poplars 56% and willows 62%) (King et al. 2006).

With increasing metal concentration chlorosis, necrosis and leaf rolling are found. Simultaneously, shoot and root biomass production decreases. As shown for *Salix viminalis*, the general high tolerance towards metals depended on the duration of exposure, because short-term tolerance was certainly higher than long-term tolerance. In particular, the root tips revealed browning in a line with lignification and resulted in decreasing nutrient uptake (Cosio et al. 2005).

Because of their relatively strong resistance to air pollution willows can also be afforested at dryer industrial areas (brownfields) for phytoremediation purposes. Along with birches willows are frequently found in industrialised areas with severe air pollution (Kuzovkina and Quigley 2005). Moreover, from the physical point of view the continuous vegetation cover results in optimised site stabilisation in accordance with reduced erosion and contaminant leaching.

Pot experiments with moderately (Cd: 4.3 mg kg^{-1}, Zn: 220 mg kg^{-1}) and highly (Cd: 32.7 mg kg^{-1}, Zn: 1,760 mg kg^{-1}) contaminated material in Austria revealed for the first one bioconcentration factors (BCF) of 4.0–15.9 (Cd, nine *Salix* species, e.g. *Salix caprea, Salix babylonica, Salix smithiana, Salix dasyclados*), 2.7–5.3 (Cd, two *Populus* species *Populus nigra* and *Populus tremula*), 1.1–3.9 (Zn, nine *Salix* species) and 1.3–1.9 (Zn, two *Populus* species) respectively. In the highly contaminated soils the values were 2.6–13.5 (Cd, nine *Salix* species), 4.4–4.8 (Cd, two *Populus* species), 0.8–1.4 (Zn, nine *Salix* species) and 0.5 (Zn, two *Populus* species). In relation to the willows concentrations ranging from 17.4 to 70.0 mg kg^{-1} DW (Cd) and from 240 to 870 mg kg^{-1} DW (Zn) in the moderately contaminated soils were reported in leaves. The corresponding results for the highly contaminated pots were 84–440 mg kg^{-1} DW (Cd) and 640–2,410 mg kg^{-1} DW (Zn). Compared to research projects based on more realistic field trials, the phytoremediation potential must be assessed as extremely overestimated (Dos Santos Utmazian and Wenzel 2007).

The life cycle of willows can be approximately 30 years. Whether this period suffices to reduce the metal concentration to a level which conforms with regulations might be doubtful. Viewed mathematically, this aim will only be achieved, if the plant tissue concentration attains an acceptable level. As shown in Table 6.11, the number of years required to reduce the metal concentration depends on the contaminated soil depth, plant tissue concentration and the metal values in the soil targeted.

In general, the Cd concentration in leaves in field investigations will be in the range of 10 mg kg^{-1} DW rather than 100 mg kg^{-1} DW. It is a problem that plant removal only based on the coppice technique will not be sufficient, since high amounts of metals are concentrated in the leaves that should be harvested continuously in autumn (Dickinson 2006).

Inoculation with mycorrhizal fungi to the willow hosts was tested using pot experiments to discover an influence on phytoextraction capacity. The fungi indicate several advantages, since they store metals in the vacuoles or on the surface of the mycelia and their hyphal length mean increased root growth and subsequent improved nutrient supply. Accordingly, the biomass of the hosts accelerates. While the concentrations of heavy metals in stems and leaves did not show a tendency to increase after fungi inoculation to *Salix dasyclados*, an elevated phytoextraction rate still occurs due to the enhanced biomass production (Baum et al. 2006).

6.4.3 Phytodegradation

Recalcitrant compounds such as pentachlorophenol (PCP), PCB, PAH, explosives, halogenated compounds (e.g. TCE, PCE), phenols and chlorine and phosphorus-based pesticides are, in principle, interesting for degradation (Bradl 2005). Phytodegradation follows different strategies. Firstly, the plants can take up and degrade the compounds biochemically to harmless products. During the process oxidative and reductive enzymes are used in different parts of the plant. The process, also termed phytotransformation, generally takes place in the plant tissue, but a further degradation may occur after plant death in the uppermost soil horizon.

The main strategy with reference to phytodegradation, however, is located in the belowground part of the plants, in the rhizosphere. This strategy is also termed rhizodegradation. The rhizosphere is a 1–3 mm-thick region surrounding the plant's roots with high biological activity. An important characteristic of the rhizosphere is the production and release of root exudates. In relation to the biodegradation potential exudates such as carbon containing sugars and proteins are ideal nutrient sources for the microbes, whose biological activity will be stimulated to a great extent. Hence, rhizodegradation can be described as plant-assisted bioremediation (see Sect. 6.3.1).

Regarding metal contamination it should be noted that root exudates can impact phytoremediation counterproductively, since they consist of biochemicals which can facilitate precipitation within the root zone by complex formation and adsorption to clay particles and the organic matter. Accordingly, the uptake and translocation to the aboveground parts of the plants are prevented. Furthermore, the metals are sequestered into the vacuoles of the root cells, preventing subsequent transport upwardly. The contrasting behaviour of root exudates, however, is preferentially meaningful in the context of phytoextraction (see Sect. 6.4.2).

A study dealing with dredged canal sediments in Warrington, United Kingdom, which were planted with alders, poplars and willows, resulted in a strong decrease in mineral oils (TPH) over a period of 32 months, but the authors explained the decline with plant-independent degradation and volatilisation, since soil cracks caused by drying were increasingly produced, allowing greater oxygenation. Certainly, the crack development would not be generated without the influence of the shrubs. The uptake did not play an important role because there were no differences between planted and fallow land. At the same period of time the PAH concentration did not exhibit decreasing values (King et al. 2006).

Willows are used in the context of phytodegradation. They may transport oxygen to the roots through aerenchyma, whereby the growth conditions for aerobic microbes are improved to degrade organic pollutants. In relation to wetlands and sludge fields other plant species such as *Phragmites australis* providing a ventilation system for roots are also used for phytodegradation.

A high number of successful projects on rhizodegradation in association with plants has been reported. Some species are able to stimulate, for instance, oil-degrading microbes in association with their roots, as found in soils contaminated by oil spills. After the First Gulf War in Kuwait *Senecio glaucus* was grown in desert sand that was strongly polluted by mineral oils caused by the criminal attacks of Saddam Hussein. The landscape was covered to a great extent by oily lake-like surfaces. The plants, whose roots are associated with bacteria strains of *Arthrobacter*, are capable of degrading hydrocarbons. Within 4 years the TPH concentration decreased drastically (Radwan et al. 1995).

Near Fairbanks, USA, an accidental spill of an oil storage tank was treated after fertilizing and seeding grasses and compared to adjacent contaminated sites without establishment of other vegetation. Whereas in the planted area 80% of the TPH concentration was reduced within 100 days, in the unvegetated plots it was not possible to achieve this level after 238 days. The differences resulted mainly in the bacteria population density, which was much higher in the planted plots. On day 238 about 3.5×10^5 bacteria g^{-1} soil were determined in the unvegetated area but in the other area about 10^8 bacteria g^{-1} soil were analysed. Supported by the nutrients, the plant growth was stimulated and the roots enhanced the growth and activity of the microbes significantly (Bradl and Weil 2008).

Poplars, for instance, stimulated microbes in the rhizosphere which were able to degrade monoaromates such as BTEX (Trapp and Karlson 2001). With the help of the bacterial symbiont *Rhizobium galageae* and in the presence of the inoculated bacterium *Pseudomonas putida* under the nitrogen-fixing legume *Galega orientalis* toluene was effectively degraded in experiments in Northern Finland. The degrading microorganisms are metabolically stimulated by nutrient-enriched plant exudates and the fixed nitrogen. Subsequently, increased bacterial numbers and metabolic activities combined with a spread of the degraders due to the intensive root development improved the toluene degradation. Obviously, under the conditions of this northern temperate climate there are also possibilities to solve the polluted soil problem (see Sect. 6.3.4). In general, *Pseudomonas putida* is able to colonise the rhizosphere of polluted soils (e.g. TPH, toluene) but attention should be paid to plant growth, which can be negatively affected (Lindström et al. 2003).

Besides mineral oils, cyanides containing soils have been also decontaminated with the help of plants (Trapp and Karlson 2001; Kuzovkina and Quigley 2005). In Holte, Denmark, a former gasworks revealing high cyanide pollution of up to 932 mg kg^{-1} complexed CN and up to 95 mg kg^{-1} free CN was treated with willows and poplars capable of metabolising the cyanides very rapidly. The CN content was concentrated in the roots, whereas a translocation into the leaves did not take place.

Poplar roots living symbiotically with white rot fungi such as *Trametes trogii* exhibited faster and more effective degradation of organic pollutants such as BTEX, PAH, PCB, explosives and some pesticides. The compounds were oxidised into more soluble products with enhanced bioavailability and consequently a higher mineralisation rate. Affinities were found between specific white rot fungi and pollutants (e.g. *Phanerochaete chrysosporium* and BTEX) (Levin et al. 2003).

In the plant cells a large amount of organic pollutants can metabolise after being taken up. In field trials in the USA hybrid poplars were grown and exposed to the VCHC trichloroethylene (TCE) added by underground injection. During the season at least 95% of TCE was eliminated from the input water stream (uptake), but only 5% of TCE was transpired. Consequently, the main percentage was metabolised in the plant tissue to chloromethyl derivatives or entirely mineralised to harmless carbon dioxide (Gordon et al. 1998).

In order to establish vegetation remediating soils polluted with organic compounds knowledge of the maximum tolerable concentrations are of importance to exclude plant damages. Unfortunately, the concentrations recommended vary considerably and must be evaluated prior to the application in pre-test studies. For instance, based upon literature the values published were 38 mg L^{-1} perchloroethylene (hybrid poplars), 4.47 mg L^{-1} free cyanides (willows), >1,000 mg L^{-1} ferro ferricyanides (poplars) and <5 mg L^{-1} TNT (hybrid poplars) respectively (Trapp and Karlson 2001).

6.4.4 Phytovolatilisation

Phytovolatilisation is defined as gaseous stomata losses of potential volatile contaminants. Dissolved in the xylem selenium and mercury can be volatilised into the atmosphere. The oxidised form of selenium (selenate, selenite) is easily taken up by the roots of e.g. *Astragalus racemosus.* This process is enhanced by rhizosphere bacteria. Afterwards, both selenate and selenite are assimilated to selenocystein in a reductive way and subsequently transformed into a dimethylated form which can be volatilised. In the case of mercury Hg (II) is mostly biotically reduced to volatile Hg (0) or methylated by anaerobic bacteria. Apart from this mechanism, the soil bacteria of the rhizosphere may detoxify Hg (II) and organomercurial compounds enzymatically (Bradl 2005).

Furthermore, chlorinated solvents (e.g. trichloroethylene, chlorobenzene) and non-substituted monoaromates (e.g. toluene) in dissolved form are compounds which can be taken up, changed into a volatile form and ultimately transpired (Trapp and Karlson 2001).

Phytovolatilisation poses the problem of release of toxic substances into the atmosphere and is subsequently responsible for the generation of air pollution. A fast re-deposition of, for instance, Hg(0) cannot be excluded. Thus, the approach should be avoided in the vicinity of residential areas.

6.4.5 *Phytostabilisation*

Plants used for phytostabilising have to be tolerant to soil contaminants, so that excluders such as *Agrostis tenuis, Agrostis capillaris, Festuca rubra* are preferred. In field trials maximum tolerable values for effective phytostabilisation were reported at 120 mg kg^{-1} (Cd), 2,600 mg kg^{-1}(Cu), 270 mg kg^{-1}(Ni), 4,200 mg kg^{-1}(Pb) and 12,000 mg kg^{-1} (Zn) respectively (SUMATECS 2008).

The plants must be adapted to the given soil characteristics such as salinity, structure, pH value, water household, etc. In general, phytostabilisation is applicable to every soil type but is most effective for cohesive soils with a high organic matter content associated with optimised growing conditions for the vegetation. It would be optimal to achieve a fast growth, a dense rooting system, a high transpiration rate and a long life. In the case of detrimental growth conditions restoring measures are applied such as manipulating pH value, adding organic matter, improving microbial activity, reducing soil compaction and balancing soil moisture regime. In most cases it is less possible to compensate for restricted plant growth caused by unfavourable climatic or hydrologic conditions.

One stabilising effect is related to the permanent ground cover that reduces erosion and deflation. The approach is particularly applied to ore mining waste heaps endangering the adjacent areas. Stabilising grasses (e.g. *Agrostis tenuis, Agrostis capillaris, Festuca rubra*) are planted to stabilise the heaps and to prevent erosion and deflation of the material contaminated by copper, lead and zinc. Mixtures of legumes and grasses are also amended. The species used are identical to the species usually colonising the site spontaneously (dicotyledonous pioneers) but due to the slow cover of the site an additional seed mixture is preferred. Moreover, *Brassica juncea* and *Brassica napus* indicating high biomass production and fast growth are also used for phytostabilisation purposes (SUMATECS 2008).

Furthermore, plantation of waste heaps is used to reduce water percolation after uncontaminated material has been deposited. In contrast to proper soil cover systems as described in Sect. 5.1.1, many waste deposits have been covered non-professionally in the past. At least dense vegetation was planted to minimise leachate and subsequent danger to the groundwater. Close to Hamburg, Germany, a landfill consisting of ash, combustion residues, wood waste and construction rubble was covered with 0.5–1.5 m uncontaminated soil. By comparison, areas with bare soil, grass vegetation and willows were investigated with regard to the water household, in particular the leachate in the course of time. Based on a simulation model the leachate under willows and grasses was supposed to be significantly lower than on the unvegetated site (Fig. 6.23). In particular, willow areas fulfill the function of

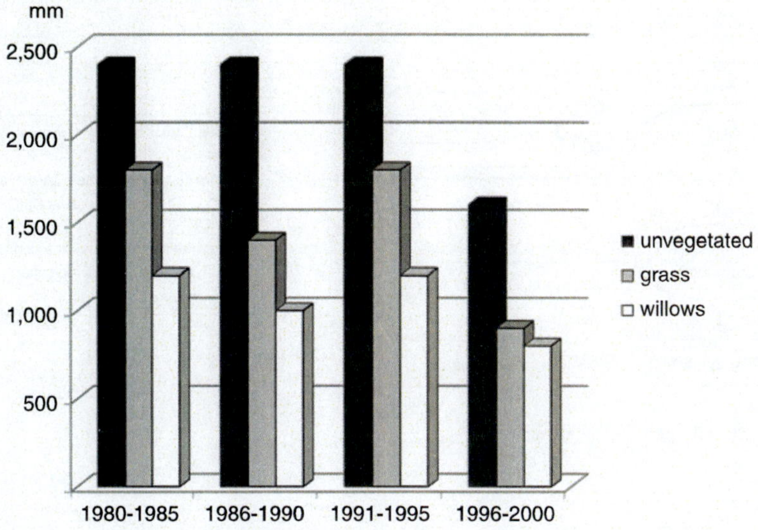

Fig. 6.23 Vegetation effects on leachate quantity (mm) at an unprofessionally covered landfill site near Hamburg, Germany

minimising leachate, since the amount was clearly reduced – by 39–54% (Kahle et al. 2002). It is possible to plant vegetation which manifests high transpiration rates in order to obtain phytostabilising effects.

6.5 Thermal Treatment

6.5.1 Technical Devices Used

The first facilities treating contaminated soil thermally have existed since the beginning of the 1980s. Most devices are used off site at installations, while some *in situ* technologies became more important in the last two decades, in particular vitrification and microwave radiation (see Sect. 5.3.3). Microwave heating shows a high degree of effectiveness for removing organic pollutants from soils with low permeability, namely clay-rich soils, since microwaves are preferentially captured by clay material. The release of the pollutants at relatively low temperatures results from an increased volatility and a decrease in viscosity, which improves the mobility simultaneously. Moreover, an alternative technique is offered with overheated water steam or air that is injected into the contaminated subsurface (see Sect. 7.2.5). Some organic pollutants and mercury can be treated in this way and because of the relatively low temperature applied (approximately 250°C) this procedure is classified as more soil-protective.

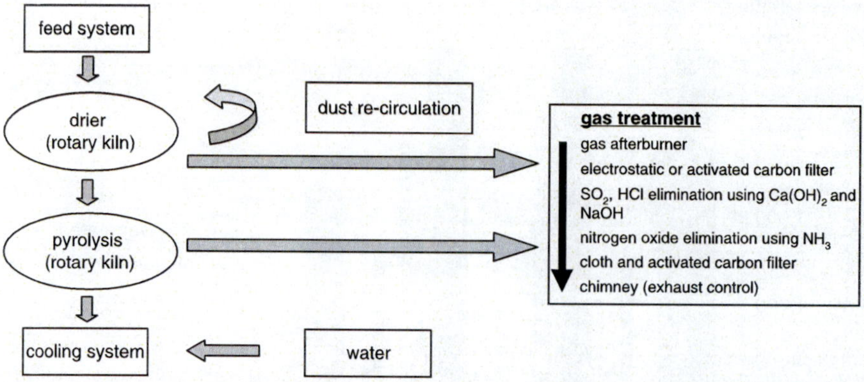

Fig. 6.24 Thermal treatment flow chart

The stationary systems mainly use rotary kilns but there are also installations based on circulating fluidised beds. Regarding the temperature there are systems operating with low temperatures ranging between 400 and 700°C, where the pollutants are desorbed and volatilised and then sent to a gas treatment chamber. They are destroyed in this chamber, which operates with a temperature of 800–1,200°C. The soil passes usually zones of increasing temperature. In a way, this technique is a physical separation system (Mirsal 2004). Although many installations operate without oxygen during the soil treatment (pyrolysis), the soil is more or less soil-like after this thermal desorption.

The alternative technique, which has been less frequently used during the last two decades, may operate with higher temperatures from 970 to 1,200°C (Mirsal 2004). Here organic pollutants are combusted and the soil matrix is burnt and subsequently strongly altered. This incineration technique, however, frequently leads to the emission of products of incomplete combustion requiring intensive air treatment. The installations usually operate in the presence of oxygen.

The combustors are differentiated into directly and indirectly heated equipment. Only the former causes a contact between the soil and the flame (incineration). This approach is predominantly used for high temperature treatment. The direct version enables a complete burning of organic substances apart from some strongly oxidisable polymer molecules. The indirect technique leads to a thermal depolymerisation to decomposition products such as CO_2, CO, H_2, CH_4 and O_2. However, it is expected that a tar-like residual component remains up to a processing temperature of about 460°C. After exceeding this temperature the organic residues increasingly tend towards graphitising. The indirect technique is applied to both versions with and without oxygen.

In Fig. 6.24 a flow chart of a widely used and commercially available incineration technique is introduced. Depending on the facility used, a first rotary kiln reaching a temperature of about 105–120°C is used to dry the contaminated soil after passing the soil conditioning processes generally needed (see Sect. 6.1) and the feed system processing between 15 and 40 t h^{-1}. This process takes 5–7 min and the targeted residual moisture is usually reduced to about 3%. Dust produced from the drying procedure is re-circulated and treated again.

This procedure follows a second rotary kiln indicating distinct temperatures depending on the pollutant type and usually acting as pyrolysis. The operation needs a varying amount of time, due to the pollutant character. A retention time between 15 and 60 min is possible but usually 15–25 min are allotted. The treated material must be cooled down in a cooling system with water. Finally, the water content aimed for ranges from 3 to 10 vol%.

During the pyrolysis dust, which can be re-circulated and treated once again, is also emitted. The gaseous components that are emitted by the incineration technique must be cleaned up in different ways:

- Gas afterburner: the temperatures range between 970 and 1,200°C and subsequently the development of NO_x is strongly reduced. The afterburning lasts a few seconds only and occurs in the presence of oxygen. To prevent dioxin formation the gases should be cooled down quickly, in particular in the range from 600 to 300°C.
- Cloth filters, activated carbon filters and electrostatic precipitators: these filter systems adsorb nearly all pollutants after cooling down of the hot gases by means of the so-called quencher. The operation temperature is targeted at approximately 240°C but some filter systems require values between 140 and 190°C. The extremely polluted filter ash must be collected and is deposited for instance in mining shafts.
- Sulphur oxide elimination with CaO; this results in:

$$CaO + SO_3 \rightarrow CaSO_4 \text{ (gypsum)} \tag{6.3}$$

- Hydrochloric acid elimination with CaO; this results in:

$$CaO + 2HCl \rightarrow CaCl_2 + H_2O \tag{6.4}$$

- Nitrogen oxide elimination using NH_3 based on the formula:

$$NO_2 + NH_3 \rightarrow N_2 + H_2O + OH^- \tag{6.5}$$

Before releasing the treated gases to the atmosphere via the chimney an emission control gauging station observed the gas continuously. With reference to mercury the problematical physical vapour deposition requires a comprehensive technique which is based on the condensation of Hg-laden gases; for this reason, many providers are not willing to accept soils contaminated with Hg. Figure 6.25 shows a thermal treatment facility located in Breda, The Netherlands.

Fig. 6.25 Thermal treatment facility in Breda, The Netherlands

6.5.2 Required Soil Properties and Treatable Contaminants

The material handling opportunities depend on some soil characteristics which may detract from the success of the process. Construction debris, rock fragments, stones and bulky refuse are normally removed during the soil conditioning process (see Sect. 6.1). In the case of sludges to be treated or soils with high water content a dewatering is needed. Problems are well-known with reference to clay-rich soils containing aggregates and tightly packed substrates which do not allow the penetration of heat. In principle, a high water content >40 vol% and cohesive soils are detrimental in association with the drying process because they take up considerably more time and need co-incineration with fuels.

Depending on the contaminants of concern the temperature of the rotary kiln can be regulated more or less exactly to reduce the energy consumption during the decontaminating operation. Table 6.12 lists the boiling points of several contaminants. For instance, in the case of soils polluted by TPH an operating temperature of approximately 240°C can be supposed to be sufficient to decontaminate the soil. For volatile chlorinated hydrocarbons (VCHC) a temperature of approximately 90–320°C is recommended. If soils containing monocyclic aromatic compounds such as benzene are ascertained to be decontaminated, the operating temperature can be reduced to about 150°C. Pesticides are effectively treated with temperatures between approximately 320 and 540°C (NJDEP 1998). Thus, knowledge about the pollutants cleaned up plays a decisive role for temperature regulation. It should be taken into account that halogenated compounds require the gas afterburner technique to exclude the unwanted dioxin formation. With reference to the organic pollutants

Table 6.12 Boiling points of different contaminants

Parameter	°C
Mineral oil hydrocarbons (TPH)	40–240
BTEX aromates	80–144
Volatile chlorinated hydrocarbons (VCHC)	90–320
Polycyclic Aromatic Hydrocarbons (PAH, three benzene rings)	265–340
Polycyclic Aromatic Hydrocarbons (PAH, four benzene rings)	375–400
Polycyclic Aromatic Hydrocarbons (PAH, five and six benzene rings)	481–542
Polychlorinated Biphenyls (PCB)	Less than 600
Cyanides	450–600
Arsenic (As)	613
Cadmium (Cd)	767
Chromium (Cr)	2,599
Copper (Cu)	2,395
Mercury (Hg)	357
Nickel (Ni)	2,732
Lead (Pb)	1,740
Zinc (Zn)	907

the total content can be up to 10% of the soil mass. Some vendors may accept concentrations up to 30%.

Apart from the organic pollutants listed, the cyanides can also be examined in the thermal treatment equipment. Complex cyanides are processed at a temperature of 450–600°C. However, released hydrogen cyanide (HCN) has to be aftertreated in the gas afterburner at a temperature of approximately 1,100°C according to:

$$2HCN + 2O_2 \rightarrow N_2 + H_2O + 2CO_2 \tag{6.6}$$

Most of the metals, however, are hardly treatable due to the defined high boiling point. Since the pyrolysis technology is applied to operating temperatures less than 750°C merely the arsenic and cadmium concentration might be reduced but a complete volatilisation of these metals does not occur. The application of the high temperature technique associated with combustor technology causes vitrification of the metals (see Sect. 5.3.3). Accordingly, the metals are included in the matrix and are at least temporarily immobile. Nevertheless, most of the facilities dealing with thermal treatment only accept soil with maximum concentrations in relation to the metals, e.g. for As 50 mg kg^{-1}, for Cr 800 mg kg^{-1}, for Cu 500 mg kg^{-1}, for Ni 500 mg kg^{-1}, for Pb 600 mg $kg^{-1}$and for Zn 3,000 mg kg^{-1}respectively.

In particular, the combustors operating with oxygen are supposed to enhance the mobility of some metals. For instance, in oxidative and alkaline conditions during the combustion it is possible to generate toxic and mobile chromate (CrVI). In alkaline conditions usual for contaminated soils containing technogenic substrates the metalloid arsenic can be mobilised as well after reacting to $H_2AsO_4^-$. Mercury tends to volatilise rapidly, causing complex after-treatment, as mentioned before.

Table 6.13 Results from examples of thermal treatment (low-temperature pyrolysis) in Herne, Germany (Data from SITA 2009, unpublished)

Contaminant	Concentration before thermal treatment (mg kg^{-1})	Concentration after thermal treatment (mg kg^{-1})	Decontamination rate (%)
TPH	37,800	8.0	99.9
PAH	31,450	3.3	99.9
PCB	712	0.1	99.9
Cyanides	942	8.0	99.2
Mercury	410	0.5	99.9

Apart from soil contaminated with inorganic or organic pollutants, asbestos containing soil can also be treated, if the processing temperature exceeds 900°C (Nathanail and Bardos 2004).

The thermal treatment normally achieves high decontamination results. An example of a German remediation facility is given in Table 6.13. It is possible to achieve values exceeding 99%. Nevertheless, in relation to the results from thermal treatment only the total concentration after treatment should be taken into account, since the decisive factor is the quality standard that must be achieved in all cases.

6.5.3 *Re-use of the Treated Material*

The soil conditions after passing the rotary kiln are less favourable regarding the re-use of the material. As investigated in a field trial in Essen, Germany, dealing with deposited material originating from loess that had been treated in the pyrolysis, some properties changed significantly. An influence on the texture was not found and clay minerals were not seriously damaged, since the temperature fell below 650°C. The bulk density indicated a somewhat lower value (1.37 g cm^{-3}) compared with the value usually found in loess soils. The soil structure was degraded because gully erosion occurred, although the slope gradient measured values <2.5% (Fig. 6.26).

In spite of this, soil mineralogy was influenced because some chemical properties reacted measurably. For instance, the salinity was initially enhanced but decreased quickly within months (Fig. 6.27). A visible white salt crust appeared at the surface (Fig. 6.26). The pH value revealed elevated results as well with decreasing tendencies within the same period of time (from 7.9–8.4 to 7.5–8.0). Calcium carbonate was detected as a result of the presence of technogenic substrates such as construction debris which were excavated and treated together with the contaminated soil. The macronutrients phosphorus, magnesium and potassium did not show any clear change.

The nitrogen content was reduced since organic nitrogen was supposed to be volatilised during the thermal treatment (Table 6.14). However, the carbon content

Fig. 6.26 Deposit of decontaminated soil obtained from the thermal treatment (pyrolysis) in Essen, Germany; gully erosion in spite of a low slope gradient and salt crust formation (*white colour*) are visible

ranged from 2.8 to 6.1% with an average value of 3.3%, although the humus percentage was eliminated simultaneously with the nitrogen. The C content stemmed from inorganic tar-like residues of the pyrolysis (soot), while the organic matter was certainly decomposed. A very low carbon content of <1% is only expected in association with an oxidative treatment. According to the carbon alterations, the soil colour changed from brownish yellow (loess material) to black.

Of course, the fauna was destroyed leading to a more or less missing biological activity. According to the impact on the edaphon the enzymatic activity showed low values. For instance, the dehydrogenase activity of the field was nearly not detectable, apart from the plot fertilised with compost. This variation reached up to 55 μg TPF g^{-1} 24 h^{-1} within the first 10 months after establishment of the trial.

A surface-near backfill of the decontaminated material is usually not required but it can only be guaranteed, if the revitalisation of the site takes place. Regarding the field trial mentioned it was feasible to revitalise the soil by the application of organic fertilizers such as compost and by the use of legumes such as clover and *Lupinus*. After amendment of organic fertilizers and legume cultivation the soil was almost

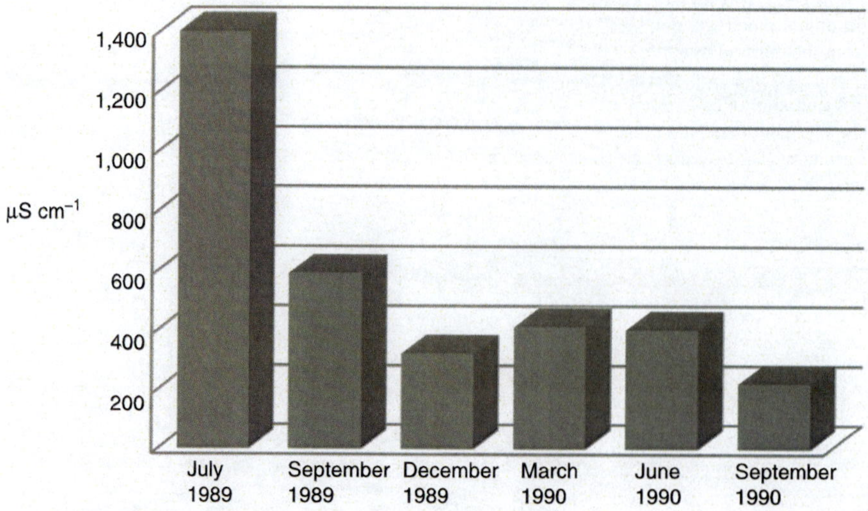

Fig. 6.27 Development of the soil salinity (electrical conductivity in μS cm^{-1}) after depositing of thermally treated soil (pyrolysis) in Essen, Germany (Data from Burghardt et al. 1991)

completely covered, leading to the reduction of soil erosion (Fig. 6.28). Even within a few months the microorganism activity was restored and differences between the field trial site and the adjacent park areas were not possible to find.

The chemical characteristics of the soil changed significantly as well. For example, within 4 years the plot fertilised by compost (3,000 kg ha^{-1}) and planted with a grass-clover-mixture reached an N content of 0.18–0.20% and a cation exchange capacity (CEC) of 91–99 $mmol_c$ kg^{-1}, while the plot seeded with grass and clover resulted in 0.07% N and 28–41 $mmol_c$ kg^{-1} CEC with mineral fertilizing or without fertilizing. The CEC parameter was chosen because the increasing CEC might be correlated with the increasing humus content, which was not possible to detect separately due to the presence of inorganic carbon sources (soot development). The N content increased in the course of time in all plots because organic substance was accumulated due to root residues. Taking the C/N ratio into consideration the composted variation was 28, while the plot not amended by compost showed higher results, varying from 30 to 50.

In addition, an impact on the physical characteristics was found. The difference of the composted option compared to the untreated plot was 9–17 vol% to 8–11 vol% in relation to the air capacity and 29–31 vol% to 33–37 vol% related to the available field capacity. In conclusion, the compost fertilizing improved the air capacity, while the available water content did not show any beneficial effects (Burghardt et al. 1991).

In a field experiment in Homecourt, France, the revitalisation of thermally treated soil from a former coking plant was investigated. The material also revealed organic residues of a heating process (9.4%), coupled with a wide C/N ratio of 97, an alkaline pH value (9.0) and a relatively high calcium carbonate content amounting to

Table 6.14 Soil and plant characteristics of a field trial in Essen, Germany, dealing with deposited thermally treated soil (Data unpublished)

	Zero option	Grass-clover mixture without fertilizing	Grass-clover mixture with mineral manure	Grass-clover mixture with compost fertilizing
C (%)	3.3	na	3.5	5.2
N (%)	0.11	0.07	0.07	0.18
C/N	30	–	50	28
N (%), 5 years later	0.14	0.14	0.12	0.20
CEC ($mmol_c$ kg^{-1})	41	41	28	99
CEC ($mmol_c$ kg^{-1}), 5 years later	61	79	70	91
P-DL (mg 100 g^{-1})	na	8.5	10.8	8.9
Dehydrogenase activity (μg TPF g^{-1} 24 h^{-1}), 7 months after sowing	0	0	1.1	53.4
Dehydrogenase activity (μg TPF g^{-1} 24 h^{-1}), 10 months after sowing	0	0	0.6	55.4
Soil cover of the vegetation (%), 1 month after sowing	15	na	8	48
Soil cover of the vegetation (%), 2 months after sowing	32	na	48	57
Soil cover of the vegetation (%), 3 months after sowing	42	na	49	68

Date of sowing and fertilizing: 16.06.1989
Grass-clover-mixture: 30 g m^{-2}
Mineral manure: 50 kg ha^{-1} N (ammonia-nitrate-manure), 90 kg ha^{-1} P
Compost application: 3,000 t ha^{-1} equivalent to 45 kg ha^{-1} N and 12 kg ha^{-1} P
Investigated soil depth: 0–40 cm
na not analysed

Fig. 6.28 Deposit of revitalised decontaminated soil obtained from the thermal treatment in Essen, Germany; legume growth reduced the bare soil surfaces and subsequently reduced soil erosion

24.5%. After thermal treatment the material still indicated low but detectable PAH concentration (e.g. benzo(a)pyrene 2.0 mg kg^{-1}, naphthalene 0.8 mg kg^{-1}) and enhanced Pb (496 mg kg^{-1}) and Zn (760 mg kg^{-1}) concentrations. The field trial worked with one option where 15 cm of compost overlaid the 60 cm-thick thermally treated material and an alternative plot where, below the 15 cm of compost, the material was intensively mixed with paper mill sludge. Apart from unvegetated plots displaying natural succession, rye-grass (*Lolium perenne*) and alfalfa (*Medicago sativa*) were afterwards sown.

With reference to the heavy metals the organic fertilizer partly initiated accelerated metal mobility due to the formation of organo-metalic complexes. The high values occurred sporadically, otherwise the high pH value might have reduced the metal mobility. This result should generally be taken into consideration, since thermally treated soil might frequently contain continuous metal contamination. In spite of this, the addition of the paper mill sludge led to improved plant growth. The rye-grass showed a higher total biomass production on the paper mill plot than on the other plot, whereas the alfalfa plants adapted to N_2 fixation did not reveal

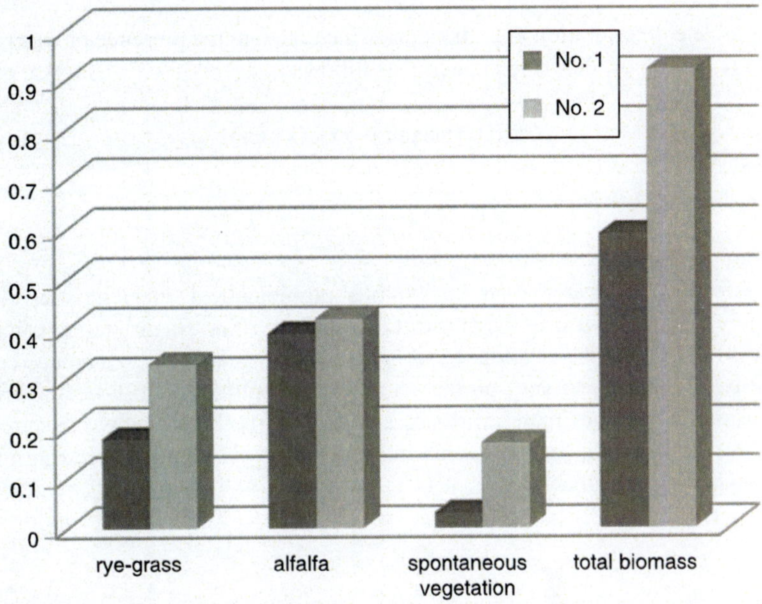

Fig. 6.29 Biomass production (kg m^{-2}) of plants growing in pure thermally treated material (*No. 1*) and thermally treated material after mixing with paper mill sludge (*No. 2*); both options were overlain with 15 cm compost (Data from Sere et al. 2008)

any differences (Fig. 6.29). In summary, the total biomass production was 0.92 kg m^{-2} in the paper mill plot and only 0.59 kg m^{-2} in the alternative plot. Similarly, the root development of the paper mill plot reached a depth of 70 cm, while the roots growing in purely thermally treated material only overlaid by 15 cm compost were limited to the uppermost 25 cm of soil depth. The plots used for spontaneous vegetation exhibited a higher species diversity and biomass production in the profile with paper mill sludge. In conclusion, the paper mill sludge variation allowed a faster ecological restoration due to advanced physical and chemical properties (Sere et al. 2008).

The thermal treatment of contaminated soils usually occurs at low temperatures. In the past some facilities operated at a high temperature level, exceeding 650°C and reaching even 1,200°C in some case studies. The material was completely changed after the burning process and did not bear resemblance to natural soils.

Apart from the humus and nitrogen losses and the destruction of the soil, there was an extreme change in the chemical properties. Clay minerals were damaged and the structure changed to a porous, partly glassy, pellet structure with a rough surface previously unknown. The metals were incorporated into the new glassy and crystalline structure and hence the solubilisation of metals was strongly limited. An exception was chromium, which was mobilised under the alkaline and aerobic conditions present after the treatment.

Furthermore, the pH value rose to 11.8 till 12.0. It has been found that the reason for the extreme alkalinisation is the high temperature exceeding 600°C and resulting in a complete deterioration of carbonates, which react in the presence of water used for the rewetting of the treated material as follows:

$$CaCO_3\,(\text{heated}) \rightarrow CaO + CO_2 \tag{6.7}$$

$$CaO + H_2O \rightarrow Ca^{2+} + 2OH^- \tag{6.8}$$

Physically, the treated material exhibits problematical characteristics which definitely exclude re-use in the construction industry. The single grains indicate a rough surface due to the developed carbonate crusts, which leads to a reduced compatibility. They are also very porous and thus susceptible to frost weathering. In conclusion, the lack of material re-use as much as the high energy consumption have been responsible for the decreasing number of case studies referring to the high temperature approach.

6.6 Electrokinetic Remediation

6.6.1 Technical Devices Used

For this treatment electrodes are placed in the soil and an electrical current is applied, which leads to the movement of ions to the respective electrode. Cations migrate to the negatively charged cathode, anions to the positively charged anode (Fig. 6.30). Not only are metals of interest, polar organic substances are moved towards the respective electrode as well. Non-polar substances do not migrate on the basis of the electrical current but nevertheless they are moved to the cathode by the pumping effect of the pore fluid as described below.

The electrical field at the cathode causes production of a high pH up to 12.0, while at the anode a low pH value up to 2.0 occurs. The process is conducted by different redox potentials. At the cathode reductive conditions prevail, which cause, generated by hydrolysis, the establishment of hydroxyl groups, indicating alkaline reaction

$$4H_2O + 4e^- \rightarrow 2H_2 + 4OH^- \tag{6.9}$$

At the anode oxidation is dominant in line with the creation of acidic conditions

$$2H_2O - 4e^- \rightarrow O_2 + 4H^+ \tag{6.10}$$

Subsequently, the front of the solution of high pH produced at the cathode and the front of low pH produced at the anode move towards the opposite electrode and at the border between both a sharp jump in pH value is observed. The pH gradient

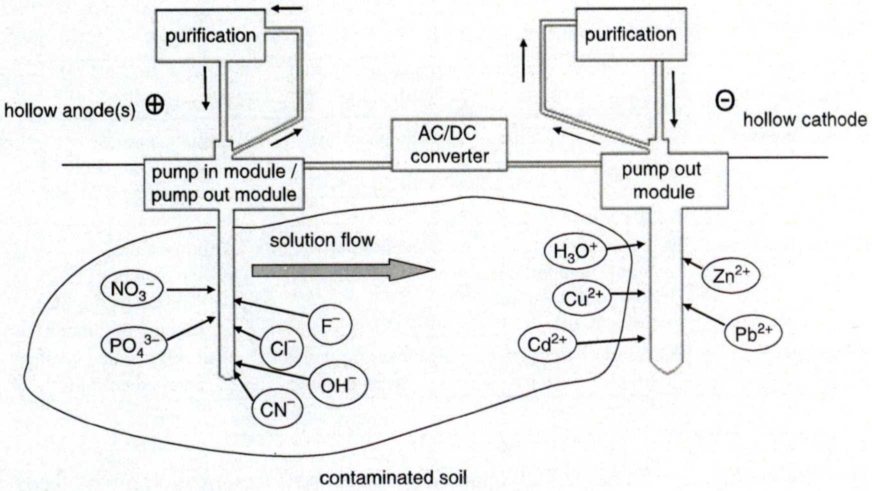

Fig. 6.30 Diagram of the electrokinetic treatment for contaminated soils

influences the movement of the ions. In general, the acid front moves faster than the base front, since the H^+ mobility exceeds the mobility of OH^-. Hence, the soil becomes predominantly acidic throughout its volume, except for the surrounding area of the cathode. After current application a liquid pumping effect is recognisable, in which the movement from the anode to the cathode, where the ion enriched liquid is pumped off the soil, increases. To prevent shrinkage and drying of the soil the water must be pumped back using the anode or fresh water must be added at the same electrode. The soil drying is a result of the heating effect of the electric current. It must be avoided due to the development of uneven flow paths and consequently cessation of the fluid flow.

The acidic front (extremely high H^+ concentration) processed at the anode is much larger than the number of other cations adsorbed. Thus, the H^+ ions can replace the cationic elements at the sorptive sites, so that a huge movement of cations towards the cathode takes place. Apart from the metal cations, contaminants exhibiting cation behaviour such as amphoteric hydrocarbons and organic bases also flow towards the cathode. Subsequently, most of the cations will be electrodeposited on the cathode (e.g. in the case of cadmium and lead 75–95% have been reached in each case). A high portion of the soluble cations will precipitate at the cathode as oxides, hydroxides and carbonates because the alkaline reaction favours the precipitation but a successful decontamination depends on permanent maintenance of the metal solubility. Thus, a high precipitation rate may reduce the effectiveness of the electro-remediation, since clogging of the pores may occur, resulting in a cessation of the flow (Marks et al. 1994; Page and Page 2002).

Electrokinetic remediation offers three options, namely the electroosmosis, electromigration and electrophoresis (Probstein 1994; Voss et al. 2001; Simon et al. 2002) (Table 6.15). The first one follows the principle of a cation and water film movement through the diffuse layers of clay minerals. Clay minerals provide dif-

Table 6.15 Electrokinetic remediation strategies

Type	Principle	Favoured soil application	Pollutants
Electroosmosis	Cation and water film movement through the diffuse double layer of clay minerals	Silty and clayey soils, fine-grained sludges	Cations, anions, charged soluble organic pollutants
Electromigration	Transport of soluble cations and anions and non-polar substances (mass flow)	Coarse-grained soils (gravel, sand)	Cations, anions, charged and non-polar, soluble organic pollutants
Electrophoresis	Transport of charged colloids (>10 μ)	Coarse-grained soils (gravel, sand)	Positively or negatively charged colloids

fuse double layers with negatively charged surfaces. After application of an electrical field forces are exerted on the charges which lead to the movement of the ions, if the forces exceed the ion attraction to the surfaces. When water is added to the diffuse double layer of clay minerals, the cations are no longer held and diffusion from the bulk phase starts equalising the concentration gradient in the double layer. The cation bulk concentration gradient influenced by the acid front will disintegrate the double layers and ultimately release the adsorbed cations to the bulk solution to a great extent. The movement, which is irrespective of pore size distribution, occurs towards the electrodes installed. The velocity of the ion transport can reach several centimetres per day (Bradl 2005). The electroosmosis treatment is preferentially applied to silty and clayey soils because the presence of clay minerals is required. For this reason, at least the electroosmosis option is advantageous in soils with low permeability which are normally very difficult to treat.

Electromigration is based on the mass flow (advection) of ions dissolved in the pore fluid and accordingly it is feasible in different texture classes but most effective in coarse-grained soils. Electromigration is predominant in these soils, since electroosmosis requires the presence of clay minerals which are quantitatively reduced. The ion velocities of electromigration are 5–40 times higher than those of electroosmosis (Bradl 2005). If the contaminants are included in macromolecules, the ion migration becomes more problematical. After possible desorption from attracting constituents like clay minerals and the organic matter macromolecules must be divided into micromolecules and the latter finally into the mobile ions. Hence, the process of transformation into soluble ions takes up a lot of time.

Electrophoresis means the transport of charged colloids mostly present in sandy and pebbly soils as well as sludges. The charged particles exhibit distinct sizes and the larger ones are less movable, since they are partly larger than the pore sizes. Therefore, electrophoresis is predominantly relevant for coarse-grained soils exhibiting a lot of macropores.

Deposited heterogeneous soils containing technogenic substrates >10 cm show limitations to an advantageous electrokinetic remediation. In particular, metal-enriched deposits reduce the possibilities of the electroremediation to a great extent, because they consume the electric energy and disrupt the normal current path. Typical materials disturbing the electroremediation but often found are construction steel and iron barrels. In general, the high cost-intensive energy input ranging from 40 to 400 kWh t^{-1} might be a reason for the failure of this technique in a number of contaminated sites (Genske 2003).

Irrespective of the electrokinetic principle, water is required to implement the technique. Electromigration and electrophoresis are based on mass flow and subsequently dependent on a relatively high water content. Electroosmosis, however, can also not be carried out in dry soil conditions, because the ion migration depends on the water film movement through the diffuse double layers of clay minerals. Consequently, the water content generally influences the solubilisation and migration and ultimately the effectiveness of the procedure considerably. Optimised conditions for electromigration and electrophoresis prevail at 50–70% of the field capacity, while electroosmosis needs a water content of approximately 80% of the field capacity.

The electroremediation technique is applied *in situ* and at great soil depths. It is even possible to use it below buildings. Heterogeneous soil properties are not preferred, but the technique is generally applicable in soils with alternating textures. In most cases one central cathode is installed surrounded by a number of external anode electrodes. The configuration of the installed electrodes results alternatively in hexagonal, triangular or quadratic patterns. The maximum distance between the electrodes should be restricted to 10 m, the maximum depth is 12 m and the tensions (voltage) applied vary from 20 to 100 V m^{-1}, exceptionally 500 V m^{-1} (Acar and Alshawabkeh 1993; Simon et al. 2002). Cathode and anodes are housed in a permeable electrode casing. Accordingly, the water containing soluble ions can be pumped out of the ground and afterwards cleaned and conditioned aboveground (see Sect. 7.1.4).

In situ electroremediation, which is more commonly applied than *ex situ* approaches, has a detrimental impact on soil microorganisms. The one-dimensional flow from the anode to the cathode may sweep the microorganisms towards the cathode. At the anode base-tolerant species are killed, because the pH value ranges from 2 to 3. On the other hand, acid-tolerant species are destroyed near the cathode, which has a pH value ranging from 8 to 12. Consequently, the optimum pH value between 5 and 7 can never be achieved in the electrical field. Hence, the soil will not be suitable for a metabolic destruction of organic pollutants administered at a later date (see Sect. 6.3.1). Furthermore, electrokinetic remediation can enhance the soil temperature, reaching sometimes more than 60°C. Thus, it is expected that the edaphon will be strongly disturbed (Marks et al. 1994).

Electroremediation can be combined with other clean-up techniques. For instance, after application of an electric field upstream of a reactive wall the amount of constituents touching the barrier system is reduced, whereby the function of the wall is improved, since increasing clogging of the pore system in the barrier has not taken place (see Sect. 7.1.7).

6.6.2 *Treatable Contaminants*

The following contaminants have been successfully remediated by the different options of electroremediation:

- Cations: heavy metals such as Cd^{2+}, Cu^{2+}, Ni^{2+}, Pb^{2+}, Zn^{2+}, ammonia (NH_4^+)
- Anions: cyanides (CN^-), arsenate ($H_2AsO_4^-$), chromate (CrO_4^{2-}), nitrate (NO_3^-), phosphate (PO_4^{3-}), fluoride (F^-), chloride (Cl^-), sulphate (SO_4^{2-})
- Polar organic compounds: phenols, benzene, trichloroethylene.

Metals do not migrate only to the cathode, since some compounds such as chromates and arsenates are negatively charged. The element chromium, for instance, may flow to the cathode, appearing as Cr^{3+} and may flow to the anode, if the chemical compound is CrO_4^{2-}. Apart from the contaminants of concern, some nutrients are mobilised as well, e.g. nitrate and sulphate. Hence, the leaching of anionic nutrients must be carefully monitored.

Phenol is an organic pollutant which is successfully treated by the electroremediation in the field. Under acidic conditions phenol is undissociated (C_6H_5OH), leading to transportation towards the cathode, while in the presence of higher pH values in the soil the anionic phenol proportion ($C_6H_5O^-$) stemming from the phenol dissociation

$$C_6H_5OH \leftrightarrow C_6H_5O^- + H^+ \tag{6.11}$$

tends to move towards the anode (Page and Page 2002).

The organic pollutants are less effective in association with the electroremediation approach, since desorption from clay minerals and the organic matter is strongly reduced. Therefore, surfactants such as detergents are additionally used to improve the solubility of some organic pollutants, e.g. phenols. Heavy metals can be solubilised additionally after amendment of solubilising agents such as acetic acid and chelating agents, in particular EDTA (see Sect. 6.2.3). Organic phase contaminants (LNAPL, DNAPL), which are frequently soil constituents relating to tar oils and petroleum hydrocarbons, cannot be mobilised as a result of electroremediation.

A number of case studies in the USA starting in 1991 reveal the potential adequacy of the electroremediation approach. In Europe there were examples of successful application of the technique in the past. For instance, an average reduction of 74% for copper and lead has been obtained at a site of a former paint factory in the Netherlands. A 2-year project in Woensdrecht, The Netherlands, which treated 3,500 m^3 of contaminated soil, exhibited a strong reduction of heavy metals, achieving results ranging from 87 to 91% of the initial concentration, e.g. chromium from 7,300 to 755 mg kg^{-1}, nickel from 860 to 80 mg kg^{-1}, lead from 730 to 108 mg kg^{-1} and zinc from 2,600 to 860 mg kg^{-1}respectively (Lageman 2007). Even radioactive isotopes were treated by electroremediation, for example ^{90}Sr in the USA and ^{137}Cs in Russia (Page and Page 2002). A case study dealing with a former gasworks displayed a cyanide reduction of 96% and a phenol reduction of 73% (Lageman 2007).

Electroremediation has also been performed *ex situ* based on a commercially available flow cell used for electrolysis. This technique is aimed at the reductive dechlorination of chlorinated compounds to an extent that would enable aerobic biodegradation at a later date (see Sect. 6.3.2). Apart from chlorinated monoaromates, even more complex structures such as PCB and PCDD/F were involved in this technical approach. As explained above, at the cathode reductive conditions are produced, resulting in completely anaerobic conditions which make the reductive de-chlorination possible.

In relation to a mixture of polychlorinated benzenes including hexachlorobenzene (HCB) the reduction led to 92% monochlorobenzene and 8% benzene. Mixtures of dichlorotoluene and trichlorotoluene were reduced to 54–84% monochlorotoluene and 4–13% toluene. Moreover, compounds consisting of dichlorobiphenyls and trichlorobiphenyls were transformed to biphenyls, yielding up to 99% and some traces of monochlorobiphenyls. Another example dealing with chlorophenols revealed similar results. Dichlorophenols and trichlorophenols were converted to monochlorophenols to approximately 80%, and even pentachlorophenols (PCP) resulted in 44% dichlorophenols, 43% monochlorophenols and 4% phenols. The treatment was successfully applied to technical waste products as well. In a case treating industrially produced PCB the electroreduction resulted in the formation of 92% biphenyls and only small amounts of chlorinated biphenyls. With reference to the contaminant concentration in another example dealing with a high concentration of polychlorinated dioxins in an oily suspension (12,200 $\mu g\ kg^{-1}$) the electroreduction was able to eliminate the toxic substance to a value of about 10 $\mu g\ kg^{-1}$. A dioxin containing copper slag termed kieselrot, which has been used for the construction of sports fields in Germany and the Netherlands, initially indicated a dioxin concentration of 500 $\mu g\ kg^{-1}$ but after treatment the chlorinated compounds were completely de-chlorinated (Voss et al. 2001).

6.7 Treatment Centres

Measures requiring a lot of transport from the contaminated site to different off-site facilities are environmentally unfriendly. Hence, the best method of environmentally protective handling is the establishment of soil treatment centres offering combined approaches including soil conditioning, soil washing, bioremediation and thermal treatment. They are adapted to the usually complex contamination of the excavated soil and deal with different soil materials and qualities, so that a rapid backhaul of clean material used for the backfilling operations can be realised. Since most of the remediation cases are connected to building measures, quick soil delivery on a specific date might be of importance.

Moreover, remediation centres probably enable a complete decontamination and are suitable for different contaminated materials in addition to soil such as dredged material, solid waste and road litter. Consequently, the material finally disposed of is strongly minimised, disburdening landfills and other types of dumpsites. Some clean material can be re-used at other sites and in this way money can be made.

The centres mentioned must be linked up to the infrastructure, namely motorways for transport by lorry, waterways for transport by ship as well as railways. A large area is used providing a location for storage and interim storage facilities and the treatment technology. Some beneficial effects can be coupled, e.g. use of the waste heat produced during the thermal treatment for bioremediation purposes (regeneration pits). With regard to the transport and possible negative effects such as dust emissions and development, noise and vibration, the centres must be accepted by the people living in the neighbourhood.

References

Acar, Y. B., & Alshawabkeh, A. N. (1993). Principles of electrokinetic remediation. *Environmental Science and Technology, 27*, 2638–2648.

Alexander, M. (2001). Bioavailability of organic compounds sequestered in soils. In R. Stegmann, G. Brunner, W. Calmano, & G. Matz (Eds.), *Treatment of contaminated soil*. Berlin/New York: Springer.

Anonymus. (2008). *Soil washing on the Olympic Park*. www.london2012.com/blog/2008/07/soi-washing-on-the-olympic-park.php. Accessed 15 Sept 2011.

Baker, A. J. M., & Brooks, R. R. (1989). Terrestrial higher plants which hyperaccumulate metallic elements. A review of their distribution, ecology and phytochemistry. *Biorecovery, 1*, 81–126.

Bandyopadhyay, S., Bhattacharya, S. K., & Majumdar, P. (1994). Engineering aspects of bioremediation. In D. L. Wise & D. J. Trantolo (Eds.), *Remediation of hazardous waste contaminated soils*. New York: Marcel Dekker.

Baum, C., Hrynkiewicz, K., Leinweber, P., & Meißner, R. (2006). Heavy-metal mobilization and uptake by mycorrhizal and nonmycorrhizal willows (*Salix x dasyclados*). *Journal of Plant Nutrition and Soil Science, 169*, 516–522.

Black, H. (1995). Absorbing possibilities: Phytoremediation. *Environmental Health Report, 103*, 1106–1108.

Bradl, H. B. (2005). *Heavy metals in the environment*. Amsterdam: Elsevier.

Bradl, N. C., & Weil, R. R. (2008). *The nature and properties of soils* (Pearson international edition). Upper Saddle River: Pearson Prentice Hall.

Burghardt, W., Hiller, D. A., Hintzke, M., Meuser, H., & Wessel, R. (1991). Abiotic and biotic properties of a thermally treated soil. *Notes of the German Soil Society, 66*, 609–612 (in German).

Chu, W. (2003). Remediation of contaminated soils by surfactant-aided soil washing. *Practice Periodical of Hazardous, Toxic and Radioactive Waste Management, 7*, 19–24.

CL:AIRE. (2007). Understanding soil washing. *Technical bulletin, TB 13*. Contaminated land: Applications in Real Environments. www.claire.co.uk. Accessed 10 Sept 2011.

Cosio, C., Vollenweider, P., & Keller, C. (2005). Localization and effects of cadmium in leaves of a cadmium-tolerant willow (*Salix viminalis* L.). *Environmental and Experimental Botany, 58*, 64–74.

DEC (DEME Environmental Contractors). (2011). *Mobile soil washing plant*. Antwerpen, Belgium.

Dennis, R. M., Dworkin, D., & Zupko, A. J. (1994). Soil-washing processes for site remediation. In D. L. Wise & D. J. Trantolo (Eds.), *Remediation of hazardous waste contaminated soils*. New York: Marcel Dekker.

Dickinson, N. M. (2006). Phytoremediation of industrially-contaminated sites using trees. *NATO Series IV: Earth and Environmental Sciences, 68*, 229–240.

Dos Santos Utmazian, M., & Wenzel, W. W. (2007). Cadmium and zinc accumulation in willow and poplar species grown on polluted soils. *Journal of Plant Nutrition and Soil Science, 170*, 265–272.

EPA (Environmental Protection Agency USA). (1990). *Engineering bulletin: Solvent extraction treatment* (EPA/540/2-90/013), Cincinnati.

Eschenbach, A., Mescher, H., Wienberg, R., & Mahro, B. (2001). Humification of PAH and TNT during bioremediation – Evaluation of long term risk and sustainability. In R. Stegmann, G. Brunner, W. Calmano, & G. Matz (Eds.), *Treatment of contaminated soil*. Berlin/New York: Springer.

Fan, C. Y., & Tafuni, A. N. (1994). Engineering application of biooxidation process for treating petroleum-contaminated soil. In D. L. Wise & D. J. Trantolo (Eds.), *Remediation of hazardous waste contaminated soils*. New York: Marcel Dekker.

Feitkenhauer, H., Hebenbrock, S., Deppe, U., Märkl, H., & Antranikian, G. (2001). Degradation of xenobiotica at elevated temperatures. In R. Stegmann, G. Brunner, W. Calmano, & G. Matz (Eds.), *Treatment of contaminated soil*. Berlin/New York: Springer.

Fletcher, R. D. (1994). Practical considerations during bioremediation. In D. L. Wise & D. J. Trantolo (Eds.), *Remediation of hazardous waste contaminated soils*. New York: Marcel Dekker.

Garcia, G., Conesa, H. M., & Faz, A. (2002). *Phytoremediation of zinc polluted soils by Mediterranean pant species: Usefulness of bioaccumulation and tolerance capabilities*. 4th WG2 workshop, Bordeaux, France.

Genske, D. D. (2003). *Urban land – Degradation, investigation, remediation*. Berlin: Springer.

Gordon, M., Choe, N., Duffy, J., Ekuan, G., Heilman, P., Muiznieks, I., Ruszaj, M., Shurtleff, B. B., Strand, S., Wilmoth, J., & Newman, L. A. (1998). Phytoremediation of trichloroethylene with hybrid poplars. *Environmental Health Perspectives, 106*, 1001–1004.

Grotenhuis, T. J., & Rijnaarts, H. H. (2011). In situ remediation technologies. In F. A. Swartjes (Ed.), *Dealing with contaminated sites*. Dordrecht: Springer.

Heely, D. A., Werk, E. S., & Kowalski, R. G. (1994). Bioremediation and reuse of soils containing No. 5 fuel oil in New England using an aboveground treatment cell: A case study. In D. L. Wise & D. J. Trantolo (Eds.), *Remediation of hazardous waste contaminated soils*. New York: Marcel Dekker.

Hupe, K., Koning, M., Lüth, J. C., Heerenklage, J., & Stegmann, R. (2001). Optimisation of microbial soil treatment. In R. Stegmann, G. Brunner, W. Calmano, & G. Matz (Eds.), *Treatment of contaminated soil*. Berlin/New York: Springer.

Kahle, P., Criegee, C., & Lennartz, B. (2002). Willow stands as an alternative method for the reduction of leachate at contaminated sites – Numerical investigations. *Journal of Plant Nutrition and Soil Science, 165*, 501–505.

Kanaly, R. A., & Harayama, S. (2000). Biodegradation of high-molecular-weight polycyclic aromatic hydrocarbons. *Journal of Bacteriology, 182*, 2059–2067.

Karthikeyan, R., & Kulakow, P. A. (2003). Soil plant microbe interactions in phytoremediation. In D. T. Tsao et al. (Eds.), *Phytoremediation*. Berlin: Springer.

Karvounis, Y., & Kelepertsis, A. (2000). *Chemical clean up and properties of soils polluted with heavy metals near mining areas*. Proceedings Vol. 3. Paper presented at 1st international conference on soils of urban, industrial, traffic and mining areas, Essen, Germany.

Kästner, M., & Richnow, H. H. (2001). Formation of residues of organic pollutants within the soil matrix – Mechanisms and stability. In R. Stegmann, G. Brunner, W. Calmano, & G. Matz (Eds.), *Treatment of contaminated soil*. Berlin/New York: Springer.

Kästner, M., Streibich, S., Beyrer, M., Richnow, H. H., & Fritsche, W. (1999). Formation of bound residues during microbial degradation of ^{14}C anthracene in soil. *Applied and Environmental Microbiology, 65*, 1834–1842.

Khan, F. I., Husain, T., & Hejazi, R. (2004). An overview and analysis of site remediation technologies. *Journal of Environmental Management, 71*, 95–122.

Kiene, A., Miehlich, G., & Gröngröft, A. (2001). Influence of oil-contamination on N-mineralisation in soils. In R. Stegmann, G. Brunner, W. Calmano, & G. Matz (Eds.), *Treatment of contaminated soil*. Berlin/New York: Springer.

King, R. F., Royle, A., Putwain, P. D., & Dickinson, N. M. (2006). Changing contaminant mobility in a dredged canal sediment during a three-year phytoremediation trial. *Environmental Pollution, 143*, 318–326.

Koning, M., Cohrs, I., & Stegmann, R. (2001). Development and application of an oxygen-controlled high-pressure aeration system for the treatment of TPH-contaminated soils in high biopiles (a case study). In R. Stegmann, G. Brunner, W. Calmano, & G. Matz (Eds.), *Treatment of contaminated soil*. Berlin/New York: Springer.

Kuzovkina, Y. A., & Quigley, M. F. (2005). Willows beyond wetlands: Uses of *Salix* L. species for environmental projects. *Water, Air, and Soil Pollution, 162*, 183–204.

Lageman, R. (2007, December 9). *Applications of electro-remediation and (electro) bioremediation*. Presentation at the Environmental Symposium Tel Aviv, Israel.

Lajoie, C. A., & Strom, P. F. (1994). Biodegradation of polynuclear aromatic hydrocarbons in coal tar oil contaminated soil. In D. L. Wise & D. J. Trantolo (Eds.), *Remediation of hazardous waste contaminated soils*. New York: Marcel Dekker.

Levin, L., Viale, A., & Forchiassin, A. (2003). Degradation of organic pollutants by the white rot basidiomycete *Trametes trogii*. *International Biodeterioration and Biodegradation, 52*, 1–5.

Lindström, K., Jussila, M. M., Hintsa, H., Kaksonen, A., Mokelke, L., Mäkeläinen, K., Pitkäjärvi, J., & Suominen, L. (2003). Potential of the *Galega – Rhizobium galegae* system for bioremediation of oil-contaminated soil. *Biotechnology, 41*, 11–16.

Liphadzi, M. S., & Kirkham, M. B. (2006). Heavy-metal displacement in chelate-treated soil with sludge during phytoremediation. *Journal of Plant Nutrition and Soil Science, 169*, 737–744.

Lotter, S., Heerenklage, J., & Stegmann, R. (2001). Carbon balance and modelling of oil degradation in soil bioreactors. In R. Stegmann, G. Brunner, W. Calmano, & G. Matz (Eds.), *Treatment of contaminated soil*. Berlin/New York: Springer.

Mahro, B., Müller, R., & Kasche, V. (2001). Bioavailability – The key factor of soil bioremediation. In R. Stegmann, G. Brunner, W. Calmano, & G. Matz (Eds.), *Treatment of contaminated soil*. Berlin/New York: Springer.

Marks, R. E., Acar, Y. B., & Gale, R. J. (1994). In situ remediation of contaminated soils containing hazardous mixed wastes by electrokinetic remediation and other competitive technologies. In D. L. Wise & D. J. Trantolo (Eds.), *Remediation of hazardous waste contaminated soils*. New York: Marcel Dekker.

Meers, E., Vervaeke, P., Tack, F. M. G., Lust, N., Verloo, M. G., & Lesage, E. (2003). Field trial experiment: phytoremediation with *Salix* sp. on a dredged sediment disposal site in Flanders, Belgium. *Remediation Journal, 12*, 115–120.

Meuser, H. (2010). *Contaminated urban soils*. Dordrecht: Springer.

Meuser, H., & Makowsky, L. (2009). *Feasibility study on phytoextraction of heavy metals in a sludge field derived from lake dredging nearby Hanover, Germany*. Presentation SUITMA conference, New York.

Middeldorp, P., & Langenhoff, A. (2002). In situ clean up technologies. In S. N. Agathos & W. Reineke (Eds.), *Biotechnology for the environment: Soil bioremediation*. Dordrecht: Kluwer Academic Publisher.

Milner, M. J., & Kochian, L. V. (2008). Investigating heavy-metal hyperaccumulation using *Thlaspi caerulescens* as a model system. *Annals of Botany, 102*, 3–13.

Mirsal, I. A. (2004). *Soil Pollution – Origin, monitoring & remediation*. Berlin: Springer.

Müller, R., & Mahro, B. (2001). Bioaugmentation: Advantages and problems using micro organisms with special abilities in soil decontamination. In R. Stegmann, G. Brunner, W. Calmano, & G. Matz (Eds.), *Treatment of contaminated soil*. Berlin/New York: Springer.

Nathanail, C. P., & Bardos, R. P. (2004). *Reclamation of contaminated land*. Chichester: Wiley.

Nathanail, J. F., Bardos, P., & Nathanail, C. P. (2002). *Contaminated land management ready reference*. Richmond: EPP Publications/Land Quality Press.

Neeße, T. (2001). New developments in soil washing technology. In R. Stegmann, G. Brunner, W. Calmano, & G. Matz (Eds.), *Treatment of contaminated soil*. Berlin/New York: Springer.

NJDEP. (1998). *Revised guidance document for the remediation of contaminated soils*. Department of Environmental Protection State of New Jersey, USA.

Norris, R. D. (1994). *Handbook of bioremediation*. Boca Raton: Lewis Publisher.

Page, M. M., & Page, C. L. (2002). Electroremediation of contaminated soils. *Journal of Environmental Engineering, 128*, 208–219.

Paul, E., & Clark, F. (1996). *Soil microbiology and biochemistry*. San Diego: Academic.

Pearl, M., Pruijn, M., & Bovendeur, J. (2006). The application of soil washing to the remediation of contaminated soils. *Land Contamination and Reclamation, 14*, 713–726.

Probstein, R. F. (1994). *Physicochemical hydrodynamics – An introduction*. New York: Wiley.

Radwan, S., Sorkhon, N., & El-Nemri, I. (1995). Oil biodegradation around roots. *Nature, 376*, 302–308.

Sere, G., Schwartz, C., Ouvrard, S., Sauvage, C., Renat, J. C., & Morel, J. L. (2008). Soil construction: A step for ecological reclamation of derelict lands. *Journal of Soils and Sediments, 7*, 1–7.

Siegel, F. R. (2002). *Environmental geochemistry of potentially toxic metals*. Heidelberg: Springer.

Simon, F. G., Meggyes, T., & McDonald, C. (2002). *Advanced groundwater remediation: Active and passive technologies*. London: Thomas Telford.

SITA. (2009). *SITA remediation*. Information brochure, Herne, Germany. www.sitaremediation.com. Accessed 1 Sept 2011.

SUMATECS. (2008). *Sustainable management of trace elements contaminated soils – Development of a decision tool system and its evaluation for practical application* (Final Research Report, No. SN-01/20), Vienna, Austria.

Tamura, H., Honda, M., Sato, T., & Kamachi, H. (2005). Pb hyperaccumulation and tolerance in common buckwheat (*Fagopyrum esculentum*). *Journal of Plant Research, 118*, 355–359.

Tiehm, A., & Stieber, M. (2001). Strategies to improve PAH bioavailability: Additions of surfactants, ozonation and application of ultrasound. In R. Stegmann, G. Brunner, W. Calmano, & G. Matz (Eds.), *Treatment of contaminated soil*. Berlin/New York: Springer.

Tiehm, A., Stieber, M., Werner, P., & Frimmel, F. H. (1997). Surfactant-enhanced mobilization and biodegradation of polycyclic hydrocarbons in manufactured gas plant soil. *Environmental Science and Technology, 31*, 2570–2576.

Trapp, S., & Karlson, U. (2001). Aspects of phytoremediation of organic pollutants. *Journal of Soils and Sediments, 1*, 1–7.

Vandecasteele, B., De Vos, B., & Tack, F. M. G. (2002). Cadmium and zinc uptake by volunteer willow species and elder rooting in polluted dredged sediment disposal sites. *The Science of the Total Environment, 299*, 191–205.

Vandecasteele, B., Meers, E., Vervaeke, P., De Vos, B., Quataert, P., & Tack, F. M. G. (2005). Growth and trace metal accumulation of two *Salix* clones on sediment-derived soils with increasing contamination levels. *Chemosphere, 58*, 995–1,002.

Vipulanandan, C., Wang, S., & Krishnan, S. (1994). Biodegradation of phenol. In D. L. Wise & D. J. Trantolo (Eds.), *Remediation of hazardous waste contaminated soils*. New York: Marcel Dekker.

Voss, J., Altrogge, M., Goliske, D., Kranz, O., Nünnecke, D., Petersen, D., & Waller, E. (2001). Degradation of chlorinated arenes by electroreduction. In R. Stegmann, G. Brunner, W. Calmano, & G. Matz (Eds.), *Treatment of contaminated soil*. Berlin: Springer.

Werther, J., Malerius, O., & Schmidt, J. (2001). Ways to improve the efficiency of soil washing. In R. Stegmann, G. Brunner, W. Calmano, & G. Matz (Eds.), *Treatment of contaminated soil*. Berlin/New York: Springer.

Wick, L. Y., Springael, D., & Harms, H. (2001). Bacterial strategies to improve the bioavailability of hydrophobic organic pollutants. In R. Stegmann, G. Brunner, W. Calmano, & G. Matz (Eds.), *Treatment of contaminated soil*. Berlin/New York: Springer.

Wilichowski, M. (2001). Remediation of soils by washing processes – An historical overview. In R. Stegmann, G. Brunner, W. Calmano, & G. Matz (Eds.), *Treatment of contaminated soil*. Berlin/New York: Springer.

Winterberg, R. (2001). Humification as a remediation strategy for TNT contaminations: Applications and limitations. In R. Stegmann, G. Brunner, W. Calmano, & G. Matz (Eds.), *Treatment of contaminated soil*. Berlin/New York: Springer.

Zayed, A. M., & Terry, N. (2003). Chromium in the environment: Factors affecting biological remediation. *Plant and Soil, 249*, 139–156.

Chapter 7
Groundwater, Soil Vapour and Surface Water Treatment

Abstract Contamination is not only related to the soil matrix – the media groundwater and soil vapour are also in danger of being contaminated. Hence, in the context of remediation strategies, these media are mostly taken into consideration. For groundwater treatment purposes, a number of effective clean-up possibilities, which are introduced in this chapter in detail, are offered. Apart from the widely used pump-and-treat approach (P&T), which is described together with the aboveground water purification methods, more comprehensive and in part relatively new treatment options are mentioned. Regarding P&T, different agents to improve the solubilisation and degradation of organic pollutants are discussed, for example, tenside flushing, use of hydrogen peroxide and manganese peroxide and in situ chemical oxidation. Relevant advantages and disadvantages and the interdependence with aquifer parameters such as texture and hydraulic conductivity are weighed up. Furthermore, the permeable reactive barrier (PRB) and funnel-and-gate (F&G) technologies are introduced. In relation to these treatment options, relevant reactive reagents and their impact on contaminated groundwater are chemically described. In a similar way, the approaches dealing with contaminated soil vapour are characterised. Apart from the introduction of the frequently applied soil vapour extraction (SVE), which includes the description of the aboveground vapour purification, measures like bioventing and steam-enhanced extraction (SEE) are taken into consideration. Moreover, some techniques which relate to both groundwater and soil vapour, such as air sparging, biosparging and the use of multiphase extraction wells, complement these topics. Surface water bodies are sometimes influenced by soil and groundwater contamination in different ways. Thus, in this chapter, methods of surface water restoration (lakes, rivers) are additionally mentioned.

Keywords Agent infiltration • Funnel-and-gate (F&G) • Lake restoration • Permeable reactive barrier (PRB) • Pump-and-treat (P&T) • River restoration • Soil vapour treatment (SVE)

H. Meuser, *Soil Remediation and Rehabilitation: Treatment of Contaminated and Disturbed Land*, Environmental Pollution 23, DOI 10.1007/978-94-007-5751-6_7,

7.1 Groundwater Treatment

7.1.1 Relevant Contaminants in Groundwater

In general, the danger to the environment associated with groundwater must be assessed as more problematical than the danger to the environment as a result of soil contamination, which is spatially limited. In contrast, the performance of a groundwater remediation appears to be applied in an easier way because soil does not need to be excavated, the current utilisation of the site is hardly disturbed, the future use is not impeded and the remediation can be carried out using existing structures. However, the allocation of the contaminant source to the contaminant dispersion (plume) is sometimes difficult, for instance due to altering adsorption-desorption processes between solid phase and soluble phase. An additional problem is associated with the migration of pollutants, which occurs in different directions:

- The horizontal migration depends on the hydraulic conductivity in the horizontal direction, which results from the geology and, in particular, the pore system of the geological formation and the groundwater inclination.
- The vertical migration depends on the same geological features in the vertical direction but also on the specific gravity of the contaminants.

The most-treated pollutants with reference to the groundwater problem are volatile chlorinated hydrocarbons (VCHC), monocyclic aromatic compounds such as benzene, toluene, ethyl benzene, xylene (BTEX) and phenols, as well as inorganic anions such as cyanides, arsenate and chromate.

Heavy metals are usually of less importance because their high adsorption potential accounts for the relatively low mobility of the metals. Theoretically, it is feasible to enhance the mobility in a contaminated layer based on the addition of acids to the unsaturated zone. Hence, the leaching can accelerate significantly. However, this technique stemming from the so-called solution mining (see Sect. 3.3.3), where it is applied in aerobic conditions to mobilise e.g. copper, gold, silver and uranium, is rarely used in the context of soil and groundwater remediation.

After downward percolation of the contaminated leachate, the contaminated water reaches the groundwater surface and afterwards it might migrate laterally in groundwater flow direction. In the vicinity of the source the concentration of the plume shows enhanced values, which decrease with increasing distance to the source. Lengths and dispersion of the plume differ enormously, depending on the hydrogeological conditions present. The plan view in Fig. 7.1 illustrates the contaminated plume schematically. The groundwater flow direction follows the isopiestic lines.

With regard to different contaminants introduced to the aquifer, the specific gravity of the liquid appears to be important. Liquids exceeding 1.0 g cm^{-3} such as chlorinated hydrocarbons tend to sink to the bottom of the aquifer in the course of time, leading to an accumulation in the deeper part of the aquifer that can be several

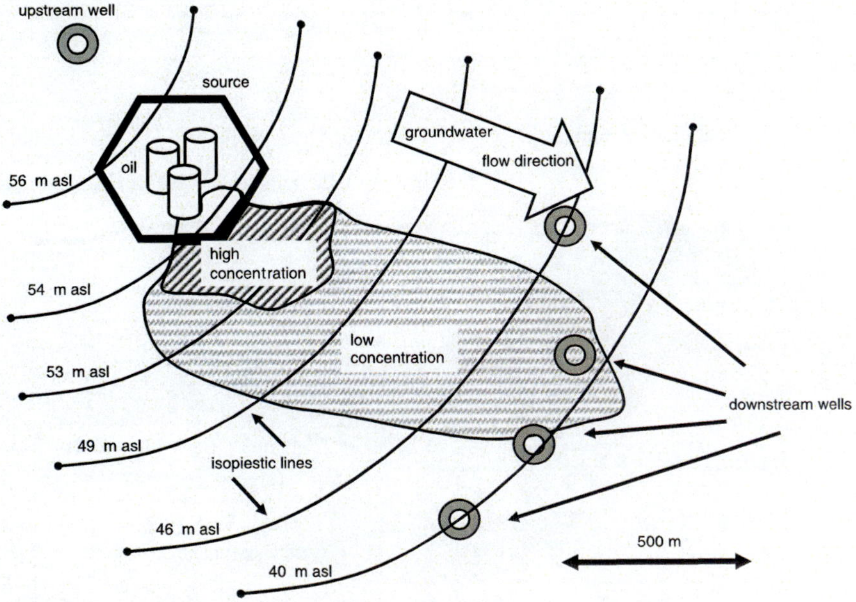

Fig. 7.1 Dispersion of a contaminated groundwater plume

metres thick. In contrast, liquids falling below 1.0 g cm^{-3} such as mineral oil might rise to the surface of the groundwater.

Furthermore, the contaminated liquids can have two different physico-chemical states. On the one hand, the contaminated liquid can dissolve in the aquifer, forming the water-soluble fraction. On the other hand, the liquid does not react with the water, leading to the presence of an independent phase of the contaminant called the non-aqueous phase liquids (NAPL). Due to different viscosities contaminated NAPL and groundwater do not come into contact with each other and the downstream migration of both occurs separately. In an analogous way to the soluble contaminants, the non-aqueous phase liquids can be differentiated into phase liquids with a specific density higher than water (dense non-aqueous phase liquids – DNAPL) and with a specific density less than water (light non-aqueous phase liquids – LNAPL). The effects are illustrated in Fig. 7.2.

Irrespective of the remediation technique applied, the general success of remediation is affected by a number of contaminant and groundwater properties, as summarised in Table 7.1. The likelihood of success depends predominantly on the interaction between the homogeneity of the aquifer and the contaminant chemistry (Table 7.2). Homogeneous aquifers consisting of only one layer are relatively easy to treat as long as DNAPL are components in small quantities. In contrast, heterogeneous aquifers with multiple layers and fractured or karst aquifers might be generally difficult to treat, in particular in the presence of non-aqueous phase liquids.

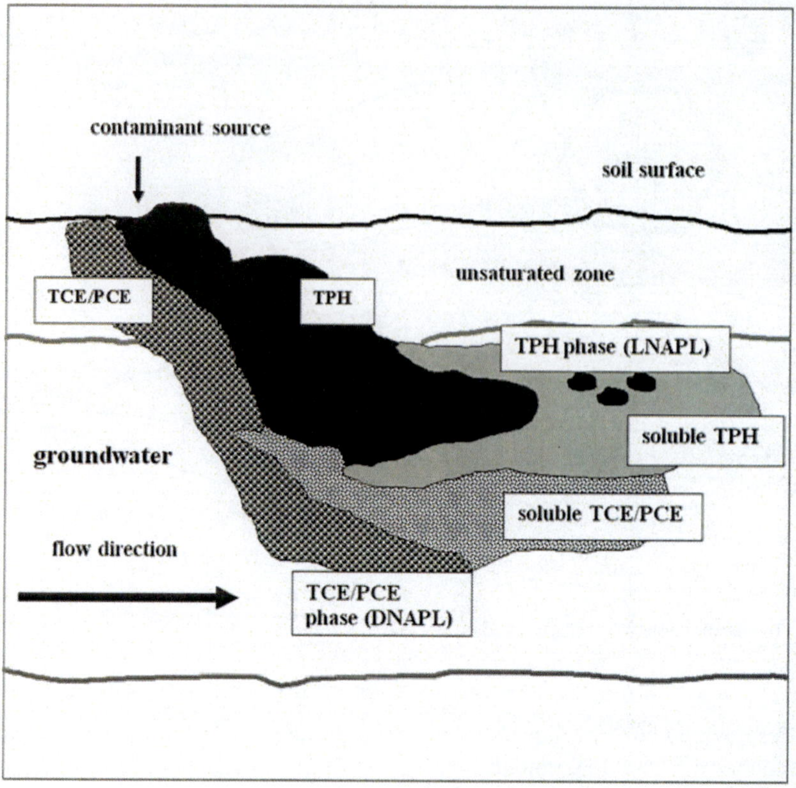

Fig. 7.2 Different distribution of water-soluble contaminants, light non-aqueous phase liquids (*LNAPL*) and dense non-aqueous phase liquids (*DNAPL*) (Data from Meuser 2010, modified)

7.1.2 Passive Groundwater Remediation

Apart from a collection and remediation of the contaminated water, the groundwater quality can be influenced in a passive way, particularly if extraction wells for drinking-water purposes are located in the direction of the contaminated plume. There are different solutions which are applied to the contaminated groundwater without subsequent aboveground decontamination:

- Construction of infiltration wells which are used for the infiltration of fresh water in a great quantity, so that the groundwater flow direction is changed away from the source of contamination
- Construction of guide walls (see Sect. 5.2.1) by means of which the groundwater may flow without coming into contact with the contaminated source
- Groundwater lowering, avoiding direct contact between the basis of the contaminated source in the unsaturated zone and the groundwater below.

Table 7.1 Factors affecting groundwater remediation effectiveness (Data from Delleur 2007, modified)

Characteristic	High effectiveness	Low effectiveness
Contaminants		
Volume	Small	High
Duration of release	Short	Long
Biodegradation potential	High	Low
Sorption potential	Low	High
Contaminant phase	Aqueous	LNAPL, DNAPL
Aquifer		
Volume of contaminated groundwater	Small	Large
Depth of contaminated groundwater	Shallow	Deep
Stratigraphy	Pore groundwater	Discontinuous strata, fractural and karst groundwater
Texture	Gravel, sand	Cohesive textures
Hydraulic conductivity	High (>10^{-2} m s^{-1})	Low (<10^{-6} m s^{-1})
Heterogeneity	Homogeneous	Heterogeneous
Vertical flow	Little	Large downward flow

7.1.3 *Pump-and-Treat System (P&T)*

The pump-and-treat system is still the most applied groundwater clean-up technology. To install a pump-and-treat system exact knowledge about the water table contour is required in order to site the extraction wells. The recovery well network is either designed to capture contaminated water from the centre of the plume, where a high contaminant concentration is expected, or from the leading edge of the plume, where the concentration might be lower but where the spread of the plume can be minimised in a better way. For the latter option the wells are sited in a semi-circle in proximity to the plume but the numbers of extraction wells needed vary considerably due to the dispersion of the plume and hydrological features such as hydraulic conductivity and aquifer thickness (Fig. 7.1). The different kinds of migration termed mass flow (convection), diffusion and hydrodynamic dispersion must be taken into consideration. The plume lengths usually range from 300 to 4,000 m and the number of extraction wells amounts to 3–7 on average and up to a maximum of 25. An optimum configuration of the wells should be based upon hydrogeological simulation models. It should be taken into account that the construction of the wells can cause a downward transfer of contaminants to the groundwater.

Apart from the downstream wells, at least one upstream well should be constructed to get information about the groundwater quality beyond the source of contamination. In urban and industrialised areas one can expect that the upstream groundwater consists not only of pure water. Indeed, there are a number of examples indicating contaminated groundwater upstream, since other sources are present which influence the contaminated plume concerned.

The different specific gravities of the contaminants should be taken into consideration with reference to the planning and localisation of the extraction well

Table 7.2 Interaction between hydrogeological conditions and the contaminant characteristics (Data from Delleur 2007, modified)

Hydrogeology			Contaminants			
Type	Homogeneity	Groundwater layers	Mobile, dissolved	Strongly adsorbed, slowly dissolved	LNAPL	DNAPL
Pore aquifer	Homogeneous	One	F	U-V	U-V	V
Pore aquifer	Homogeneous	Multiple	F	U-V	U-V	V
Pore aquifer	Heterogeneous	One	U	V	V	I
Pore aquifer	Heterogeneous	Multiple	U	V	V	I
Fractured aquifer, karst aquifer	Heterogeneous	–	V	V	I	I

F remediation technically feasible, *U* remediation uncertain, *V* remediation very uncertain, *I* remediation technically infeasible

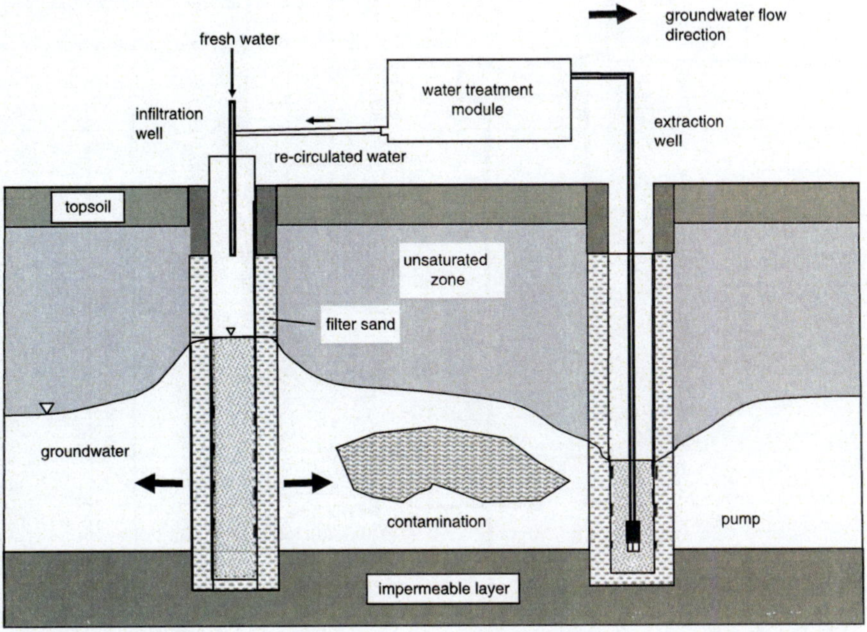

Fig. 7.3 Groundwater pump-and-treat system (*P&T*) combined with a re-circulation process

vperforation, thus avoiding the pumping of only less contaminated water, since the perforated filtering should collect the strongly polluted groundwater.

Pump-and-treat systems are focused on the plume rather than the source decontamination. As shown in Fig. 7.3, an extraction well, or usually a series of extraction wells, is installed. This pumps water continuously out of the aquifer, which is afterwards treated aboveground. The effectiveness of the system can be improved by re-infiltration of the treated water with injection wells. In relation to surface-near aquifers the infiltration can occur by means of ditches. The permanently re-circulated water might accelerate the solubilisation of the contaminants, in particular if the groundwater flows directly through the contamination (*in situ* flushing). In spite of the circulating system a complete removal of the contaminants is rarely achieved even over decades but the spread of the contaminants might be stabilised in any case. Hence, the prevention of plume spreading aimed for might be evaluated as a containment strategy.

Infiltration and extraction wells have a diameter of 100–200 mm, while observation wells mostly measure 50 mm in diameter only. The latter are used for the measuring of the groundwater table by light plummet or well whistle. The wells are made of hard-density polyethylene (PEHD), which is preferred to PVC, since PVC can possibly be damaged by the soluble pollutants. After drilling of the borehole and the setting up of the perforated plastic tube, the distance between borehole and tube must be backfilled with filter sand, which is adapted to the surrounding soil material. Sandy soils require fine gravelly filter material, while more cohesive soils

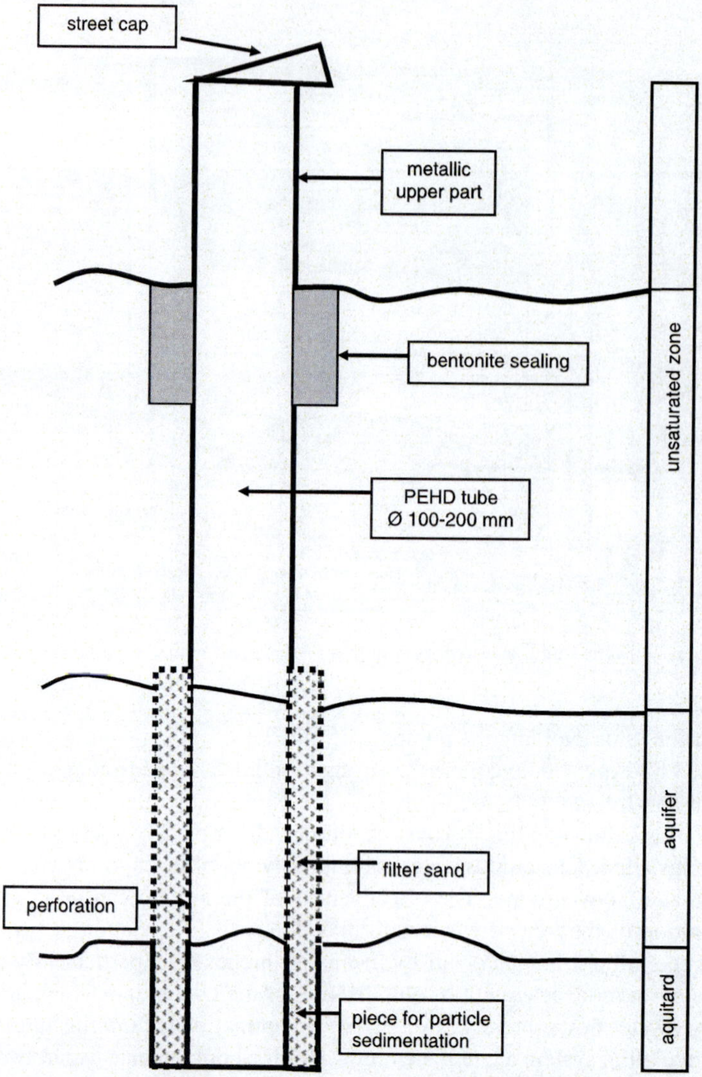

Fig. 7.4 Design of a groundwater extraction well

are backfilled with sandy material. At the top the surface surrounding the well is waterproofed using e.g. bentonite in order to prevent downward percolation of rainwater alongside the tube. Furthermore, the tube should be capped (street cap) for the same reason. The visible aboveground parts of the wells are usually made of metal. Nickel-based commercial alloys should be used to minimise corrosion (Fig. 7.4).

Beneath buildings and other aboveground structures, where the installation of vertical wells is not possible due to the limited accessibility caused by underground

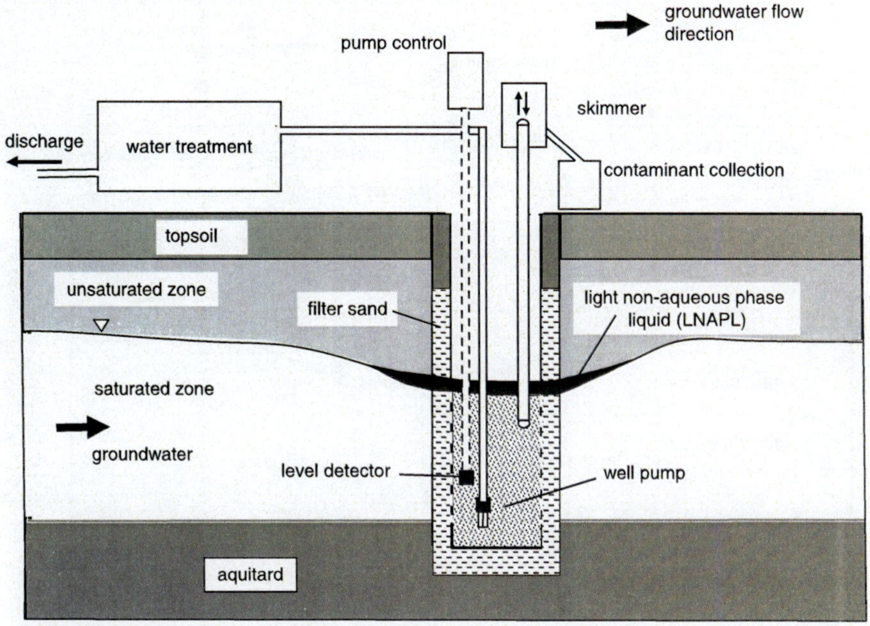

Fig. 7.5 Well system for separate extraction of contaminated water (pump system) and LNAPL (skimmer technology)

utilities and subsurface obstructions, the horizontal well technology can be an alternative. The wells are drilled at a sloped angle until they reach the aquifer.

The pump-and-treat approach is limited in the presence of non-aqueous phase liquids, particularly after the pumps are switched off. For this reason, in the case of LNAPL a separate extraction is feasible by using specialised technologies. Firstly, the LNAPL can be removed with the metallic skimmer technology, which comes into contact with the floating phase, adsorbs it and moves it upwards (Fig. 7.5). Alternatively, a diaphragm pump can be used, which directly pumps the LNAPL upwards (Fig. 7.6). The pump technique can be combined with the biological treatment of the vadose zone. Vacuum-enhanced pumping enables LNAPL to be lifted off the water table and the simultaneously drawn air improves the bioremediation of the unsaturated zone above analogously to the bioventing approach (see Sect. 7.2.4). Therefore, the recovery of phase products is associated with an enhanced bioremediation process. The combinated technique is turned bioslurping.

In the case of a simultaneous pumping of groundwater and LNAPL a depression cone develops, in which the non-aqueous phase liquids merge. During the pumping operation, when the LNAPL level decreases in response to pumping, the slurp may extract vapour. The extracted liquids (oil and water) and vapour must be differentiated aboveground using the oil-water separator and liquid-vapour separator technique, which is also required for the bioslurping technique (see Fig. 7.21).

In case studies it has observed that the LNAPL may reach enormous thicknesses. In an example in Berlin, Germany, dealing with the remediation of a former

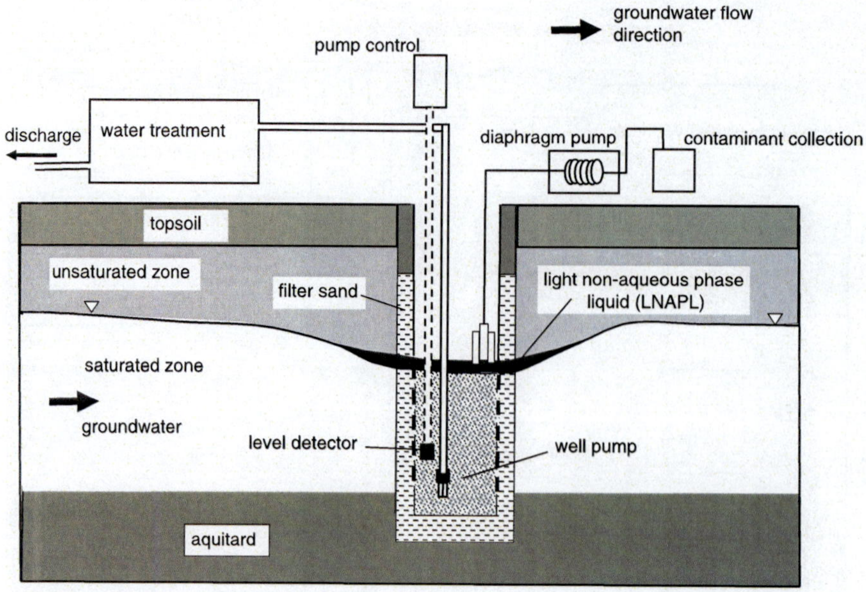

Fig. 7.6 Well system for separate extraction of contaminated water (pump system) and LNAPL (diaphragm pump)

transformer factory site, which started production in 1897 and closed in 1990, thicknesses of up to 111 cm consisting of PCB-containing oils (LNAPL) have been found. Two separate LNAPL plumes covered 2,600 and 3,300 m^2 and the oil phases fluctuated in the course of time but thicknesses of more than 40 cm have frequently been measured. Between 1996 and 2001 52,250 l of oil was extracted with the help of a submerged pumping device. After extraction the LNAPL were pumped into a sludge trap for sedimentation purposes. Afterwards, an oil-water separator was used, followed by a filter device based on activated carbon. To prevent damage to the activated carbon filter additional adsorber equipment (two sand filters) was previously included (Anonymous 2001).

During the pump-and-treat operation the contaminant concentration should decrease continuously until the clean-up value aimed for is achieved. The decrease, however, does not occur gradually, since adsorption and desorption processes as well as inhomogeneous groundwater flow influence the concentration of the contaminants in the groundwater extraction wells. In Fig. 7.7 the BTEX concentration of three extraction wells placed in the plume in groundwater flow direction at a former petrol station in Brondby, Denmark, is presented. The contaminant development did not exhibit a continuous decrease and the effectiveness of well No. 3 appeared to be higher than that of well No. 2, although both wells indicated were located at a comparable distance to the source. After 7 years the yielded concentrations achieved clearly exceeded the ascertained quality standard of 500 µg L^{-}.

Unfortunately, the slow and long-term pollutant reduction and the fluctuating values have often been observed in the execution of the pump-and-treat method.

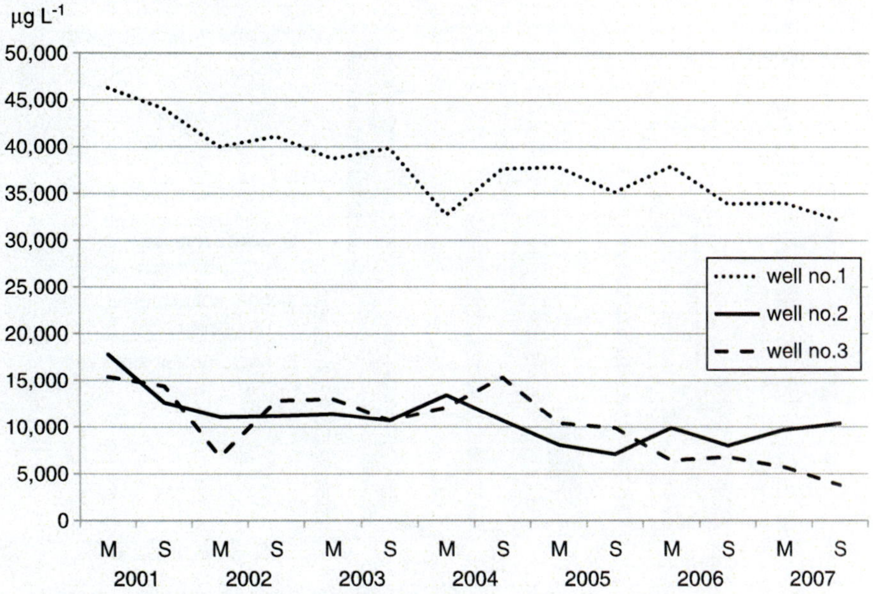

Fig. 7.7 Development of the BTEX concentration in groundwater extraction wells at a former petrol station in Brondby, Denmark (unpublished data). *Well no. 1*: placed at a distance of 20 m to the source (160 m^2), *well nos. 2 and 3*: placed at a distance of 70 m to the source Aquifer: Quaternary till, hydraulic conductivity 5×10^{-6} m s^{-1}, thickness 4 m, *M* March, *S* September

In many case studies, in spite of a long-term application, it was not possible to achieve the clean-up goal concentrations (Fig. 7.8). Within approximately the first 15 years it is normally feasible to reduce the contaminant concentration by 80% but a further reduction to, for instance, 99%, which is often required in order to achieve the quality standards decided, is very time-consuming and may increase the costs of a conventional pump-and-treat system to an unacceptable amount (Fig. 7.9). Therefore, in many projects the remediation success only referred to a reduced spreading of the contaminants (containment) (Delleur 2007). Consequently, the simple pump-and-treat approach has been modified in different ways, as introduced in Sect. 7.1.6.

In association with the planning and execution procedures of the groundwater treatment a lot of factors have to be taken into consideration, as listed below. Firstly, the pump-and-treat applications are mostly connected with a lowered groundwater table or at least altered groundwater flow direction. In particular, in areas of potential aridity, the reduced water availability to the plant roots can cause ecological damages to wet biotopes. In agriculturally used areas the lack of water resulting from missing capillarity due to the lowered groundwater table is possibly responsible for reduced crop yield. In proximity to the extraction wells depression cones caused by pumping and by nearby injection wells an elevated groundwater table resulting from the water infiltration will become visible.

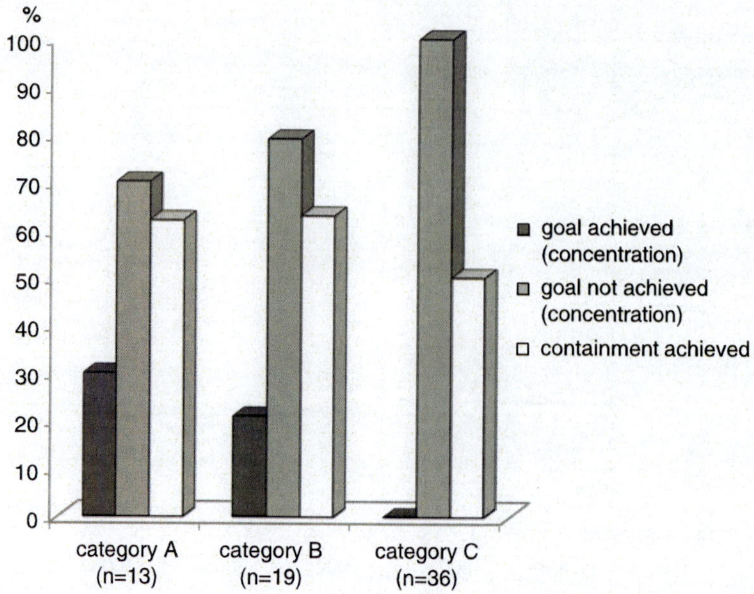

Fig. 7.8 Percentage of achieved clean-up goals with respect to differently contaminated aquifers treated by the pump-and-treat approach in the USA (Data from NRC 1994 [cited in Delleur 2007], modified). *Category A*: homogeneous and heterogeneous aquifers/mobile and dissolved contaminants. *Category B*: homogeneous, heterogeneous or fractured aquifers/potentially adsorbed but dissolved contaminants or LNAPL. *Category C*: heterogeneous or fractured aquifers/DNAPL

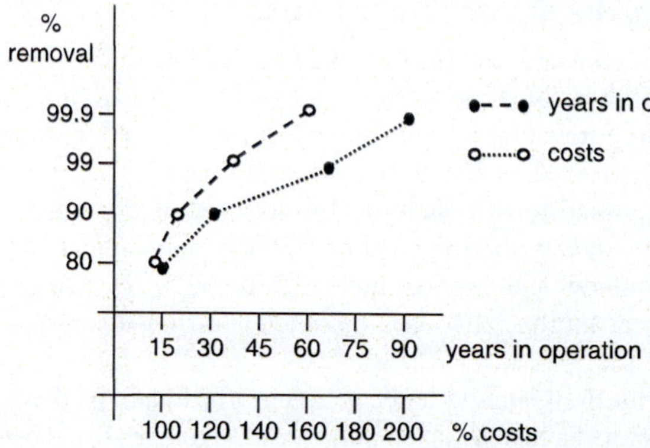

Fig. 7.9 Interaction between contaminant removal (%), costs (%) and calculated remediation time (years) for a conventional pump-and-treat system (Data from Delleur 2007, modified)

The width of the zone of influence which determined the depression cones can be calculated as follows:

$$W = Q / Ti \tag{7.1}$$

where W is the width of the influenced zone (m), Q is the pumping rate (m^3 s^{-1}), T the transmissivity of the aquifer (m s^{-1}) and i the dimensionless hydraulic gradient (Delleur 2007).

Moreover, the planning of the groundwater treatment requires exact knowledge about the unsaturated as well as the saturated zone. Based on hydrogeological studies or available maps in the run-up to the treatment, soil type and porosity of the unsaturated zone, but also aquifer parameters, must be estimated. The latter include depth, storage, hydraulic conductivity, thickness, flow direction and heterogeneity of the aquifer. In particular, the texture of the water-bearing stratum has a decisive significance, since the efficiency of the extraction well system located in a loamy, silty or clayey texture is strongly reduced. Aquifers consisting of an enhanced silt or loam content are only treatable when the flushing model (extraction-injection well network) is preferred. Furthermore, detailed knowledge must be gained about the properties of the contaminants of concern, in particular about sorption potential, water solubility, distribution rate and the presence of phase liquids. If one wants to plan a pump-and-treat system as well as an extraction-injection well network, the parameters mentioned must be well-known in detail.

Finally, in this regard the long-term treatment usually necessary should be taken into account, because the technical system might be negatively affected with an increasing tendency in the course of time. In particular, damage to the well systems caused by the precipitation of iron and manganese cannot be excluded in aquifers naturally containing a high iron (Fe^{2+}) and manganese (Mn^{3+}) concentration. Groundwater treatment in association with infiltration of oxygen-containing chemicals (see Sect. 7.1.6) might lead to accumulated iron and manganese precipitation, clogging and blocking the well equipment. In iron-rich groundwater the well systems are endangered by precipitation of $Fe(HCO_3)_2$.

7.1.4 Aboveground Groundwater Purification

If extracted and subsequently treated groundwater is provided for reintroduction into the public water supply, it is expected that certain groundwater standards and criteria are complied with. For this reason, special attention must be paid to the aboveground module dealing with the decontamination of the extracted groundwater. In spite of the techniques available, in most cases it is not possible to achieve drinking water standards.

With reference to the iron and manganese precipitation as mentioned above the extracted groundwater enriched with soluble iron usually present as $Fe(HCO_3)_2$ should implement a deferrisation module. The continuous removal of $Fe(HCO_3)_2$ occurs by means of oxygen and can be described as:

$$4Fe(HCO_3)_2 + O_2 + 2H_2O \rightarrow 4Fe(OH)_3 + 8CO_2 \tag{7.2}$$

Fig. 7.10 Ion exchanger for selective adsorption of heavy metals, located at a remediation site in Utrecht, The Netherlands

The contaminated groundwater extracted is treated on the basis of different technologies normally well-known for wastewater management. In the first instance, the use of activated carbon with reactive surface where the organic pollutants are preferentially adsorbed is widespread. Granular activated carbon is suitable for most of the organic pollutants including explosives and pesticides but the best results have been obtained in association with volatile organic compounds (Mirsal 2004). Disadvantageously, the monitored regeneration of the activated carbon must be carried out.

Also frequently applied is the principle of precipitation, where a phase transfer from soluble ions into a non-soluble phase occurs. The precipitation of metal sulphides has been proven to be very successful. The reduction by sodium sulphide (Na_2S) or hydrogen sulphide (H_2S) is exclusively applied to contamination by heavy metals such as cadmium, chromium and lead. Some metals that are less efficient with regard to precipitation such as copper, mercury and zinc are additionally removed by activated carbon filters, which are used in combination with the precipitating agents (Förstner 1998).

The alternatively applied exchangers are based on a resin laden with dissociable counter ions. The selective ion exchangers can be applied to various heavy metals and organic pollutants (Fig. 7.10). The resins are made of polystyrene, polyacrylate, phenol and formaldehyde. In general, a number of exchangers are available. There are resins which increase the cation exchange capacity and thus enhance the binding of negatively charged contaminants on the positive surfaces. Furthermore, basic macroporous resins for chromates and cyanides are used and, in addition, selective resins using carbic acid and chelate exchangers for the elements Cd, Co, Cu, Ni and Zn are distinguished. Non-ionic resins made of styrene are taken for organic pollutants. The process usually occurs in a counter-current way and the exchange proceeds from below. The exchanger must be regenerated from time to time, using, for instance, acids and alkaline solutions like NaOH (Förstner 1998).

Flocculation and filtration, which are usually important components of wastewater works, are very seldom considered in relation to the treatment of extracted groundwater. Notwithstanding, flocculation (coagulation), a process in which non-soluble small contaminated particles are aggregated into larger flocs, is another suitable approach for purifying extracted contaminated water. Afterwards, the flocks are separated mechanically. This can be done by flotation, where air bubbles which are attached to the flocs, are injected into the water. The flocs are carried to the surface, where a surface skimming device is used. The coagulation is promoted with the help of different agents such as calcium hydroxide and iron hydroxide (at high pH values) as well as FeIII salts and Al salts at low pH values. Some additives like activated alumina or bentonite may enhance the agglomeration significantly. Moreover, polymer flocculants are added later on to enable the establishment of macro-flocs, which are easier to separate. Organic flocculants may preferentially form bridge-like structures between the particles. Gravity filtration and downstream or upstream filtration are also taken into account with reference to the remediation of extracted contaminated groundwater (Hahn and Klute 1990; Bradl 2005). Moreover, suspended particles can be separated by filter beds consisting of sand or diatomaceous earth.

Additionally, for particles varying between 0.1 μm and 1.0 mm, membrane filters made of different materials such as polyamide, polyester, cellulose acetate, etc. are used. The pressure filtration is the most widely applied technique. A distinction is made between microfiltration working with 0.5–3 bar pressure, ultrafiltration working with 1–10 bar pressure and reverse osmosis operating with 20–100 bar pressure. The latter technique works with external pressure across a semi-permeable membrane to reverse the osmotic flow (Hahn and Klute 1990; Bradl 2005).

In association with organic pollutants oxidation by hydrogen peroxide (H_2O_2) or ozone (O_3) is also frequently applied aboveground, particularly for petroleum hydrocarbons (TPH) and chlorinated hydrocarbons (CHC). The chlorinated compounds frequently need further treatment of the soluble chloride anions. It should be noted that most of the aboveground techniques for groundwater purification produce waste that must be removed or additionally treated (Table 7.3) (Delleur 2007).

Mixtures of water, LNAPL and gaseous constituents are also collected and treated aboveground. Accordingly, a liquid separator is used to separate the gaseous components of the collected mixture. This can be implemented by a blower system disassociating the exhaust air, which is later purified. Moreover, an oil-water separator is necessary to differentiate oil and water. The oil is collected in an oil tank, which must be treated, for instance, by the activated carbon absorber (see Sect. 7.2.6).

Table 7.3 Aboveground treatment technologies for extracted contaminated groundwater and their residual waste produced during the operation (Data from Delleur 2007)

Treatment technology	Residual waste
Carbon adsorption	Replaced activated carbon that must be regenerated
Precipitation	Hazardous sludge
Ion exchange	Solution produced during regeneration
Coagulation (flocculation)	Hazardous sludge
Filtration	Backwash waste
Ultra-filtration (membranes)	Concentrated brine
Reverse osmosis	Concentrated liquid waste
Oxidation (H_2O_2, O_3)	Mostly none

7.1.5 Air Stripping

As shown in Fig. 7.11, for volatile chlorinated hydrocarbons (VCHC) an alternative method is applied based on the contaminant adsorption onto air bubbles. The extracted and contaminated groundwater is pumped to the top of an up to 4 m high tower filled with ceramic or plastic ingredients. While the water slowly trickles downward, fresh air is pumped under pressure into the tower in the opposite direction from the bottom to the top, causing the development of air bubbles to a great extent. The ratio between water and air ranges from 1:10 to 1:20. The chlorinated hydrocarbons tend to be adsorbed onto the surface of the bubbles, which may rise upward. After seeping through at the top the air is collected and treated by activated carbon (see Sect. 7.2.3). The water treatment capacity is up to 120 $m^3\ h^{-1}$. This common technique is termed air stripping.

The contamination potential of the groundwater is considerably reduced. In the case of insufficient decontamination, however, the process can be applied once again. A long-term operation causes precipitation of iron and manganese hydroxides, which have to be removed occasionally. The costs for the disposal or treatment of the iron-manganese sludge must be taken into consideration.

7.1.6 Enhanced Groundwater Remediation (Agent Infiltration)

The simplest way of making an impact on contaminants in groundwater might be the change of the water temperature but the artificial temperature rise to make organic pollutants more mobile is seldom applied due to the high energy expenditure. Temperature changes are of more importance with reference to the improvement of the soil vapour extraction method. Therefore, this topic is discussed in more detail in Sect. 7.2.

The main problem of the non-aqueous phase liquid removal is their differing viscosity, which is responsible for the general inertia compared to water. For this reason, some agents such as alcohol and tensides can be applied to improve the

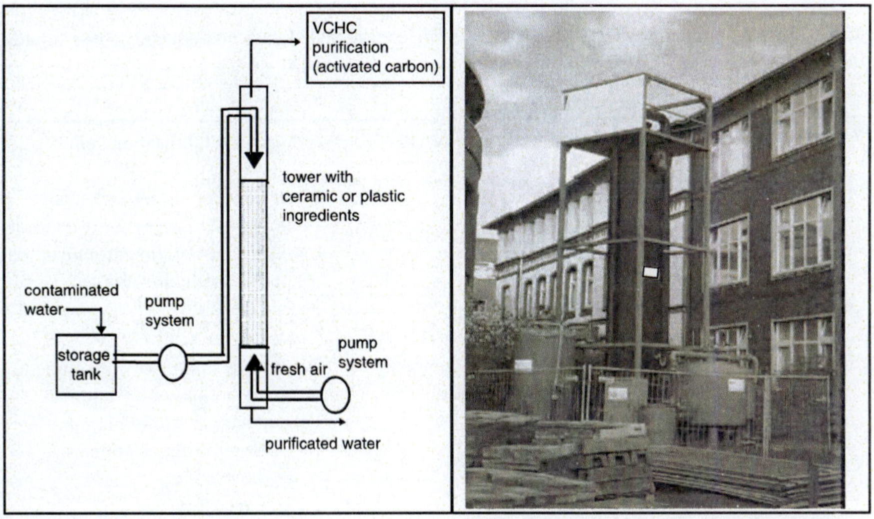

Fig. 7.11 Air stripping tower shown as drawing (*left*) and photograph (*right*)

liquid movement in the aquifer, which is contaminated predominantly with VCHC and petroleum hydrocarbons (TPH). In Table 7.4 the alcohol and tenside flushing is explained in detail. Both agents cause an enhanced solubilisation of the NAPL and are possible for both light and dense non-aqueous phase liquids. The application is limited to gravelly and sandy aquifers, and it is focused on the source clean-up. Attention should be paid to some detrimental impacts, e.g. the possible elevated downward sinking of the contaminants, tube blocking after amendment of tensides and the problematical recovery of the agents used. In some cases an additional flushing well is installed, which is located close to the contaminated hot spots. In groundwater flow direction extraction wells collect the water influenced by the flushing well. In order to control undesired spreading of the agents a series of observation wells must be constructed sideways. Alcohol and tenside flushing looks back on only a few case studies outside of the USA, where this technology has been frequently tested since the 1990s.

Because of accelerated solubilisation the possibility to biodegrade organic pollutants such as VCHC is improving. In a field investigation in Rheine, Germany, the injection of 27 m^3 (1st injection) and 63 m^3 (2nd injection) solution containing 0.5–1% tensides into a sandy Quaternary aquifer indicating a hydraulic conductivity of 10^{-5}–10^{-7} m s^{-1} led to a decrease in the less degradable component tetrachloroethylene, whereas biological degradation products like trichloroethylene and dichloroethylene tended to accumulate (Fig. 7.12) (Anonymous 2010). The tetrachloroethylene degradation took place in reductive conditions promoted by the bacterium *Dehalococcoides ethenogenes.* This was after previous injection of molasses, the biodegradation of which consumed the oxygen completely (see below).

Organic pollutants, which, in principle, are biodegradable, can also be treated in the aquifer biologically. The degradation is mainly promoted by aerobic or

Table 7.4 Alcohol and tenside flushing in aquifers contaminated with non-aqueous phase liquids (*NAPL*)

	Alcohol flushing	Tenside flushing
Principle	Enhanced solubilisation of NAPL in an alcohol-water mixture	Enhanced solubilisation and flowability of NAPL
Means	Alcohol cocktail (lipophilic, hydrophilic)	Water concentration of tensides 0.5–2%
Contaminants	LNAPL, DNAPL	LNAPL, DNAPL (in particular, petroleum hydrocarbons)
Location	Source clean-up	Source clean-up
Texture	Gravel to fine sand	Gravel to fine sand
Problems	Downward sinking of contaminants to the bottom of the aquifer Alcohol recovery	Downward sinking of contaminants to the bottom of the aquifer Tube blocking/aquifer clogging (biological degradation of the tensides) Tenside recovery

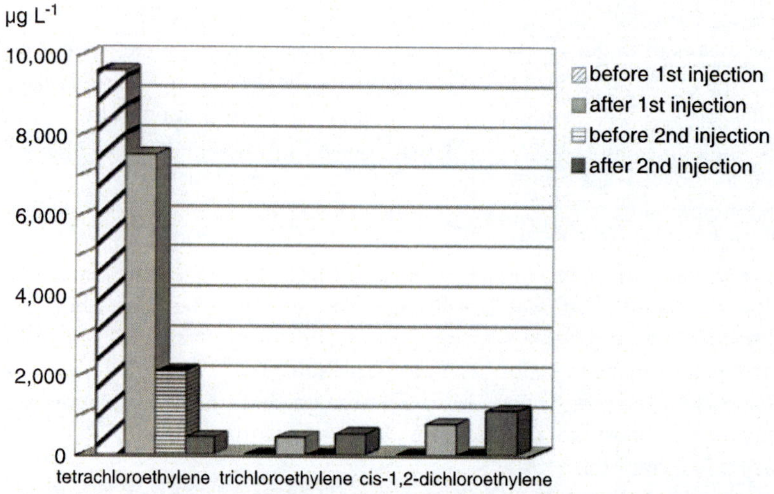

Fig. 7.12 Alteration of the volatile chlorinated hydrocarbon (*VCHC*) concentration (µg L^{-1}) in a downstream groundwater well after tenside flushing in Rheine, Germany (Data from Anonymous 2010)

methanogenic bacteria. With reference to the biological treatment some agents must be amended. The latter are fed carefully into the groundwater via wells. The agents generally help to stimulate the biodegradation process in the groundwater plume. For contaminants such as petroleum hydrocarbons (TPH), monoaromates (BTEX) and Polycyclic Aromatic Hydrocarbons (PAH) the degradation process must be conducted aerobically (Neilson and Allard 2008). Accordingly, pure oxygen or oxygen containing reactive agents (e.g. nitrate, iron hydroxide, sulphate) which act as electron acceptors are used.

In a similar way to the pump-and-treat approach introduced in Sect. 7.1.3 the amendment of liquid agents can be combined with an infiltration-extraction network. Infiltration galleries may re-inject previously extracted groundwater to generate a circulation of oxygen-amended groundwater through the contaminated zone. It is supposed that the remediation time will be shorter than a pump-and-treat system, which avoids the amendment of agents.

In the following formulae the benzene (C_6H_6) degradation is exemplified:

$$\text{Aerobic respiration: } 7.5O_2 + C_6H_6 \rightarrow 6CO_2 + 3H_2O \tag{7.3}$$

$$\text{Denitrification: } 6NO_3^- + 6H^+ + C_6H_6 \rightarrow 6CO_2 + 6H_2O + 3N_2 \tag{7.4}$$

$$\text{Iron reduction: } 60H^+ + 30Fe(OH)_3 + C_6H_6 \rightarrow 6CO_2 + 30Fe^{2+} + 78H_2O \tag{7.5}$$

$$\text{Sulphate reduction: } 7.5H^+ + 3.75SO_4^{2-} + C_6H_6 \rightarrow 6CO_2 + 3.75H_2S + 3H_2O \tag{7.6}$$

In urbanised and industrialised areas the sulphate concentration in the groundwater is mostly high in contrast to nitrate, which is more typical for agriculturally used areas. The accelerated sulphate concentration is associated with the deposits of technogenic substrates, in particular construction debris (gypsum, mortar), slag, ashes and coal mining waste. Consequently, a high sulphate capacity reacting with organic pollutants in the aquifer is generally assumed to be present. Thus, the opportunity for an effective biodegradation in the presence of sulphate is basically provided in a high number of contaminated areas (Meuser 2010).

Furthermore, the benzene degradation can be carried out methanogenically. The methanogenic degradation of benzene can be described according to the formula:

$$4.5H_2O + C_6H_6 \rightarrow 2.25CO_2 + 3.75CH_4 \tag{7.7}$$

Regarding CH_4 a methane-air-mixture is also infiltrated to stimulate the biological degradation of short-chain chlorinated hydrocarbons such as dichloroethylene and trichloroethylene in aerobic conditions. The methane concentration varies between 0.5 and 2.2% so that it falls below the explosion limit. The treatment is applied to homogeneous and permeable aquifers. Because of stripping effects a soil vapour extraction module is usually combined. In a similar way to the infiltration of other agents previously mentioned it is necessary to observe the groundwater flow downstream in order to prevent uncontrolled dispersion of the agent concerned.

In general, the organic parameters are correlated to the agents differently. In Table 7.5 the agents functioning as electron acceptors are listed in association with the targeted organic pollutants. Most of the parameters apart from some volatile chlorinated hydrocarbons can be treated by oxygen injection into the aquifer, while nitrate, iron and manganese as oxidising agents are mainly restricted to some monoaromates like benzene and toluene. Positive results have been clearly observed

Table 7.5 Relationship between oxidising agents injected to the aquifer and organic pollutants that must be degraded (Data from van Agteren et al. 1998)

Parameter	O_2	NO_3^-	Fe^{3+}	Mn^{4+}	SO_4^{2-}	CO_2
Benzene	+	**+co**	**+co**	–	+	+
Toluene	+	+	+	**+co**	+	+
Ethyl benzene	+	**+co**	–	–	+	**+co**
Xylene	+	+	–	–	+	**+co**
Vinyl chloride	+	–	–	–	**+co**	**+co**
cis-1,2-dichlorotheylene	**+co**	–	–	–	**+co**	**+co**
Trichloroethylene	**+co**	–	–	–	**+co**	**+co**
Tetrachloroethylene	–	–	–	–	**+co**	**+co**
Dichloromethane	+	+	–	–	–	+
Tetrachloromethane	**+co**	–	**+co**	–	**+co**	**+co**
Naphthalene	+	**+co**	–	–	**+co**	**+co**
Benzo(a)pyrene	**+co**	–	–	–	–	–

+ beneficial
+co beneficial with co-substrates
– detrimental

in the context of sulphate, in particular with respect to VCHC and monoaromates. Moreover, induced carbon dioxide succeeded as well based on the same contaminants.

For a number of contaminants it was only possible to guarantee the beneficial treatment in the presence of co-substrates serving as an additional carbon source. The agents are only successfully applied in homogeneous aquifers with moderate to high permeability (>10^{-5} m s^{-1}) and missing phase liquids. Optimised biodegradation is expected at a pH value between 5 and 9 and a nutrient supply in the order of 100:10:1:1 (C:N:P:K) (EPA 2004). Aquifers containing sulfide might not be adequate media for treatment due to the possible sulphide oxidation in line with increasing acidification.

The aerobic biodegradation of monoaromates occurs enzymatically and is predominantly carried out by the bacterial strains *Acinetobacter* sp. and *Pseudomonas* sp. in the groundwater. Close relationships have been found between some monoaromates such as toluene and specific bacteria (*Azoarcus tolulyticus*). With regard to the PAH the aerobic degradation may play the major role. PAH consisting of two, three or four rings are aerobically degradable with the help of the bacterial strains *Acinetobacter* sp., *Bacillus* sp. and *Pseudomonas* sp. (van Agteren et al. 1998).

Affinities between bacteria and the pollutant group VCHC have also been discovered. For instance, the aerobic degradation of dichloroethylene and vinyl chloride is associated with the bacterium *Methylosinus trichosporium* and tetrachloroethylene was decomposed by *Dehalospirillium multivorans*. However, it should be noted that VCHC are relatively well biodegradable in anaerobic conditions as well. Consequently, most of the bacteria are able to biodegrade some VCHC in both anaerobic and aerobic conditions. For example, chloromethanes were treated by *Hyphomicrobium* sp. irrespective of the oxidative or reductive environment.

As already mentioned, particularly the volatile chlorinated hydrocarbons such as trichloroethylene and tetrachloroethylene can be degraded in the presence of co-substrates. In general, the degradation process of these pollutants occurs anaerobically. The co-substrates, namely molasses, alcohol and lactic acid are used for this purpose. They produce protons (H^+) which are oxidised, while the VCHC are gradually reduced. The degradation is driven by the bacterial strain *Dehalococcoides ethenogenes*. The treatment has often been carried out in homogeneous and permeable aquifers to clean up predominantly the pollutant plume. During VCHC decomposition attention should be paid to possible toxic metabolites and dead-end-products such as vinyl chloride, which in a number of case studies did not take part in a continuing degradation process.

The insufficiently degradable compounds such as tetrachloroethylene and tetrachloromethane, which must only be decomposed anaerobically, are sometimes treated in the presence of lactic acid polymer added to the infiltration well. The degradation occurs in a co-metabolic way and in reductive conditions. Whereas the chlorinated compounds serve as electron acceptors, the degradation of lactic acid leads to the release of hydrogen, which is oxidised (electron donator). The chloride atoms are substituted successively by hydrogen atoms stemming from the co-substrate decomposition. The degradable and hydrogen-producing lactic acid is named hydrogen releasing compound (HRC®). In homogeneous and permeable aquifers the HRC® approach has been applied in the USA, in particular, for plume decontamination purposes. Again, this reductive de-chlorination process is promoted by *Dehalococcoides ethenogenes* and the establishment of undesirable dead-end products should be taken into consideration. More information about the biological decontamination of organic pollutants – related to terrestrial ecosystems – can be found in Sect. 6.3.

Furthermore, other agents are taken in order to stimulate the aerobic degradation process significantly (Table 7.6). They are linked to a chemical decomposition, if they are applied in great quantities, since they tend to be toxic to microorganisms (H_2O_2, MgO_2). For this reason, the proportioning of the agent is very important to prevent toxicity to microorganisms and subsequently extreme reduction of the effectiveness. The infiltration of the agents is aimed at a relatively rapid chemical reaction to oxygen, which may improve the biological decontamination.

The main effect of the use of hydrogen peroxide is associated with the improved aerobic conditions. In a field experiment in Berlin, Germany, it has been found that the redox potential increased significantly after adding H_2O_2 to a sandy aquifer of 7 m thickness over a period of 7 days in contrast to the addition of nitrate, which did not show any considerable alteration (Anonymous 2006). The achieved values of up to approximately 600 mV ensure the oxidative conditions required for an acceptable aerobic decomposition of organic pollutants such as VCHC.

For hydrogen peroxide it is well-known that the formation of oxygen occurs rapidly, while manganese peroxide needs more time to be transformed into oxygen. Generally speaking, the latter is advantageous because it is considered to be a

Table 7.6 Agents used for stimulating the aerobic degradation process of organic pollutants in the groundwater

	H_2O_2 (adequate, low concentration)	MgO_2 (manganese peroxide)	O_2 (bubble-free)
Process	Rapid reaction	Long-term source of O_2	Oxygen injection (analogical to air sparging, but using microporous hollow fibers (iSOC™))
	$H_2O_2 \rightarrow H_2O + 0{,}5\ O_2$	$MgO_2 + H_2O \rightarrow 0{,}5\ O_2 + Mg(OH)_2$	
Requirement	Permeable, homogeneous aquifer	Permeable, homogeneous aquifer	Permeable, homogeneous aquifer
Location	Plume clean-up (also applied to the source area to improve pump-and-treat)	Plume clean-up (also applied to the source area to improve pump-and-treat)	Plume clean-up (also applied to the source area to improve pump-and-treat)
Contaminants	BTEX, TPH, PAH, VCHC	BTEX, TPH, PAH	BTEX, TPH, PAH

long-term source of O_2. Magnesium peroxide belongs to the so-called oxygen releasing compounds (ORC®). Hydrogen peroxide is applied in a liquid form and magnesium peroxide is injected as slurry phase or placed within the wells in a solid form (filter socks). Both agents require homogeneous and permeable aquifers, are located at the plume decontamination and are generally applied in the absence of phase liquids in the case of BTEX, petroleum hydrocarbons and PAH pollution in the groundwater. Besides, the agent H_2O_2 is frequently applied to aquifers that are contaminated with volatile chlorinated hydrocarbons (VHCH). For example, the chemical reaction of H_2O_2 and trichloroethylene is described as:

$$3H_2O_2 + C_2HCl_3 \rightarrow 2CO_2 + 2H_2O + 3HCl \tag{7.8}$$

In general, before application of every kind of agent pre-tests should be involved, e.g. microcosm studies, to evaluate the effectiveness and proportioning of the most adequate agent (Wiedemeier et al. 1998). The bench-scale tests should find answers to the microbial population present, oxygen and nutrient requirement and the biodegradability of the pollutants. For instance, a microcosm investigation was carried out for an industrial area in the German Ruhr district which looks back on a long and intensive industrial epoch. The 50 ha site of concern was used as a coking plant, colliery and chemical plant, beginning in 1928. Consequently, the aquifer geologically derived from the Cretaceous Period was highly contaminated with BTEX, PAH and VCHC. Because of the limited remediation success of a pump-and-treat approach a microcosm study over a period of 4 weeks with a mixture of extracted groundwater and sediment was prepared. The results for the same organic pollutants are presented in Fig. 7.13. The degradation took place rapidly in aerobic conditions which were adjusted with H_2O_2, while the poisoned option (sodium azide) did not reveal degradation due to the complete killing of the microorganisms. Obviously, the only amendment of hydrogen peroxide did not lead to a reduction of the biological degradation, since the concentration (18–30 mg H_2O_2 L^{-1}) appeared to be compatible with the living conditions of the microbes (Anonymous 2008). They achieve the biochemical degradation potential and to prevent the disturbance of the microbial activity a concentration ranging from 100 to 200 mg L^{-1} is recommended (EPA 2004).

Similar to the air sparging technique (see Sect. 7.1.8) oxygen can also be pumped directly into the aquifer under pressure. The creation of air bubbles, however, should be avoided, because otherwise the contaminants adsorbed to the bubbles might migrate upward to the unsaturated zone after leaving the aquifer. Accordingly, a soil vapour extraction system must additionally be installed (see Sect. 7.2.2). The technique presented in this context, however, is aimed at the stimulation of only the biological degradation. For this reason, the oxygen injection should occur without any establishment of air bubbles. This can be done successfully by microporous hollow fibres (according to the *in situ* submerged oxygen curtain called iSOC™).

To decompose organic pollutants chemically some more or less aggressive agents can be used, which might lead to a quick chemical reaction in the aquifer called *in situ* chemical oxidation (ISCO) (ITRC 2005b). For this purpose, Fenton's agents

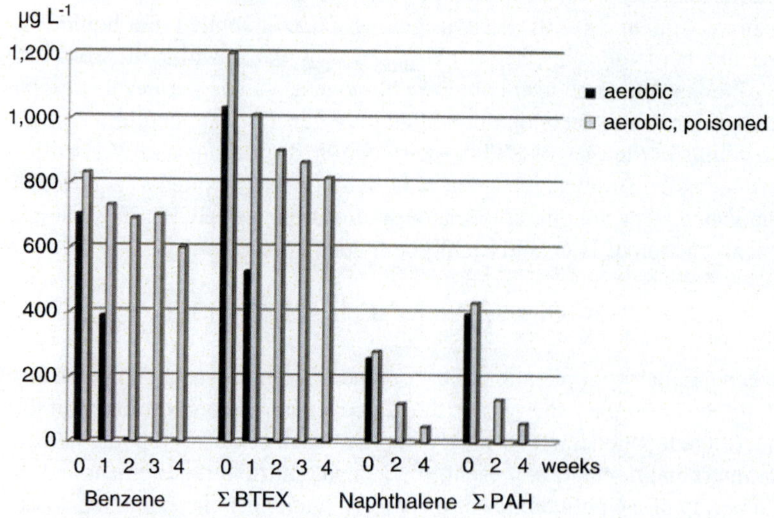

Fig. 7.13 Biodegradation of some organic pollutants (µg L^{-1}) in oxidative conditions after addition of hydrogen peroxide in a microcosm study (Data from Anonymous 2008)

consisting of H_2O_2 and $FeSO_4$ as well as ozone (O_3) and permanganate ($NaMnO_4$, $KMnO_4$) are injected into the contamination source. The addition of liquid permanganate has been frequently applied, especially in the USA, and was successfully evaluated several times. The more problematical gaseous ozone injection (Black 2001) has rarely been used (only USA and The Netherlands) up to now. In the case of Fenton's agent liquid hydrogen peroxide together with $FeSO_4$ acting as a catalyst are used. These release free hydroxyl radicals (OH group), which are well-known for an effective decomposition of the pollutants. In the USA and the Netherlands a lot of experience in treatment with Fenton's agent has been gained.

In Table 7.7 the treatable contaminants are summarised. Case studies showed successful rapid oxidation of unsaturated aliphatic compounds, aromatic compounds such as benzene and VCHC. By way of example, the treatment of trichloroethylene (C_2HCl_3) is demonstrated on the basis of the following formula (Mirsal 2004):

- Oxidation with ozone

$$O_3 + H_2O + C_2HCl_3 \rightarrow 2CO_2 + 3HCl \tag{7.9}$$

- Oxidation with potassium permanganate.

$$2KMnO_4 + C_2HCl_3 \rightarrow 2CO_2 + 2MnO_2 + 2KCl + HCl \tag{7.10}$$

Organic pollutants are partly volatilised, because ISCO creates heat. Hence, an exhaust collecting system (see Sect. 7.2.2) is mostly associated with this groundwater

Table 7.7 Methods of the *in-situ* chemical oxidation (*ISCO*)

	Fenton's agent ($H_2O_2 + FeSO_4$)	Ozone (O_3)	Permanganate ($NaMnO_4$, $KMnO_4$)	Persulfate ($Na_2S_2O_8$, $K_2S_2O_8$)
Contaminants	BTEX Short-chained TPH 2 or 3 ring PAH Free cyanides VCHC (LNAPL)	BTEX 2 or 3 ring PAH VCHC Phenols	Toluene, xylene, ethyl benzene VCHC	BTEX 2- or 3-ring PAH VCHC Short-chain TPH
Requirement	pH < 6 (Fe-stabilisation) Low organic matter content Homogeneous, permeable aquifer	pH < 6 Low organic matter content Homogeneous, permeable aquifer Necessary vapour extraction	Homogeneous, permeable aquifer	Homogeneous, permeable aquifer
Location	Source clean-up			
Problems	Short-term effectiveness Exothermal reaction	Low operating distance	MnO_2 precipitation CrIII oxidation Disturbing effects of phase liquids	Damage to concrete foundations

remediation approach. Furthermore, some drawbacks have to be taken into consideration and these require careful and safe handling (see Sect. 4.4). The agents are extremely aggressive compounds which endanger human health during the application process. Furthermore, problems can be caused due to the persistence of some agents like potassium permanganate. In any case, the reagents must be prevented from migrating out of the treatment area in an uncontrolled manner.

Apart from the aquifer conditions relating to the homogeneity and permeability, more requirements such as a low organic matter content and an iron (Fe^{2+}) stabilising pH value of <6 must be fulfilled. It should be noted that some negative impacts must be taken into account after injection of the agents mentioned. The low pH value tends to be responsible for an accelerated mobility of heavy metals in the aquifer. Furthermore, the chemical reaction of Fenton's agent produces heat (exothermal reaction) and ozone displays a reduced migration movement, leading to a low operating distance. Permanganate can result in MnO_2 precipitation, blocking the injection tubes, and it is not recommended in the case of chromium contamination in the groundwater because of CrIII oxidation to the more toxic CrVI.

The infiltration of persulfate solution also results in fast decomposition of organic pollutants such as VCHC, BTEX, PAH, petroleum hydrocarbons and even PCB at the source area. The impact of free hydroxyl radicals is stronger in comparison with Fenton's agent. The permeability of the aquifer must be relatively high. There may be problems associated with the release of substances such as heavy metals and organic pollutants which cannot normally be oxidised sufficiently. The approach in many case studies in the Netherlands dealt normally with $Na_2S_2O_8$ and $K_2S_2O_8$. The chemical reaction (example trichloroethylene C_2HCl_3) occurs as follows:

$$3Na_2S_2O_8 + C_2HCl_3 + 4H_2O \rightarrow 2CO_2 + 9H^+ + 3Cl^- + 6Na^+ + 6SO_4^{2-} \quad (7.11)$$

The formation of sulfate in high quantities, however, requires careful observation with regard to the sulfate plume development. Moreover, a high sulfate concentration in the aquifer shows negative impacts on building foundations, since the solution might cause danger to the belowground structures made of concrete.

7.1.7 Permeable Reactive Barriers (PRB) and Funnel-and-Gate Systems (F&G)

Practical experience showed a lack of efficiency when treating groundwater with the pump-and-treat technology, in particular with regard to chlorinated hydrocarbons, BTEX aromates and PAH. The relatively low success has mostly been evident in the underground inhomogeneity and inhomogeneous groundwater flow combined with irregular contaminant distribution. Soluble contaminants are transported predominantly by advection, but dispersion and diffusion even into low permeability regions occur simultaneously, leading to impacts on the contaminant transport by convection and to an extension of the transport time. Diffusion and temporary retardation of the contaminants exacerbate the planning of a certain time schedule aimed at a complete remediation of the contaminated plume. For this reason, the pump-and-treat technique might take more time than calculated on the basis of only advection and dispersion.

Alternatively to the pump-and-treat techniques the installation of reactive walls termed permeable reactive barriers (PRB) can decontaminate the groundwater effectively. In groundwater flow direction a wall is constructed to the maximum depth where the aquifer comes into contact with the aquitard below. It is necessary to construct the walls into a layer of low permeability to avoid contaminant migration underneath the wall. The wall must include the width and height of the entire contamination plume. Impermeable side barriers (see Sect. 5.2.1) can be constructed additionally at the sides to the right and left to prevent the migration of the contaminated plume laterally. The wall is filled with reactive substances for plume clean-up purposes.

The construction of the barrier system is achieved by standard machinery, as already explained in Sect. 5.2.1. If the maximum depth does not exceed 10 m, the construction process might be relatively simple. The wall is executed in an open construction pit with reinforced steel walls during the construction to prevent collapse of the trenches. In the case of deep-reaching construction up to approximately 25 m the methods applied include technologies such as slurry trenching, jet grouting and sheet-pile wall installation. However, in the context of the construction attention should be paid to machinery compaction in proximity to the barrier system, since a constant groundwater flow through the barrier must be assured during the later operation. Furthermore, the permeability of the wall must not be reduced when using slurry components as a reactive medium.

According to the technique applied the thickness of the reactive wall varies depending on the contaminant concentration and reaches sometimes more than 1 m. Furthermore, the thickness of the barrier increases with increasing groundwater velocity in order to generate an acceptable contact time between groundwater and the reactive material. In general, heterogeneous aquifers such as karst and insufficiently weathered parent material with fluctuating velocity complicate the thickness calculations considerably. Apart from the reactive material, an inert sand-gravel mixture is additionally filled into the wall to complete the filling.

Alternatively, deep mixing reaching maximum depths of 30 m and conducted by augers can be applied. A number of augers, which mix the soil locally with the reactive substance (percentage 40–60%) in the porous permeable aquifer, is used. In this way an artificial reactive zone is established in which the chemical or biological reaction can occur. The augers must be placed exactly so that a barrier of up to 1 m width is created (Gavaskar et al. 2000).

The dimension of the reactive wall is of importance because the best management is achieved when the reactive material does not have to be replaced during the remediation time. This objective, however, is not achieved in a great number of case studies. Thus, after some time of operation the reactive material must be replaced (exchange of reagents). This is facilitated by the use of filter cassettes that are

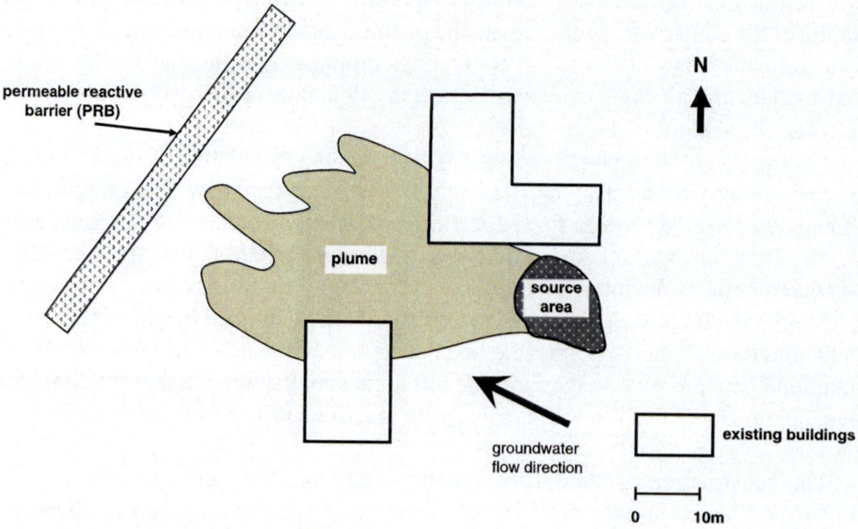

Fig. 7.14 Plan view of a permeable reactive wall construction (*PRB*)

exchanged from the top. The walls are surrounded by gravel filters to prevent washing in of fine earth (silt, loam, clay) which is responsible for clogging of the pores and accordingly preventing a continuous water flow through the barrier system.

Figure 7.14 illustrates the reactive wall installation as plan view. The hydraulic conductivity of the wall is comparably high or even higher than the hydraulic conductivity of the aquifer. The contaminated water flows continuously through the permeable wall, reacting with the material filled in. Down-gradient the contaminant concentration must be significantly lower in accordance with the quality standards or the concentration level required. It should be noted that the planning of reactive walls requires exact hydro-geological knowledge. Considering the three-dimensional heterogeneous flow system stochastic flow and transport models must be used in order to optimise the design of the walls. In principle, the hydraulic conductivity of the reactive barrier should exceed the average conductivity of the aquifer. The groundwater may flow continuously through the reactive barrier, if the hydraulic conductivity is comparably high or even higher than the hydraulic conductivity of the aquifer. Basically, apart from the soluble contaminants, phase liquids are also treatable.

There are a number of reactive substrates which produce favourable results in reactive wall technology. Volatile chlorinated hydrocarbons are treated with microscale or nanoscale zero-valent iron or ferric oxides mixed with sand or pea gravel (Zhang 2003; Phenrat et al. 2007). The reacting surface of Fe^0, in particular, is of great magnitude (35 $m^2\ g^{-1}$). This iron-bearing solid has the ability to decontaminate spontaneously the VCHC described as X*Cl following the formula:

$$2Fe^0 + 3H_2O + X*Cl \rightarrow 2Fe^{2+} + 3OH^- + H_2 + X*H + Cl^- \qquad (7.12)$$

The toxic hexavalent chromium occurring as an anion ($HCrO_4^-$, $Cr_2O_7^{2-}$) is more soluble than the CrIII compounds. The iron-bearing solid is also used to reduce toxic chromium to CrIII as follows:

$$Fe^0 + 2H_2O + HCrO_4^- + H^+ \rightarrow Fe(OH)_3 + Cr(OH)_3 \tag{7.13}$$

Furthermore, arsenate and selenate might also be treatable in this way and radioactive substances like uranium have also been successfully remediated in the presence of zero-valent iron:

$$Fe^0 + UO_2^{2+} \rightarrow Fe^{2+} + UO_2 \tag{7.14}$$

UO_2 is an amorphous or crystalline uranium oxide that precipitates in the reductive conditions but the reduced uranium can be rapidly oxidised and dissolved into the groundwater in the case of a return to oxidative conditions. The zero-valent iron amendment has been frequently used in case studies in different countries such as Canada, Italy and the Czech Republic (Mace et al. 2006). One problem related to the zero-valent iron application is the limited dispersion of the suspension resulting from a high adsorption potential of the iron-containing suspension. For this reason, organic acids (e.g. polycarboxylic acid) are added to establish a more negative charge of the iron particles. Particular safety measures are required to prevent direct contact between humans and the strongly alkaline suspension during operation time (see Sect. 4.4).

The velocity of the reaction with zero-valent iron can be accelerated by means of catalyst bimetals. The bimetals are plated onto the surface of the iron particles, speeding up the de-chlorination process of VCHC at least ten times. Recommended combinations between iron and bimetals are palladium (Pd/Fe), platinum (Pt/Fe) and silver (Ag/Fe), whereas Ni/Fe and Cu/Fe are more problematical because nickel and copper have their own contaminant character (Muftikian et al. 1995).

Alternatively, a precipitation of cationic and anionic metal compounds can be carried out, e.g. chromium is treatable by using sulphides as a precipitating agent in reactive walls:

$$Cr_2O_7^{2-} + 2FeS + 7H_2O \rightarrow 2Cr(OH)_3 + 2Fe(OH)_3 + 2S + 2OH \tag{7.15}$$

In general, metals can be removed by chemical precipitation where a soluble metal is transformed into an insoluble form. For instance, to eliminate lead hydroxides are recommended:

$$Pb^{2+} + 2OH \rightarrow Pb(OH)_2 \tag{7.16}$$

Furthermore, a high level of efficiency is guaranteed with the help of sulphides:

$$Pb^{2+} + S \rightarrow PbS \tag{7.17}$$

For instance, in Canada a PRB operating with sulphate-reducing bacteria as reactive media was installed at an industrial site, so that a quick heavy metal precipitation in the form of sulphides occurred. After 21 months of operation a clear decrease in soluble metals was detected. The copper concentration was reduced from 3,600 to 10.5 μg L^{-1}. Cadmium dropped from 15.3 to 0.2 μg L^{-1}, cobalt from 5.3 to 1.1 μg L^{-1}, nickel from 131 to 33.0 μg L^{-1} and zinc from 2,410 to 136 μg L^{-1} respectively (Ludwig et al. 2002).

The lowest solubility is expected by using phosphates, which may react immediately:

$$10Pb^{2+} + 6H_2PO_4^- + 2H_2O \rightarrow Pb_{10}(PO_4)_6(OH)_2 + 14H^+ \tag{7.18}$$

It is also possible to remove metals by adsorption onto inorganic sorbents such as ion-exchange resins, zeolites and ferric oxides and oxyhydroxides, since these materials have a high cation exchange capacity. The adsorption possibility, however, is restricted to cationic metals, whereas anions are not adsorbed electrostatically. There are further agents such as amine hexadecyltrimethylammonium adsorbed onto zeolites and providing a positively charged interface, which may indicate a strong sorption capacity to anions such as chromate and selenate (Haggerty and Bowman 1994; Lehmann et al. 1999). In conclusion, the reactive wall technology in the context of groundwater remediation is equivalent to solidification/stabilisation approaches conducted in terrestrial systems (see Sects. 5.3.1 and 5.4).

In mining areas the use of reactive walls is conducted as well. The acid mine drainage (AMD) resulting, for instance, from copper mines, is treated with alkaline substances such as granulated limestone, soda ash or potassium hydroxide, which neutralise the acidity, e.g. following the formula:

$$Cu^{2+} + H_2SO_4 + 4KOH \rightarrow Cu(OH)_2 + K_2SO_4 + 2K^+ + 2H_2O \tag{7.19}$$

In general, organic pollutants can be treated after the wall has been filled in with granular activated carbon, which adsorbs the pollutants such as BTEX, VCHC, phenols and some aliphatic hydrocarbons which migrate through the wall system. Even PAH-contaminated groundwater has been remediated in this way several times. However, for fuel hydrocarbons this technique appeared to be less effective in a number of case studies (Birke et al. 2003). Addition of nutrients and microorganisms leads to the possibility of an organic pollutant degradation executed in the wall. Accordingly, the wall appears to be bioreactive, which reminds one of the biological treatment methods (see Sect. 6.3.1). Autochthonous bacteria are identified, isolated and afterwards reinjected as an inoculum for the biobarrier formation. The living conditions of the bacteria are improved by addition of e.g. molasses, so that they show a fast and strong growth. Instead of molasses other organic materials are used such as compost, wood chips and cotton fibres. Theoretically, the aerobic biodegradation can be improved by adding oxygen-releasing compounds (ORC®) to the reactive barrier. The microbial biomass causes the formation of biofilms that, on the one hand, show a high biodegradation potential but, on the other hand, the water permeability might be reduced. *Pseudomonas* and *Klebsiella* are effective candidates for use in biobarrier systems (Bradl 2005).

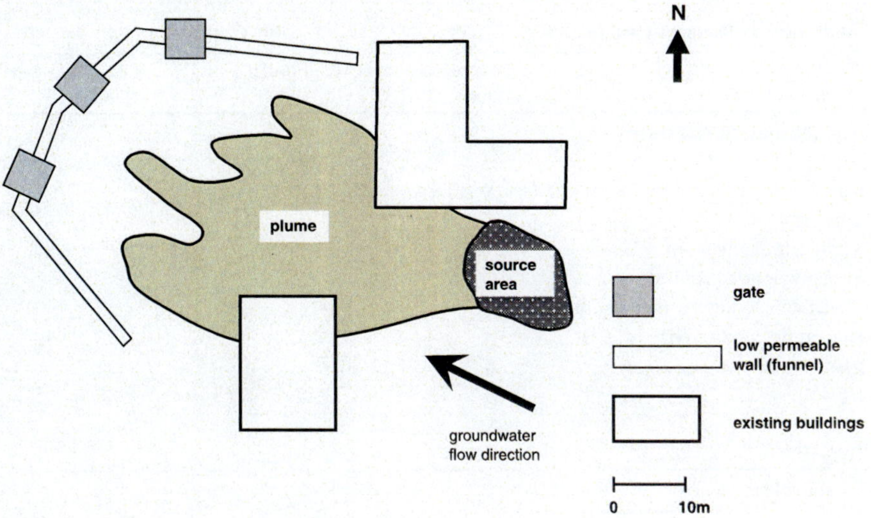

Fig. 7.15 Funnel-and-gate remediation system (*F&G*)

The clean-up of the groundwater plume can be concentrated by using small-sized gates which interrupt the side barrier system. Funnels guide the contaminated groundwater to the permeable gate, where the contamination is degraded or immobilised. This funnel-and-gate system (F&G) first developed in Canada in 1991 enables the precise decontamination of the plume more exactly. Therefore, F&G is primarily applied when the plume is too long, wide or deep so that the digging of the trench and the implementation of the reactive wall across the entire width of the plume is technically too complex and too expensive. The funnels show very low permeability in contrast to the gates, indicating only 1/1,000 to1/10,000 of the gate permeability (Genske 2003). As shown in Fig. 7.15, the set-up of the system includes side barriers with some gates in main groundwater flow direction that are filled with the reactive substances. It is only possible for the groundwater to flow through the open gates combined with the chemical reaction of the ingredients or with the possible biodegradation, while the funnel system does not allow a groundwater migration. An optimum streaming into and through the gate is only certain, if the gate permeability exceeds the natural aquifer permeability by a factor of 10 (Bradl 2005).

The funnel-and-gate system can be modified by the addition of drain systems (tube arrangement) called drain-and-gate system (D&G) that collect the contaminated groundwater and subsequently transport it to the gates. This technique might be an efficiently controllable reactive barrier, since the entire groundwater is collected and guided to the treatment location. Due to the limited area where the groundwater can flow at all and subsequently react with the chemicals the amount of the reactive material needed in F&G and D&G is smaller than in the PRB approach. A subsequent decision can be taken whether the water treated in the gates is acceptable for pumping directly into surface waters using a tube network. Otherwise, it must be forwarded to distributors.

Table 7.8 Hydro-geological properties affecting the choice of distinct types of reactive barriers

Parameter	Permeable reactive barrier (PRB)	Funnel-and-gate (F&G)	Drain-and-gate (D&G)
Aquifer material (texture)			
Gravel	X	X	
Sand	X	X	
Silty sand	X	X	X
Sandy silt, sandy loam			X
Strongly fissured aquifer			X
Moderately to low fissured aquifer	X		X
Hydraulic conductivity (m s^{-1})			
$>10^{-2}$	X		
10^{-4}–10^{-2}	X	X	
10^{-6}–10^{-4}	X	X	X
10^{-8}–10^{-6}			X

X = suitable approach

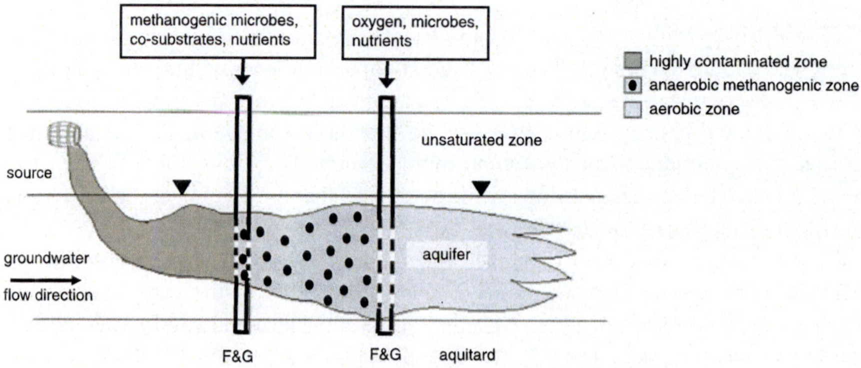

Fig. 7.16 Performance of a sequential permeable reactive barrier system

The funnel-and-gate approach is mostly applied to moderately to highly permeable aquifers, whereas the drain-and-gate system is chosen in less permeable saturated zones. In comparison, the material conditions and hydraulic conductivity of the aquifer is decisive for the choice between PRB, F&G and D&G (Table 7.8) (ITRC 2005a)

It is possible to place several reactive walls successively, resulting in the treatment of different contaminants one by one. With reference to the different pollutants, which are aerobically and anaerobically biodegradable, the set-up of the reactive barrier system can theoretically treat different pollutant groups successively (Fig. 7.16). For instance, a first series of gates enriched with methanogenic microbes and filled with nutrients and co-substrates is focused on the degradation of CHC pollutants like volatile chlorinated hydrocarbons, which are usually anaerobically biodegradable. In this way, firstly a controlled redox zone across the aquifer is created.

After this, a second series of gates is installed which are filled with carbon sources (e.g. compost, sawdust, carbohydrate containing substances), nutrients and aerobically living microbes, and which are additionally injected with e.g. oxygen. Solid oxygen-releasing (ORC®) and hydrogen-releasing (HCR®) compounds can be used. There, the degradation of pollutants such as remaining monoaromates or other contaminants additionally present such as petroleum hydrocarbons (TPH) and PAH can occur. While the first series causes an anaerobic methanogenic zone, the second one is classified as aerobic zone. Accordingly, a mixture of distinct organic contaminants can be treated, if alternating reactive walls or gate series are constructed in groundwater flow direction. Theoretically, a reductive and oxidative zone can be created by a series of wells, too, but the effectiveness might be more limited.

Generally, the treatment effect is mostly not completed within the reactive wall system, but it continues as the groundwater migrates onwards. Ultimately, the groundwater quality standards have to be achieved when reaching the extraction wells of the water supply. Apart from that, the reactive wall technology is associated with some detrimental effects. The creation of precipitate products such as carbonates or sulfides may lead to a reduced hydraulic conductivity within the wall in the course of time. But there are further reasons for the so-called clogging, e.g. products from microbial biomass degradation, products of corrosion which may coat the reactive material, air bubbles generated during the chemical reactions, etc.

Furthermore, the construction of the walls is difficult to realise in the presence of a hydraulic heterogeneity of the aquifer, in particular in aquifers indicating preferential flow situations caused naturally or by F&G. The groundwater flow conditions can be changed significantly, resulting in a groundwater rise upstream of the funnel. The continuous flow might be also influenced by precipitation of iron and manganese. In relation to the microbial degradation in bioreactive walls unknown degradation products can be generated (biofouling), clogging the permeable wall. Contrary to the reactive wall technology, the exchange of the reagents is rather easy, since in principle the gates may encompass removable treatment cassettes.

In general, there is a lack of long-term experience in this technology, because the first case studies date from only 1991. On the one hand, reactive walls avoid active pumping and the volume to be treated is a lower one. If vertical barriers can be keyed into the less permeable underlying strata, the total amount of groundwater to be pumped might be considerably lower. On the other hand, the cost aspect is hard to estimate, since the operation time of the reactive wall is preferentially of relevance. In relation to the reactive walls additional costs must be considered because of the need to replace the reagents every 5–10 years (Nathanail and Bardos 2004). In comparison with the pump-and-treat solution the reactive wall technology appears in all probability to be a cost-saving approach after 3–5 years. In the beginning the investment costs are relatively high compared to the pump-and-treat idea, but the operation (energy expenditure) and maintenance costs might be decisively lower.

Preliminary investigations should be carried out in the context of the planning of the reactive barrier system. Based on hydraulic modelling of the exact location, configuration and width must be estimated. The planning should involve the construction of various observation wells (up-gradient and down-gradient) to monitor

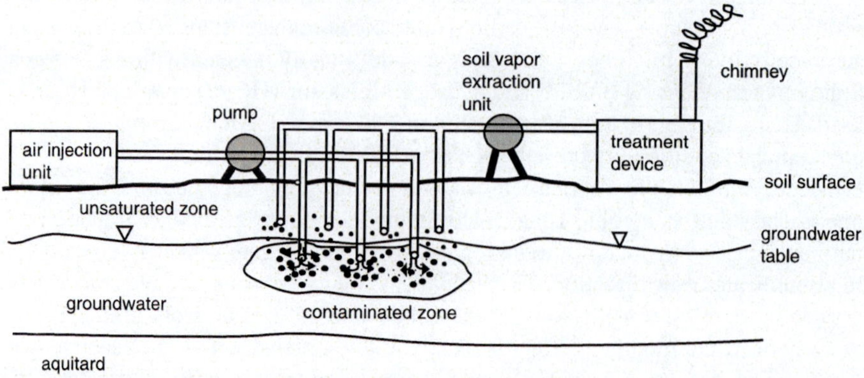

Fig. 7.17 Air sparging technology

the effectiveness of the procedure. With regard to the reactive substances bench tests should be performed to enable optimum contaminant removal in the groundwater.

7.1.8 Air Sparging and Biosparging

Without doubt the techniques mentioned are based on the decontamination of the groundwater itself. One should usually differentiate between groundwater treatment and soil vapour treatment (see Sect. 7.2), which is located in the unsaturated zone only. In the case of the so-called air sparging technology, also termed *in situ* air stripping and *in situ* volatilisation, both groundwater plume and soil air are included, so that it is not possible to draw a clear definition line. Fresh air or oxygen is injected under pressure to the aquifer, reaching the groundwater table below. Bubbles adsorbing volatile chlorinated hydrocarbons (VCHC) are created and rise upward. The volatilisation of trapped and adsorbed contaminants present below the groundwater table occurs simultaneously in the capillary fringe (*in situ* volatilisation). Subsequently, all gaseous contaminants will penetrate into the unsaturated zone, where soil vapour extraction wells are drilled, ultimately pumping the gases out of the ground (Fig. 7.17).

Phase liquids and heterogeneous aquifers with low permeability are limiting factors. The main problem is associated with the limited accessibility of all contaminants to the treatment regime. In the course of time contaminants may migrate from less accessible spots, re-contaminating the groundwater and ultimately the soil vapour. Accordingly, the effectiveness of air sparging is hard to calculate because of the difficulty to predict exactly the pathways, especially in the saturated zone (Fields et al. 2002). In particular, the efficiency of the technology is reduced if the vertical air passage becomes disturbed while the lateral air movement is being increased (Khan et al. 2004). Under optimum conditions, meaning sandy and not stratified soils, the treatment time is relatively short – usually less than 1–3 years (NJDEP 1998).

Air sparging differentiates between groundwater in which the air is injected and vapour that is extracted from the unsaturated zone. This differentiation can be avoided, if gas and groundwater extraction occurs from a single well, termed dual-phase extraction system (DPE) (see Sect. 7.2.6). Buildings with basements or cellars should not be located in proximity to the extraction wells, since volatile organic compounds tend to migrate into the belowground structures.

The treatment of air sparging can be modified by taking the biodegradation potential into consideration. Apart from air pumped into the aquifer, microorganisms and nutrients are inoculated simultaneously, resulting in an additional microbiological decontamination in the unsaturated zone. Nutrients and microbes are alternatively added to ditches cut at the soil surface. The further development of the biodegradation potential to the establishment of the air sparging is called biosparging. This technique focused on plume decontamination can treat aquifers contaminated with BTEX, (light) petroleum hydrocarbons (TPH) and even 2-ring or 3-ring PAH as long as phase liquids are absent. The direct injection of oxygen instead of air appears to be more beneficial (EPA 2007).

The techniques introduced are more favourable in soils with high and homogeneous permeability $>5 \times 10^{-4}$ m s^{-1} (e.g. sandy soils). Otherwise the airflow is only concentrated in fissures. High organic matter content in the unsaturated zone is detrimental due to the adsorption potential for organic pollutants (Nathanail and Bardos 2004). There are limitations on biosparging if basements or other underground structures are present because of dangerous vapours which can be generated during the operation.

7.1.9 Fracturing Technology

To facilitate the decontamination above and below the groundwater table fracturing technology can be added. Therefore, this technology refers to both groundwater and soil vapour remediation. The main target is to provide enhanced access for the applied agents (liquids, gases) by enlarging already existing fractures or by the establishment of new fractures in the saturated as well as unsaturated zone. There are distinct technologies for achieving this target. Pneumatic fracturing deals with bursts of high pressure air, whereas hydraulic fracturing resorts to liquids under high pressure. Moreover, fractures can be created by explosives but this technique is rarely applied.

In general, the fracturing improves the accessibility for soil vapour extraction (see Sect. 7.2.2), pump-and-treat technology and treatments dealing with agents which have to be introduced to the aquifer (e.g. nutrients, reactive materials such as zero-valent iron or potassium permanganate). Ultimately, the soil permeability and the radius of injection and extraction wells increase and the dispersion of added substances occurs to a greater and more homogeneous extent. Thus, fracturing is mostly used at contaminated sites consisting of cohesive soil such as clay and loamy clay, but also less weathered shales, sandstones, etc. Attention should be paid to uncontrolled dispersion and movement of liquid contaminants after fracturing and damage to buildings and other structures caused by blast fracturing and subsurface elevation (Nathanail and Bardos 2004).

7.2 Soil Vapour Treatment

7.2.1 *Relevant Contaminants*

Vapour treatment is defined as the extraction of the volatile components and subsequent aboveground treatment. Soil vapour treatment deals with volatile compounds, namely the volatile chlorinated hydrocarbons (VCHC), the monoaromates benzene, toluene, ethyl benzene and xylene (BTEX), naphthalene, petroleum hydrocarbons (belonging to the TPH), methane (CH_4), hydrogen sulphide (H_2S) and ammonia (NH_3). With regard to the TPH heavier fuels such as diesel fuel, heating oil and kerosene are not readily removed by the technique. The treatable contaminants reveal a high vapour pressure exceeding approximately 66 Pa. In the past especially landfill gases consisting of a mixture of different gases including methane, carbon dioxide, carbon monoxide and hydrogen sulphide as a result of aerobic and anaerobic degradation have been of interest in the context of soil vapour treatment. Nowadays, priority is given to the pollutant group of VCHC.

With reference to the different toxic substances the contaminant adsorption increases with decreasing water solubility, increasing hydrophobicity and decreasing vapour pressure. The volatile portion of the contaminants depends on both the total and the water-soluble concentration in the soil. The gaseous percentage is derived from the interaction between the distinct states of matter. In the first instance, the water-soluble concentration (mg L^{-1}) is related to the total soil concentration (mg kg^{-1}) based on the adsorption isotherm (e.g. K_{OC}). Next, the relation between the water-soluble content and the gaseous component is defined by the Henry's Law Constant, which is well-known for all the organic pollutants and describes the quotient of the partial pressure (Pa) and the water-soluble concentration. Accordingly, it is possible to derive the volatile concentration (mg m^{-3}) regarding the measured total soil concentration as well as the calculated water-soluble concentration. In order to quantify the volatile percentage the total concentration must be multiplied with the soil volume, the water-soluble concentration with the water content defined at field capacity and the volatile concentration with the air capacity of the soil.

7.2.2 *Soil Vapour Extraction (SVE)*

The treatment of contaminated soil vapour might be preferentially necessary, if the entry of gas into habitable parts of buildings cannot be prevented or the escape of gas in e.g. multi-storey car parks cannot be stopped. Alternatively, the use of

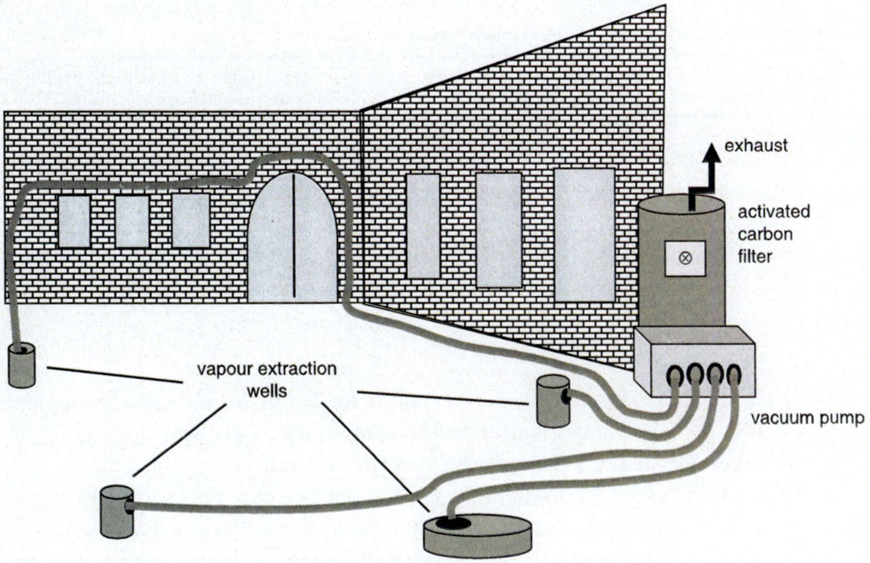

Fig. 7.18 Set-up of a soil vapour treatment facility

geotextiles as barriers to the gas movement and the construction of preferential pathways by trenches filled with gravel have previously been investigated but in the case of a failure of these approaches active pumping with extraction wells or extraction galleries might be the best decision. Nevertheless, the construction of gas-resistant membranes, sealing of cavities and active ventilation in buildings are additional measures often required.

In Fig. 7.18 a set-up of a soil vapour treatment facility is shown. Extraction wells located only in the unsaturated zone are constructed and connected by a tube system. The tubes are needed to pump the vapour out of the ground and subsequently into a container which consists of the treatment module, e.g. activated carbon absorber. The perforated extraction wells composted of polyethylene or polyvinyl chloride reveal a diameter of 50–100 mm (borehole diameter 150–200 mm). The distance between the plastic tube and the borehole wall is backfilled with a sand-gravel mixture. At the top the tube is sealed with swelling bentonite or concrete. The depth of extraction refers to the unsaturated zone below approximately 50 cm. The extraction of soil vapour is not related to the upper soil layers near to the surface, since otherwise the pumping equipment would draw too much atmospheric air. However, if the contamination is quite shallow and the area of concern large, it is possible to use horizontal piping systems or trenches (NJDEP 1998).

The set-up can be installed in built-up areas, since the space needed is limited to the drilled extraction wells only. Buildings, pavement and roads are not disturbing factors in association with soil vapour clean-up.

The extracted volume usually varies between 25 and 500 $m^3\ h^{-1}$. The maximum well distance should not exceed 50 m to enable a continuous extraction. The reach

Table 7.9 Blower and vacuum units for soil vapour extraction

Texture	Permeability (k value m s^{-1})	Type	Extraction flow (m^3 h^{-1})	Maximum negative pressure (hPa)
Gravel/coarse-grained sand	$>10^{-4}$	Low-pressure/ high-pressure fan	300–2,000	100
Medium-grained sand to silty sand	10^{-4}–10^{-6}	Side-channel blower	150–500	450
Silty and clayey classes	$<10^{-6}$	Vacuum pump	25–400	980

range can be expanded in the presence of sealing made of plastic sheets. During the vapour extraction some important air components such as oxygen (O_2), nitrogen (N_2) and carbon dioxide (CO_2) should be measured on-line.

The extraction occurs intermittently and not continuously. This is caused by the relatively slow contaminant transport from the soil matrix to the gaseous component (see Sect. 7.2.1). For this reason, the relation between operating time and pause should be in the order of magnitude of, for example, 1:3. The relation depends on the equilibrium time and subsequently it changes depending on the site. Moreover, in cold seasons the equilibrium process might generally take up more time.

Technically, there are special blowers and vacuum units for soil vapour extraction purposes (Table 7.9). The chosen type depends on the soil texture and permeability. In general, the units allow different extraction flows according to the soil conditions present. SVE is occasionally used in combination with bioremediation processes which subsequently take place, since the increased airflow stimulates the aerobic biodegradation of contaminants (see Sect. 6.3.1).

For the SVE planning a number of aspects must be evaluated and estimated in detail (USACE 2002):

- The volume influenced by the contaminated gases is difficult to estimate in the case of heterogeneous soil conditions, since the gases do not migrate continuously. Generally speaking, the higher the soil heterogeneity, the more different the migration pathways and directions.
- In general, the vapour extraction is limited to the unsaturated zone, which should not be influenced by the groundwater and its capillarity. A groundwater table of >3 m appears to be generally appropriate, while a shallow groundwater table (e.g. 1 m below the surface) is critical. The water content of the unsaturated zone, however, is also extremely interesting, because moist soil conditions demand an additional and complex soil-water separator. Accordingly, the vapour extraction effectiveness depends on the season and long-term wet periods of time are usually problematical. Due to the different viscosity and mobility of organic pollutants SVE at low soil temperatures in the wintertime is less effective.
- The physical soil properties such as air-filled pore space (vol%), air conductivity (m s^{-1}) and texture are of particular importance. There is no doubt that the most adequate texture classes are gravel and sand (Table 7.10). While silty sand can

Table 7.10 Assessment of physical soil conditions in association with soil vapour extraction methods

	++	+	–	––
Air-filled pore space (vol%)	>20	>15–20	10–15	<10
Hydraulic conductivity ($m\ s^{-1}$)	$>10^{-4}$	10^{-4}–10^{-5}	$<10^{-5}$–10^{-6}	$<10^{-6}$
Texture	Gravel/sand	Sandy to silty texture	Loam	Loamy to clayey texture

++ optimum
+ good
– hardly possible
–– impossible

also be assessed as favourable, a loamy and, in particular, clayey texture is usually the reason why it is excluded from application. In a similar way to the texture, an air-filled pore space exceeding 20 vol% and an air conductivity higher than 10^{-4} $m\ s^{-1}$ are very good soil conditions for the extraction of contaminated soil vapour but an air-filled pore space ranging from 15 to 20 vol% and a conductivity of $>10^{-5}$ $m\ s^{-1}$ are acceptable as well. In contrast, if the pore space falls below 15 vol% in line with lower conductivity of $<10^{-5}$ $m\ s^{-1}$, the effectiveness of the treatment is reduced to a great extent. Moreover, sites consisting of weathered rock indicating a lot of fissures can also be appropriate. It is noted that heterogeneous soil physics complicate vapour remediation. For instance, if sandy and loamy soil layers are present the extracted air is preferentially connected to the sandy layers, which are more exhausted than the adjacent loamy layers. The total vapour extraction is limited in the case of heterogeneous conditions with preferential airflow in highly permeable zones, whereas the airflow of less permeable zones cannot be affected by SVE.

- Furthermore, there are the chemical characteristics of the pollutants which must be remediated. In particular, the potential volatility defined as partial pressure (Pa) must be taken into consideration. Attention should be paid to the sorptive capacity of the soil and especially to the organic matter content that is predominantly responsible for a high binding capacity of the organic pollutants. Consequently, subsoils with a very low TOC content are more effective with regard to gas extraction.
- Residues of free phase products (NAPL) in the unsaturated zone may prolong SVE immensely, since contaminants are continuously delivered to the gaseous components.

Like the groundwater treatment all kinds of soil vapour treatment provide detailed bench-scale tests or field investigations to estimate the effectiveness of the procedure planned. *In situ* field investigations might be necessary to determine the number and spacing of the extraction well configuration, particularly for the estimation of the radius of influence. Bench-scale tests are associated with some modifications of SVE, for instance the bioventing technology.

Some problems and dangers are associated with the soil vapour extraction technique. Explosion risk must be taken into account with reference to some pollutants. For instance, this risk must be calculated, if the methane concentration varies between 5 and 15 vol%. (Nathanail and Bardos 2004). Furthermore, other substances may tend to cause an explosion such as benzene (1.2–8 vol%) and ammonia (15–30 vol%). The risks are not reduced to the explosion danger. Soil vapour extraction can cause terrible odour in the vicinity of the extraction wells, leading to annoyance among the neighbouring inhabitants, and, besides, subsidence can occur after long-term vapour extraction because of considerable disturbances in the soil structure. The latter may cause visible damage to buildings, roads and bridges.

7.2.3 Aboveground Soil Vapour Purification

In any case, after the vapour extraction the toxic gases must be treated aboveground, because evaporation into the atmosphere is prohibited as a result of the Clean Air Acts enacted in most of the countries. The most common technology used for the clean-up of the contaminated vapour is adsorption onto activated carbon. Although the contact time is limited, activated carbon might beneficially filter the toxic gases, since the hydrophobic substance indicates an enormous inner surface of 300–2,000 m^2 g^{-1} at a texture of 2–8 mm. The material, however, must be regenerated to guarantee its efficiency. This process, resulting in a mass loss, is done by hot steam treatment or by thermal treatment at 700–800°C.

Moreover, catalytic oxidation, biofiltration and condensation are alternatively used. The catalytic oxidation is obviously an energy-intensive technique, because a reactive temperature of 350–550°C is required. Furthermore, effective catalysers have to be used such as copper and chromium oxides or platinum. With regard to the organic pollutant groups BTEX, CHC and TPH, both carbon adsorption and catalytic oxidation are preferred irrespective of the concentration of the contaminants (Table 7.11).

In contrast, biofiltration is less effective and mostly not recommended, while the condensation is limited to some specific pollutants like mercury. Biofilters usually consist of organic materials like well-degraded compost, bark mulch, heather, brushwood, peat and finely chopped branches combined with mineral components such as volcanic ash and pumice stone or with synthetic material like polystyrene. The material, which is sometimes inoculated with microorganisms, provides the microbes with nutrients, adsorbs water and enables favourable air capacity and porosity. If strong dryness due to the elevated temperature exceeding 40°C occurs, fissures will develop, leading to a detrimental filter capacity. This problem can be solved by re-wetting, but the water content should not reach too high values, since anaerobic conditions, which might cause terrible odour, particularly H_2S and NH_3 odour, can arise.

Table 7.11 Assessment of the air treatment facilities (high: >1,000 mg m^{-3}, low: 10–1,000 mg m^{-3})

Technique	Concentration	BTEX aromates	Chlorinated hydrocarbons (CHC)	Petroleum (TPH)
Carbon adsorber	High	+	+	+
	Low	+	+	+
Catalytic oxidation	High	+	+	+
	Low	+	+	+
Biofiltration	High	–	–	–
	Low	o	–	o
Condensation	High	o	o	o
	Low	–	–	–

+ beneficial
o moderate
– detrimental

7.2.4 *Bioventing*

The vapour extraction technique can also be combined with the microbiological treatment in the unsaturated zone. In this case, the process of volatilisation is not reduced to the gaseous losses of the contaminant itself, since the microbiological degradation of the organic pollutants leads to the simultaneous generation of gaseous decomposition products such as carbon dioxide and methane, which tend to be volatilised as well and which can be successfully extracted in the soil vapour extraction well system. Advantageously, in the initial period the vapour extraction accelerates the air capacity and is responsible for the elimination of toxic gases, so that the living conditions for the microbes are increasingly beneficial. Additionally, it is possible to improve the degradation conditions using, for instance, nutrients such as nitrogen and phosphorus applied to the unsaturated zone by irrigation or by ditches.

The combined vapour extraction and biological treatment is termed bioventing. This focuses mainly on the stimulation of aerobic pollutant degradation. Normally, mid-weight petroleum products are treated in this way, because lighter compounds tend to volatilise fast (SVE preferred) and long-chained pollutants need longer time periods to be degraded (bioremediation preferred) (Khan et al. 2004).

The air flow into the unsaturated zone is conducted by injection and/or extraction wells. The injection configuration is based on the introduction of air under pressure into the contaminated zone. Extraction wells use a vacuum in order to generate a controlled airflow through the contaminated zone. Air injection is preferred to the extraction approach, since the radius of air distribution shows higher expansion and the effectiveness of the removal of toxic vapour might be stronger. For this reason, in particular in the presence of deep soil contamination, injection wells are favoured. On the other hand, because of the different radius in areas with buildings or other structures, which, generally speaking, can be negatively affected, the milder extraction approach is mostly chosen. Usually vertical wells, which can reach deep contamination up to 30 m below the surface, are installed. Shallow contamination can be treated by horizontal wells which are placed into previously excavated trenches (Leeson and Hinchee 1996; Bhandari et al. 2007).

Whereas SVE is designed to maximise the volatilisation of organic pollutants with the help of high airflow rates, bioventing is targeted at the maximisation of the biodegradation using normally lower air flow rates than SVE. In optimum conditions the bioventing appears to be a relatively rapid treatment, taking between 6 months and 2 years (EPA 1994).

The expected biodegradation in the context of bioventing depends on the more or less same physico-chemical conditions as discussed in Sect. 6.4. In summary, the key factors are an oxygen content of 5–10%, enhanced air permeability ($>10^{-5}$ m s^{-1}), soil temperatures exceeding 10°C, an optimum moisture content ranging from 40 to 70% of field capacity, slightly acid to neutral pH value, relatively low humus and clay content, balanced nutrient concentrations (C:N:P:K = 10:1:1:1) and a homogeneous soil structure. Again, heterogeneous structures, for instance alternating sandy and clayey layers, are problematical in relation to this technique. Using this example the air will preferentially flow through sandy layers, leading to wrong assumptions with regard to the air distribution and consequently possible overestimation of the radius of influence. In comparison with SVE bioventing allows the treatment of moderately permeable soils, because a reduced volume of air is required. Moreover, very high contaminant concentrations which are toxic to microorganisms are criteria for exclusion (Leeson and Hinchee 1996; EPA 1994). Irrespective of the soil properties present bench-scale tests are recommended in order to estimate the biodegradation rate, the required nutrient supply as well as the oxygen consumption before ultimately the well configuration is planned and installed.

7.2.5 *Steam Enhanced Extraction (SEE)*

To decontaminate volatile components in the unsaturated zone the hot steam technique operating at a temperature between 50 and 170°C and called steam enhanced extraction (SEE) can be applied. After distribution of the hot steam injected by an infiltration well the volatile components volatilise. The boiling of organic pollutants is combined with water evaporation (steam distillation). The principle is based on the so-called eutectic temperatures, which lead to the boiling of organic pollutants at relatively low temperatures (e.g. benzene 69°C, perchloroethylene 88°C). In spite of the low vapour pressure of the pollutant it tends to volatilise quickly. Additionally installed vapour extraction wells are responsible for the extraction of the contaminated gases.

Apart from hot steam injection heat pipes, which raise the temperature of the surrounding soil considerably, can be constructed belowground. As presented in Fig. 7.19, both the hot steam injection as well as the installation of heat pipes is

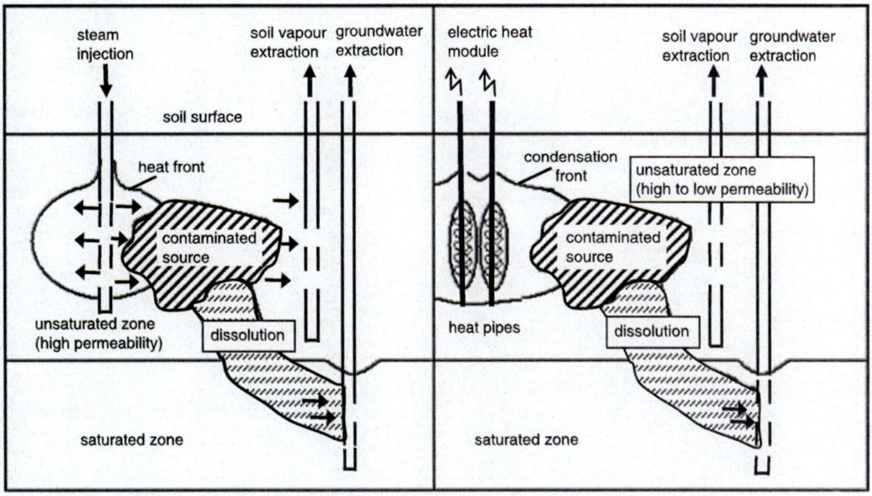

Fig. 7.19 Hot steam injection technology (*left*) and heat pipe installation technology (*right*)

useful not only for the unsaturated soil depth but also the aquifer is influenced in this manner. The hot steam condenses when entering the surrounding soil, generating a cool-water bank which may flush dissolved contaminants downward. After solubilisation, condensation and downward percolation dissolved contaminants can be extracted from the groundwater. Subsequently, extraction of contaminated groundwater using groundwater wells may occur simultaneously to the vapour extraction. The technique is advantageously applied for solubilised pollutants and LNAPL, while the presence of DNAPL reduces the effectiveness. Therefore, DNAPL are treated on the basis of a hot steam-air mixture injection because this agent acts as an inert carrier gas that can transport the pollutants upwards to the soil vapour of the unsaturated zone where a soil vapour extraction well system is constructed (Hiester and Schrenk 2005).

There are different aspects with regard to the simultaneous removal of gaseous and liquid contaminants. Firstly, the enhanced temperature generally improves the dissolution process of organic contaminants and, in particular, the reduced viscosity is responsible for the mobilisation of organic pollutants. Accordingly, even free phase residues (NAPL) are treatable, as already indicated above. Secondly, phase liquids (NAPL) are physically displaced when hot steam migrates from the injection to the extraction wells. Finally, pollutant destruction based on the chemical reaction caused by the hot steam may additionally contribute to the removal (USACE 2006).

The main advantages of the hot steam technique are the short period of treatment amounting to weeks to months compared to the cold vapour extraction and the effectiveness, as shown in Fig. 7.20. Rapid treatment, however, should always be preferred in spite of the high energy costs of SEE. Furthermore, the hot steam technique is applicable to a number of volatile compounds such as BTEX and VCHC, including non-aqueous phase liquids.

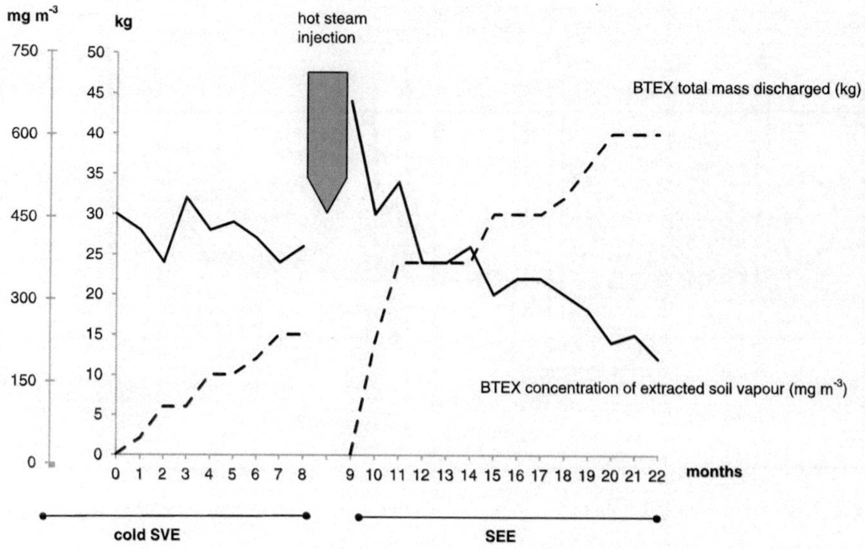

Fig. 7.20 Discharged BTEX and BTEX concentration of extracted soil vapour after alteration of the applied technique from cold SVE to hot SEE at a remediation site in Osnabrück, Germany (unpublished data)

Soils with a sandy to silty texture are of interest, if the conductivity exceeds 10^{-6} m s^{-1}. Moreover, sites located on strongly fractured bedrock are treatable in this way. In general, because of the higher conductivity of more permeable soils, the required injection pressure is lower in comparison with less permeable soils. The heat pipe configuration allows the treatment of cohesive soils as well, since the heating reaches complex geological structures completely. Silty or clayey soils indicating an air conductivity $>10^{-9}$ m s^{-1} are feasible similar to soil contamination with NAPL.

The treatable depth concerned varies between 2 and 30 m in the case of hot steam injection. In contrast, the heat pipe technology is usually reduced to the upper 4–5 m, resulting from the product-specific maximum heat pipe length. The upper limit of approximately 2 m is caused by possible soil instability in the context of the introduced pressure. Similar to the SVE approach an artificial sealed surface (e.g. plastic sheets) might compensate for this problem. The lower limit is associated with the introduced pressure generating a sufficient amount of heat, because with increasing depth the pressure requiring heat will also immensely increase.

On the other hand, some disadvantages must be taken into consideration. As already mentioned, with reference to the injection of superheated hot steam pollutant leaching cannot be excluded after condensation of the gases at the borderline between the heated and the cold zone. Consequently, in most of the case studies the groundwater must be treated additionally. In this context, the implementation of a monitoring system generally appears to be necessary. Moreover, the construction requires temperature-stable and cost-intensive material such as steel. With regard to the aboveground vapour treatment the gas must be cooled down because extremely heated gases cannot be treated by activated carbon absorber.

There are disadvantageous impacts on the microbial activity and the living conditions for the edaphon caused by the enhanced soil temperature. Apart from higher organisms, the microbes are affected and the communities are altered into mesophile and particularly thermophile ones. The vapour extraction in combination with the risen temperature is responsible for strong soil dehydration, which in turn may cause subsidence. Besides, the injection of steam under pressure causes fractures, which are responsible for structural damage to buildings. The treated soil tends to dry out during the operation time, resulting in accelerated air conductivity and an increasing lack of water influencing the vegetation and the edaphon.

Furthermore, hot steam may break out in undesired places. Thus, hot steam injection as well as hot pipe installation should definitely not be applied close to infrastructure pipes such as gas pipelines, electricity cables and communication lines. For the same reason, some contaminated areas do not appear to be suitable for this technique, e.g. contaminated petrol station sites with underground petrol storage tanks.

7.2.6 Multi-Phase Extraction (MPE)

Well techniques which are capable of extracting soil vapour, groundwater and LNAPL simultaneously in consideration of subsequent separation tools are now on offer. In the case of multi-phase-extraction (MPE) the same wells are placed and used in both the unsaturated and the saturated zone. Hence, it is possible to extract soil vapour, contaminated groundwater and the swimming light non-aqueous phase liquids (LNAPL) simultaneously. If the extraction is reduced to two phases only, e.g. groundwater and LNAPL, the term dual-phase extraction (DPE) is used. MPE and DPE have been successfully applied in the USA and United Kingdom, for instance, to remediate the source area. They focus mainly on mixed contamination like BTEX, petroleum hydrocarbons (TPH), VCHC and LNAPL and are particularly available for contaminated sites of (former) petrol stations. To apply MPE and DPE soil and aquifer should reveal high to moderate permeability. The technique can be performed up to a depth of approximately 15 m (Baldwin et al. 2009).

The basic principle of the separation tools is shown in Fig. 7.21. Firstly, a vapour-water separator is used. The gaseous component is extracted by a vacuum pump and treated with activated carbon, while the fluid phase is treated in an oil-water separator, which collects the oily phase. The water is pumped out of the aquifer and subsequently treated, for instance, in an activated carbon absorber module.

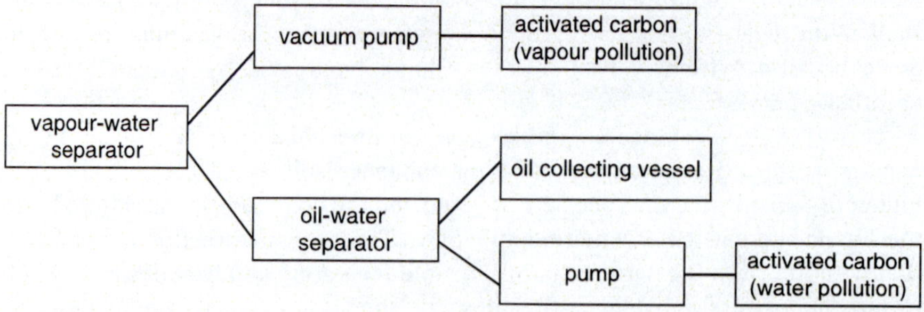

Fig. 7.21 Principle for the separation of soil vapour, groundwater and LNAPL extracted by a multi-phase extraction well

7.3 Surface Water Restoration

7.3.1 Lakes and Ponds

Intensive use of the fertilizers nitrogen and phosphorus, the growth of the population density and increasing industrialisation led to the production of organic matter in drainage basins of lakes in a lot of lake catchments. The contamination of lakes with organic matter is particularly very common in the developing countries, where wastewater treatment plants are ineffective or even absent and subsequently rapid oxygen depletion, H_2S development and fish kills occur. In contrast, in the developed countries modern technologies prevented discharges of nutrients from point sources to a great extent. For instance, in Denmark the discharges from point sources to aquatic ecosystems declined by 69% (N) and 82% (P) between 1989 and 2002 (Kronvang et al. 2005). Stagnating water bodies like lakes and ponds have a tendency towards eutrophication, if nutrient-rich wastewater is discharged or sediments are eroded in the course of time. At the bottom of the water bodies anaerobic sludge termed sapropel is accumulated, exhibiting detrimental characteristics such as odour intensity, black colour, extremely high nutrient capacity as well as contamination. As time passes by, the sediment thickness accelerates until a transition from lake to a very shallow basin or wetland occurs.

Eutrophication caused by point and diffuse sources and connected to unwanted plant growth and algae blooms is the more harmless type of lake ecosystem stress. In particular, in the presence of anthropogenic sources like sewage saprobisation can occur, usually combined with total oxygen depletion and fish kills. Moreover, the discharge of hazardous substances stemming from industrial sources may accelerate the problem significantly, so that oxygen depletion and the insufficient self-purification lead to extremely negative consequences for the lake ecosystem. The other important cause of damage to the lake ecosystems, acidification caused by acid rain, which is spread over wide areas by wind, or pyrite oxidation in lignite coal extraction areas, is discussed in Sect. 3.1.6.

The sludges must be removed or treated once in a while, in particular in proximity to residential areas. In a number of case studies this approach might be the only opportunity to solve the problem quickly, because some alternative measures, as described below, lead to a delay of 10–15 years due to the adaptation time needed by the biocenoses to the new abiotic conditions (Sondergaard et al. 2007). Moreover, dredging of lakes and reservoirs offers further advantages, such as the removal of toxic substances which have accumulated at the bottom, removal of disturbing macrophytes and contour alteration, which is important in deeper lakes for navigational purposes.

The best opportunity to remediate the water course is complete de-sludging and subsequent final disposal. De-sludging, however, is only successful, if the external loading of the lake caused by agricultural input can be interrupted or completely stopped. The excavated sludge can be dried by filter presses or it is deposited in beds for sludge drying purposes. Instead of wet dredging a number of lakes were treated on the basis of draining-down excavation (Fig. 7.22a–c).

Technically, the excavation is executed by grab excavators, endless chain buckle dredgers, barge-mounted suction dredgers or pontoon-mounted scavenging pumps. The machines are of varying design. Nevertheless, the dredged and pumped sediment is disposed of onshore or in barges. Techniques consisting of suction dredgers or scavenging pumps need long pipelines which must be controlled permanently. In order to keep the treated sediment in the working area and to prevent turbidity the dredging area is sometimes surrounded by a curtain (Cooke et al. 1993).

The dredging approach is usually associated with mechanical cleaning operations. The simultaneously applied weed control, especially the removal of the submerged vegetation with high biomass production, reduces the causes of eutrophication significantly. The mechanical cleaning of the watercourse, however, is often difficult to apply in the presence of refuse at the bottom of the lake. Disturbing artefacts such as construction debris, scrap metal and bulky refuse may cause damage to the equipment used.

If it is guaranteed that the sludge is not contaminated, re-use will be always possible. The organic matter and the sedimented fine particles, which exhibit a high sorption capacity for cationic elements, are dried and brought back to the fields where they originated from and where they can fertilize the soil again. In the case of contaminated material a mechanical separation of uncontaminated sand and the usually contaminated silty to clayey and organic fractions should take place, as normally conducted in the context of soil washing. An alternative disposal method should be avoided. Above all, different ways of sludge treatment can be applied (see Sect. 6.2.1).

If dredging cannot be executed, the problematical sediment can also be covered with polypropylene sheets impermeable by gas. The sheet must be weighted down by concrete, stones, etc. in order to suppress the so-called ballooning effect due to the gas development underneath the sheet. The period of time to interrupt the connection between sapropel and water is short, since permanent sedimentation occurs. The use of calcite or sand as sediment capping is less effective, because the barrier is quickly destroyed by bioturbation of the benthic organisms (Klapper 2003).

Because of the costs associated with sediment removal dredging is not the most frequent treatment option. An evaluation of lake restoration opportunities in

Fig. 7.22 (**a**) Lake restoration: complete emptying of the water body and excavation of the sapropel-like sediment (*black colour*). (**b**) Lake restoration: start of the re-wetting procedure after removal of the black sediment (*arrow*: poplar trees at the back for comparison purposes). (**c**) Lake restoration: continued re-wetting 3 months later (*arrow*: poplar trees at the back for comparison purposes)

Denmark (41 lakes) and the Netherlands (28 lakes) resulted in a relatively high percentage of hypolimnic oxygenation and fish manipulation, as described below. In particular, hypolimnic oxygenation in lakes exceeding 10 ha in size has been successfully applied. The percentage of lakes which were treated by sediment dredging was only 2% (Denmark) and 25% (The Netherlands) (Sondergaard et al 2007).

Apart from sediment dredging and cover, direct sediment treatment appears to be an alternative. The oxidation of the organic sediment is one approach, but it requires the careful injection of the reactive agents through tubes. Usually, nitrate is applied to react with the organic matter (CH_2O) as follows:

$$3CH_2O + 2NO_3 \rightarrow N_2 + 3CO_2 + 3H_2O \tag{7.20}$$

The oxidative biodegradation of the organic sediment can help to improve the water and sediment characteristics of the watercourse. For this reason, the mineralisation potential should be accompanied by neutralisation in the sediment conducted with lime constituents such as $Ca(OH)_2$ and $CaCO_3$. In general, the bacteria can reach the optimum biodegradation rate at slightly acid to neutral pH values. Water bodies indicating an acid pH value, however, should be treated with Na_2CO_3 to reach the targeted pH value.

It should be noted that very intensive mineralisation of the organic sediment can also mean raised mineralisation of phosphorus compounds, which, in turn, may increase the P loading of the lake water. Otherwise, except for the organic matter, the organic pollutants are also degraded by microbes, resulting in a decrease in the sediment contamination.

In the presence of high nitrogen potential oxygen can alternatively be injected using tubes. There is also a method in which plants are used to enrich the oxygen content in the sediment for biodegradation purposes, because some marsh plant species develop a so-called aerenchyma, providing the sediment with oxygen derived from the atmosphere. In most of the cases, however, oxygenation occurs technically but it must be conducted continuously over a long period of time. In Lake Sempach located in central Switzerland pure oxygen was injected from May to October (3 t day^{-1}) and from November to April pressured air was injected. In this way it was possible to maintain permanent aerobic condition (O_2>4 mg L^{-1}) and a decline in the hypolimnic phosphorus content in the water from more than 100 mg L^{-1} to approximately 60 mg L^{-1} within 20 years. The release of phosphorus from the sediment was interrupted, so that the trophic state of the lake decreased in the course of time (Gächter and Müller 2003) (see Sect. 3.3.1).

If a complete de-sludging is not feasible, there are also adequate techniques for dealing with the nutrient-rich water (Klapper 2003). Accordingly, these focus predominantly on water purification. Since during thermal stratification the highest nutrient capacity is found in the deepest water, the construction of a bottom outlet draining away the nutrient-enriched water might be a successful solution. Technically, some tubes reaching the deepest layers are installed. For instance, flexible spiral-wound plastic pipes or pressure pipelines are used. After external on-site P elimination, the water can be re-circulated and fed back into the desired lake depth.

Table 7.12 Results from different lake restoration projects in Bavarian lakes, Germany (Data from Schaumburg 1995)

Lake No.		1	2	3	4
Technique used		Bottom outlets	Deep water aeration	Continuous de-stratification	Intermittent de-stratification
Water volume (m^3)		2.2 million	1.1 million	53.1 million	0.19 million
Average depth (m)		7.0	6.1	23.9	5.7
Catchment area (km^2)		15.7	1.4	27.2	0.3
Restoration installed		1981	1981	1982	1980
P-total ($\mu g\ L^{-1}$)	1980	128	195	18 (1981)	85
	1985	80	63	16	35
	1990	64	62	18	30
	1993	35	75	17	32
Chlorophyll-a ($\mu g\ L^{-1}$)	1980	34	57	12 (1981)	53
	1985	48	28	11	33
	1990	21	21	11	34
	1993	14	18	12	44

Otherwise, the pumped water is used as irrigation water in agriculture, if it is licensed for this purpose.

In deep lakes de-stratification by mixing of deep water with high nutrient content and epilimnic water can also be a strategy to overcome the eutrophication problem. De-stratification can be achieved by the introduction of compressed air which is conducted by pipes installed in a horizontal position above the bottom of the lake. Apart from air, water oversaturated with oxygen can also be pumped into the desired depth (Klapper 2003).

In Table 7.12 results from bottom outlet technique, deep water aeration, continuous de-stratification and intermittent de-stratification are presented for four German lakes. Bottom outlets, deep water aeration and intermittent de-stratification provided a fast and effective reduction of the total phosphorus concentration, while the chlorophyll-a content, which indicates the stock of phytoplankton, remained widely unchanged (Schaumburg 1995).

Algae can be skimmed with a special shovel from the surface of algae-rich waters. Afterwards, the algae are diverted mechanically. Moreover, micro-sieving with a mesh size of <40 µm, which focuses on the removal of the algae plant cover, is applied. This approach, however, appears to be problematical, since the sieves must be cleaned continuously using the jet technique. With regard to intensive algae growth the algae removal techniques have been applied in many cases.

The main source for the eutrophication is related to the macronutrient phosphorus. Hence, an adequate means is the phosphorus precipitation using aluminium salts ($Al_2(SO_4)_3$, $AlCl_3$, $Al(OH)_2Cl$, $AlSO_4Cl$) and iron salts ($FeCl_2$, $FeCl_3$, $FeSO_4$). The metalphosphate is occluded within the flocs. In addition to the agents used the sedimentation process can be accelerated by further measures which reduce the water turbulence. For instance, the planting of wind-breaking hedges might lower the wind speed and turbulence in the water body and subsequently an improved sedimentation might be supposed.

The use of chemicals to precipitate phosphorus is often controversially discussed and sometimes not accepted. Lime compounds such as CaO, $Ca(OH)_2$ and $CaCO_3$ are environmentally more friendly. In particular, the flushing of natural calcite (lake marl) appears to be an alternative, more accepted and cheaper technology. As demonstrated in Lake Rudow in Germany, the calcite flushing approach led to a decrease of 40% in the total P concentration in the lake water and a strong biomass decrease in blue-green algae. Calcite showed a high potential for improving the water properties. After the calcite flushing to the water surface phosphorus and algae were precipitated and the generated flocs sedimented at the bottom of the lake. Afterwards, the sedimented flocs contributed to the interruption of the phosphorus re-dissolution and consequently to the danger of re-growing algae (Rönicke et al. 1995).

If iron is used as a flocculant, attention should be paid to the re-dissolution of phosphate under anaerobic conditions. Re-dissolution occurs during oxygen free periods and causes remarkably high phosphate release to the bottom layers of the water body. To avoid the re-dissolution sediment oxidation using nitrate is recommended but an additional liming is also necessary to prevent acidification initiated by the sediment oxidation process. Subsequently, a series of chemical interventions appears to be involved to minimise eutrophication. For this reason, aluminium is mostly used rather than iron (Klapper 2003). Both chemicals can be responsible for detrimental impacts, in particular fish mortality caused, for example, by aluminium toxicity at pH values varying from 5.5 to 8.5. Furthermore, nitrate itself is assessed to be the reason for eutrophication. Subsequently, denitrification might become important, if the nitrogen component is the main source of nutrient enrichment. The description of the complex lake ecosystem shows that the amendment of some chemicals can solve and cause problems simultaneously.

In principle, because of the re-dissolution problems all precipitating agents are more effective in the presence of a simultaneous oxygenation in the deep water. Hence, a number of case studies involved both chemical precipitation (e.g. with $FeClSO_4$, CaO) and deep water oxygenation. A total P reduction in the water by up to 72% (CaO) (Koschel 1990) and by up to 47% ($FeClSO_4$) (Jäeger 1994) within a few years based on enclosure experiments was reported.

Essential differences are well-known between lakes influenced by eutrophication and lakes damaged by sulphate formation and consequent sulphuric acid formation (acidification). While the re-dissolution of phosphate compounds is only expected in anaerobic conditions, reductive conditions in lakes which are potentially acidified due to the input of sulphates might be beneficial because sulphate reduction is only possible under reductive conditions due to the (iron) sulfide formation (see Sect. 3.3.1). The contrasting conditions must be taken into account, in particular in lakes which are influenced by both environmental impacts.

The nutrient decline can be accelerated by plant uptake after establishment of helophytes producing a huge biomass that must be cut yearly. Nutrient extraction by macrophytes (macrophytic biofilters), which must be harvested and removed, might be an adequate means to prevent lake eutrophication in a similar way to the phytoremediation approach (see Sect. 6.4.1). The main target of this approach is the removal of phosphorus that is stored in the biomass by plant harvest. Although the harvested phosphorus amount is surely less than the phosphorus amount in the sediment, the plant removal can reduce the phosphorus level of the lake water. The lowering of the water table, which leads to deeper light penetration to the lake bottom, the control of plant-feeding animals or the direct introduction of adequate plants support the establishment and growth of macrophytes. Improved macrophyte growth helps to prevent algae-dominated waters, which are predominantly responsible for the eutrophication tendencies. However, attention should be paid that unwanted plant species do not grow too copiously, as noticed in European countries in the context of neophytes such as *Elodea canadensis, Eichhornia crassipes, Myriophyllum spicatum* and *Salvinia natans*, because they may negatively impact the lake ecosystem in another ecological manner (Klapper 2003).

The manipulation of the fish population is also a suitable possibility to achieve dominance of macrophytes. The treatment is associated with the establishment of fish populations, which are able to consume plants. Thus, planktivores are increasingly exchanged with typical predator species. There are species like the grass carps (*Ctenopharyngodon idella*) which incorporate macrophytes and, to a less extent, the phytoplankton. Accordingly, the lake becomes a weed-free lake. Periodic fishing is necessary to regulate the fish population carefully. In general, stocking of the lake with fish should produce fewer nutrients than fishermen are able to remove. Fishermen should never enhance eutrophication caused by introduction of fish and particularly fish food (Klapper 2003).

In Denmark more than 50 lakes have been treated with regard to biomanipulation. Zooplanktivorous and benthovorous fish were substituted by stocking of piscivores. After 1–10 years tendencies towards improvement in the lake water were visible. For instance, the Secchi depth increased, while nitrogen and phosphorus did not show strong changes. In general, some properties may react at once, while other parameters do not even show long-term alterations. Although the transparency increased, the growth of submerged macrophytes showed no significant changes. Many restoration projects needed to be repeated at different time intervals to stabilise the positive effect of fish manipulation (Sondergaard et al. 2007). Nevertheless, as long as the internal phosphorus loading capacity remains in the sediment, the return of turbid conditions cannot be excluded. In Table 7.13 the most common lake restoration approaches are summarised.

Nevertheless, the water body restoration targeted at the removal of the anaerobic sediment and the prevention of eutrophication in future should be accompanied by protective measures to reduce entirely the source of the problem. These measures include different approaches. As far as possible interventions in structuring and plantation of the lake ecosystem can avoid tendencies towards eutrophication. On the lakefront shades should be formed to minimise algae growth. The impacting width is only a few meters but the shrubs and trees should not be planted close to the water body, because unfortunately fall of leaves contribute to the eutrophication.

Table 7.13 Most common lake restoration approaches and the main problems to be taken into consideration

Item	Measures	Main problems
Sediment	De-sludging	Limited re-use of the dredged material
		Waste materials damaging the equipment during operation
		Cost-intensive approach
	Sediment cover	Short-term effect (sand, calcite)
		Uncontrolled gas development (plastic sheets)
	Sediment oxygenation (nitrate injection, oxygen injection)	Long-term operation
		Unwanted nutrient mineralisation
Pelagic water	Bottom outlet construction for hypolimnic water	Aboveground water purification
	De-stratification with compressed air	Long-term, periodic operation
	Precipitation with aluminium, iron, calcite	Observation of P re-dissolution
		Possible fish toxicity
Plants	Algae skimming/micro-sieving	Intensive aboveground cleaning of the devices
	Helophyte planting	Plant biomass removal
		Establishment of undesired plant species (neophytes)
		Long-term procedure
Animals	Fish manipulation	Fish population regulation by fishermen
		Long-term procedure

In principle, planting of species with an intensive fall of leaves should be avoided. Hence, evergreen species are usually preferred in association with the plantation of lake shores and river banks.

The measures, however, achieve only limited success without additional techniques which focus on the point sources responsible for the ecological problems in lake water bodies. In the first instance, the discharge of domestic, agricultural and industrial sewage as well as of the rainwater overfall should be prevented or minimised. In general, ring sewerage systems and purification plants that involve nutrient elimination steps contribute to the reduction of the unwanted point sources considerably. At the main inlets phosphorus elimination plants (PEP) are the most effective technology. In three lakes in Berlin, Germany, the construction of PEP combined with filtration modules and the possibility to flush the lakes 1–3 times per year enabled clear reduction of the total P concentration in the lake water. Within a few years the previously high concentration of 500–800 $\mu g\ L^{-1}$ declined to 20–60 $\mu g\ L^{-1}$ (Klein and Chorus 1991).

With regard to lakes located in urban agglomerations the feeding of water birds such as ducks and swans is another source of eutrophication. To what extent prohibition signs solve the problem is highly questionable in consideration of the fact that people were observed throwing whole loaves of bread into the water.

Re-contouring of bank and lake slopes to reduce soil erosion from the adjacent areas, in particular under agricultural use, might also be of importance. Steep slopes which are strongly erosion-endangered should even be afforested. Moreover, the establishment of a water depth >100 cm in lakes and ponds is recommended. Protection zones between the intensively used agricultural land and the lake shores might help to reduce the nutrient input as well but for this reason they should be created in an environmentally friendly agricultural process. This process, for instance, includes fruit sequences with intercropping and continuous vegetation cover, fertilizing only according to the plant demand and soil handling such as ploughing and sowing parallel to the slopes. The case study Lake Schlein in South Germany showed the positive impact of diffuse nutrient load reduction. The lake had a relatively small catchment area of 45 ha, which was intensively used as grassland between 1985 and 1990 (doubling of cattle stock, manure application in liquid form). Surface runoff caused high P concentration, low transparency and permanent bloom of phytoplankton in the water. After 1990 the manure application was drastically reduced and was stopped in some parts of the catchment area. As shown in Fig. 7.23, the phosphorus concentration decreased surprisingly fast together with an increase in the water transparency (Güde et al. 1995).

Furthermore, the danger of eutrophication by inflowing rivers should also be reduced. Methods to achieve this objective are the establishment of forest belts, diversion ditches, upstream dams or screens for inflowing sediment and waste particles. Similar techniques are also applied in the case of constructed wetlands.

7.3.2 *Rivers*

Restoration measures with regard to rivers must be applied, if detrimental properties of water course and riparian zone are present. In general, in streams a continuous flow of solids occurs and eroded material is permanently sedimented in a horizontal and vertical direction. Erosion occurs also, particularly in hilly regions, where lateral runoff leads to the input of nutrients, e.g. nitrogen and phosphorus. The erosion depends on the land-use types of the adjacent areas and normally decreases in the order cropland > pasture > woodland > nature reserve. The properties of the riparian soils usually consisting of alluvial loam above terrace sand and gravel are influenced by floods and fluctuating groundwater. The water body is used for fauna and flora migrating downstream and upstream. The stream ecosystem is a functional unit that can be disturbed in different dimensions in relation to the human impact.

Harmful impact, for instance, is associated with river training structures. To compensate for level differences weirs, chutes and mills are constructed which obstruct fish passage. An extreme impact is related to the construction of bank dams, which cause complete de-watering downstream at certain times, particularly in arid

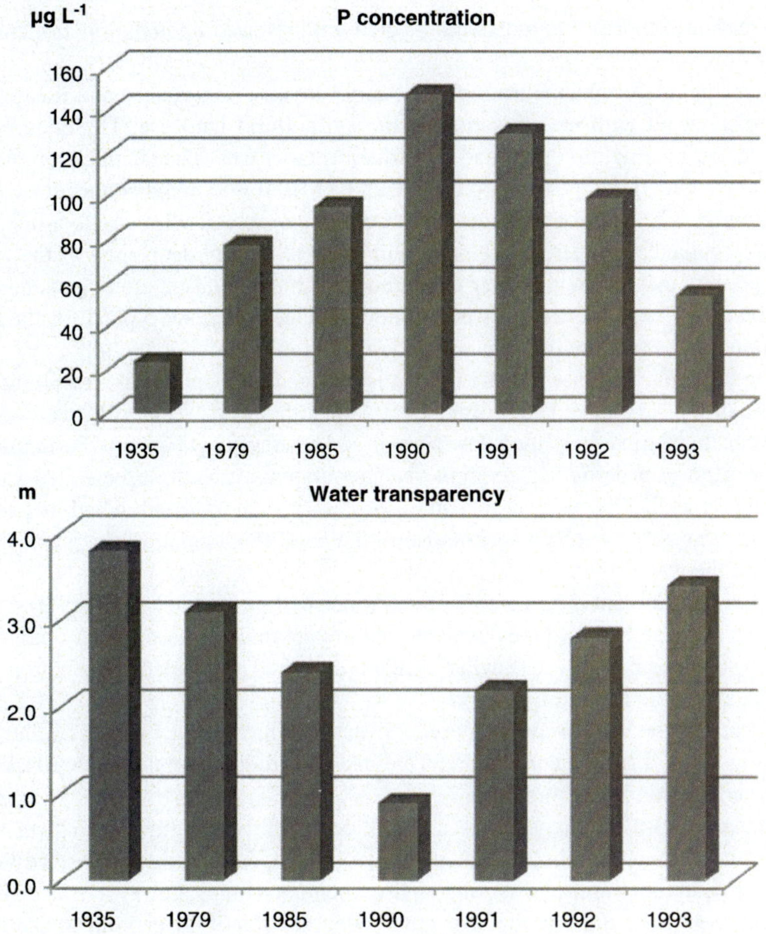

Fig. 7.23 P concentration (*above*) and water transparency (*below*) of Lake Schlein, Germany, before and after the transformation from intensively used agriculture to extensively used agriculture in the lake catchment in 1990 (Data from Güde et al. 1995)

climates. To maintain the nature of the stream ecosystem at all a minimum flow should always be guaranteed, apart from periods where flood events are required. Levees and dikes are created to regulate the stream velocity. In particular, construction of levees and dikes is used for flood control. They concentrate flood flow in channels, resulting in increased shear stress and channel erosion. The channels cause cessation of natural flooding events. Ecologically, the missing floods may change the vegetation from species adapted to wet soil conditions to more drought-tolerant species.

Detrimental effects are linked to the construction of sheet pile walls of steel, concrete shells, cemented walls and rock fills at the sides. The streamlining and diversion can stop erosion but it is responsible for the loss of natural functions. The construction activities are mostly connected with river straightening that increases water flow velocity.

Straightening or channelisation is the most adverse disturbance and it is principally connected with maximisation of the agricultural land use. The disturbance includes bulldozing, dredging and construction processes. The channel complexity is reduced, damaging the fish population, and a loss of floodplain connection can be ascertained. The lack of meandering structures combined with area confinement strongly enhances the stream velocity and leads to a rapid deepening of the gully. This process may drain the water table underneath the adjacent areas, causing conversion of the vegetation from wet-meadow species which are typical for the alluvial floodplains to more drought-adapted vegetation (Palmer et al. 2005).

For shipping purposes the stream bed is dredged once in a while, resulting in a change of the substrate. The sediment extraction should be seen critically, because it can cause sediment starvation downstream and headcutting upstream. Furthermore, ship transport, in particular, requires river regulations, causing accelerated stream velocity as well. The removal of natural obstacles such as wood, which is usually done in the case of small rivers but also in the case of some larger rivers, may also increase the water flow.

The river sediment is removed with reference to the construction of bridges, long distance pipes and sewer pipes. An extreme impact must be ascertained with reference to subsided mining areas, where streams are lifted and sealed at the bottom and thus the substrate is entirely altered.

Moreover, the water flow is altered by pumps which extract process water and drinking-water. Locally, contaminants are added to the floodplain due to discharge pipes and erosion of e.g. ore mining heaps.

Sudden land-use changes causes alteration of the water flow, which, in turn, affects the channel morphology and position on the floodplain till a new steady state is reached. In the presence of strong environmental changes, however, it is difficult to reach the steady state, so that irreversible degradation of the ecological and morphological processes of the stream system is noticeable, demanding stream restoration measures.

Basically, different types of naturalness can be distinguished. As shown in Fig. 7.24, the differentiation may involve four classes termed natural (1), semi-natural (2), artificial (3), and extremely artificial (4). The third class is frequently seen in intensively used agricultural areas, while the fourth one appears to be very common in urbanised areas. Regarding the restoration of streams and their catchment, measures applied to the third and fourth class are usually aimed at achieving class No. 2.

The differences of the water course quality and the adjacent alluvial landscape are strongly associated with the distinct land use of the catchment (Table 7.14). The man-made influence must be expected to be higher in agglomerations, since the use interests are more complex and more intensive. Alluvial floodplains serve as important traffic courses (railway, motorway) and are developing areas linked to the foundation of residential and industrial sites. While in agriculturally influenced catchments nutrient load, altered flow and habitat degradation were the most impor-

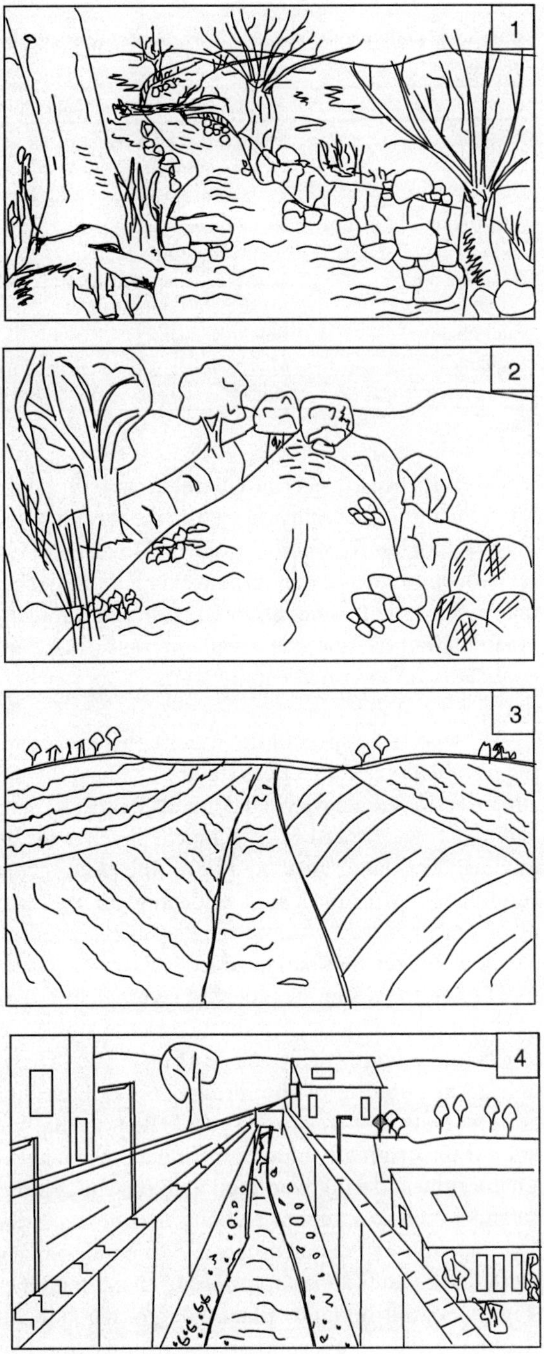

Fig. 7.24 Differentiation of the stream naturalness: *1* = natural, *2* = semi-natural, *3* = artificial, *4* = extremely artificial

Table 7.14 Differences between the agglomeration and the countryside in relation to the alluvial landscapes

	Agglomeration (urban land)	Countryside
Use interests	Many (areas for economic development, many land-use types)	Few (only agriculture and forestry)
River training structures	Many (also flood prevention)	Moderate (mostly river regulation)
Rehabilitation opportunities	Low (very few semi-natural stretches)	Moderate (more semi-natural stretches serving as stepping stones)
Water quality	Low to very low	High heterogeneity
Discharges	Many	Few
Recreation intensity	High	Low

tant disturbances, in urban environments impervious surfaces, urban settlements, infrastructural constraints as well as the introduction of contaminants were the main reasons for river restoration projects (Doyle et al. 2000; Murdock et al. 2004). Thus, rehabilitation measures in urbanised areas are technically more difficult to conduct than in the countryside. Stream restoration projects are frequently associated with rehabilitation measures in urban environments, e.g. brownfield redevelopment and application of environmentally friendly approaches to the water management (see Sects. 2.1 and 2.2).

An important aspect of the water course restoration concerns the removal of the river regulation. In practice, standards for technically and ecologically successful stream restoration have been defined (Palmer et al. 2005). Strongly deepened channels must be restored in a technical manner by gully fill or complete valley regrading. The latter method means the establishment of meandering reaches in combination with backward relocation of the dams and in consideration of the involvement of back water areas. Possible new stormwater structures should be constructed outside the riparian zone.

The width of the meandering riverbeds depends on the topography and the river dimension but with reference to the smaller rivers the meanders should not fall below a width of 20–60 m (Fig. 7.25). The measures are not targeted at a long-term flooding of the alluvial floodplain, since flood events normally last between only 2–4 weeks per year. The artificially created meanders must substitute for the originally natural meanders, because waiting for the establishment of a natural meandering would take up too much time. Furthermore, the creation of meanders reduces stream velocity and prevents the rapid deepening of the river bed. In addition, the construction of meandering river reaches offers the possibility of slack water zones appropriate to the fauna, e.g. for fish spawning purposes (Fig. 7.26). Furthermore, placing larger wood residues and boulders in the stream channel is also an appropriate opportunity to improve habitats for fish and other aquatic organisms. In general, meandering should be associated with habitat restoration for individual species and with the reconnection between stream and floodplain.

Fig. 7.25 Establishment of meanders during a river restoration project

Unfortunately, soil relocations are among the relatively cost-intensive approaches and they can be associated with soil compaction as well as shrub and tree clearing. Moreover, successful carrying out of the meander stretches must be monitored within the first 10 years at the latest.

The constructive measures should be accompanied by planting activities (Fig. 7.27). The outstanding importance of the vegetation is related to stabilisation of river banks, sediment filtering and in the case of overhanging plants supply of cover and shade for aquatic organisms. Sparse riparian vegetation is responsible for stream bank instability, excessive erosion and subsequent material slough into the stream channel. Furthermore, unstable stream banks accelerate the lateral erosion, which results in increasing stream width, decreased stream velocity and subsequent accelerated sedimentation. The eroded sediment input increases the turbidity of the stream, which adversely affects the aquatic life. This is caused, for example, by restricted light penetration and consequent reduced photosynthesis. Over-widened channels can be narrowed by gravel and cobble stemming from other riparian zones. This is important to enable the re-growth of the vegetation in association with root stabilisation of the bank soils. Biotechnical slope protection with living vegetation added by synthetic stabilising agents might be technically ambitious. Alongside the river reed and shrubs adapted to the moist soil conditions are planted in order to minimise ero-

Fig. 7.26 Implementation of a slack water zone for fish spawning

sion, which also occurs during the restoration measures as long as the constructive measures take place, since the bare soil is supposed to be particularly susceptible to erosion. The juvenile soil does not contain sufficient seeds, leading to a long period of time until the vegetation growth can be discovered. Accordingly, poorly vegetated and disturbed areas contribute particularly to the erosion potential.

Sometimes reasons for bare bank sites are runoff from the upslope area and natural occurrences such as landslide or fire. However, human disturbances resulting from poorly managed timber harvest, intensive mining activities, road and house construction near the stream bank and inadequate agricultural practice such as overgrazing in riparian areas are the main reasons. Hence, in the context of the human impact, apart from plantation and seeding, technical approaches such as slope stabilisation and the removal of poorly constructed roads and buildings must be taken (Skinner et al. 2000; Roni et al. 2002).

Under Central European conditions species chosen for stream bank vegetation are always elder (*Alnus*), willow (*Salix*) and ash (*Fraxinus excelsior*), but some tree and shrub species such as maple (*Acer*) and oak (*Quercus robur*) which do not only grow in moist ecosystems can be taken as well. Some tree species must not be recommended, although they have often been planted, e.g. the poplars. As illustrated in Fig. 7.28, the poplar roots (*Populus* hybr.) grow off the water body, while elder roots (*Alnus glutinosa*) behave conversely.

Fig. 7.27 Shrub plantation alongside the restored river with meander structure

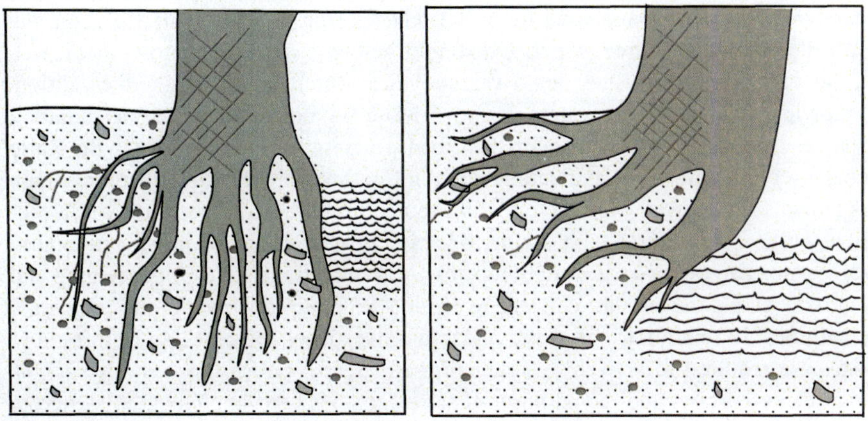

Fig. 7.28 Root growth of elder (*Alnus glutinosa*) (*left*) and poplar hybrids (*Populus hybr.*) (*right*) in river banks

In any case, riparian planting should be followed up by maintenance and long-term protection. The latter is targeted at the removal of exotic vegetation and the promotion of native vegetation. For this reason, already during the seeding and planting period local indigenous vegetation, which is preferentially capable of erosion control and bank stability, should be used. An appropriate number of species should be taken into consideration to ensure high plant diversity (Moore and Outhet 2008).

Apart from the procedures mentioned above, it makes sense to think of the dismantling of the artificial constructions in urban areas. For instance, the removal of fish barriers such as dams for water storage and hydroelectric purposes, culverts and weirs would improve the naturalness. It would enable the fish populations to reach, for instance, their traditional spawning areas and stop the limitations of upstream and downstream movement. Whereas the removal of bank dams is mostly not feasible, concrete road culverts, for example, can be replaced by bridges, embedded pipe-like culverts or open-bottom culverts. The construction of a fish pass or an alternative fish route consisting of stepped pools would be advantageous to the fish population, if the altitudinal belts have to be negotiated (Roni et al. 2002).

With reference to the non-natural substrate of the riverbed the exchange of the artificial material like concrete for natural terrace sand and gravel is one of the measures which can be carried out. This process is occasionally associated with the removal of sludgy sediments, as already discussed in relation to the sapropel treatment of lakes and ponds (see Sect. 7.3.1). Fortunately, streams rarely need to be de-sludged, since they do not tend to accumulate sediments due to the relatively high stream velocity.

The constructive measures mentioned above are targeted at the improvement of the bank properties and at the resultant biologically amended characteristics of the river ecosystem. In streams with alternating ecological conditions based on the four classes mentioned above the different habitat structures interact. Semi-natural sections have a positive impact on adjacent, structurally altered sections, causing an improvement in their conditions. This radiating effect is based upon the active and passive migration of fauna and flora. It emanates from a semi-natural section, the so-called radiation source, which is distinguished by a stable biocenosis that is rich in species. Therefore, they are river stretches which are in a good condition. Inflowing tributaries, backwaters and other water segments can also act as radiation sources for improving downstream disturbed structures or interruptions in the water course continuum. The connecting radiating pathway is the stretch upon which organisms move away from the radiation source (Table 7.15). Structurally rich sections in the water course with optimum habitat properties can be populated temporarily, thus lengthening the radiation pathway. Regional measures for enlarging and interconnecting the habitats can lead to an area-wide improvement of the ecological conditions. In relation to the water restoration it reduces the costs, since minimum sizes are required to activate the radiating effect, if supporting measures such as removal of barriers and the addition of stepping stones are implemented along the radiation pathway (DRL 2008).

The water quality will not be influenced by the technical restoration approaches. Constructive measures and water quality improvement are not interdependent.

Table 7.15 Required minimum lengths of radiation source and radiation pathway with respect to different river sediments and organisms (Data from DRL 2008)

Sediment	Organisms	Radiation source (km)	Radiation pathway in flow direction (km)
Sandy material, small stream, lowland	Macrozoobenthos	1.5	2.5
	Fish	1.5	7.5
	Macrophytes	1.5	5.0
Sandy and loamy material, river, lowland	Macrozoobenthos	2.5	3.5
	Fish	2.5	12.5
	Macrophytes	2.5	4.0
Boulders, small stream, mountainous area	Macrozoobenthos	0.5	3.0
	Fish	0.5	3.5
	Macrophytes	0.5	1.5
Fine to coarse material, river, mountainous area	Macrozoobenthos	1.5	4.0
	Fish	1.5	20.0
	Macrophytes	1.5	2.0

Hence, both points of view must be taken into account separately. In order to improve the water quality in the first instance discharge pipes must be discovered and eliminated. Moreover, regarding the agricultural influence, particularly the nutrient runoff, a 10–20 m wide buffer area that remains agriculturally unused should be established between the water body and the used land.

Moreover, the restoration will only be successful, if the entire riparian zone and its adjacent areas are also included. In particular, with regard to the water quality and the sedimentation potential, extensive farming might be essential in the bordering territories. Extensive farming means the establishment of extensive meadows and pastures in conjunction with strongly limited cropland areas. Re-wetting procedures in former drainage areas can take place but they must be carefully applied in order to avoid accommodation problems for the vegetation. The reduction of cropland in favour of meadow occasionally causes economic damage but this can be substituted with the help of land consolidation (land exchange).

In summary, the opportunities to rehabilitate negatively impacted rivers cover a wide spectrum (Table 7.16). I have reported on the theoretically possible approaches, the introduction of meanders and habitats for fish and other aquatic organisms, planting and seeding, dismantling of constructions, agricultural extensification of the catchment, sediment exchange, etc. Throughout the USA the majority of stream restoration projects dealt with the objectives of stabilising stream banks, managing the riparian zone, improving in-stream habitats, creating fish passages and enhancing water quality (Bernhardt et al 2005). A statistical evaluation in the Upper Midwest states of the USA (Michigan, Ohio, Wisconsin) revealed a dominance of in-stream habitat improvements (47%) and bank stabilisation (29%). Dam removal (6%) and channel reconfiguration (4%) did not play an important role. In particular, sand traps were introduced and large woody debris was added to improve habitat quality. Bank stabilisation was carried out by bank reshaping, bank brushing, addition of gravel and boulders, use of ripraps and grading. With reference to

Table 7.16 Most common river restoration approaches and the main problems to be taken into consideration

Item	Measures	Main problems
Stream bank	Bank stabilisation (reshaping, brushing, riprap, grading)	Soil compaction Vegetation clearing Long-term monitoring Cost-intensive approach
Course of river	Dismantling of constructions (dams, culverts, weirs)	Restricted substitution opportunities Cost-intensive approach
	Creation of meanders	Soil compaction Vegetation clearing Long-term monitoring Cost-intensive approach
	Gully fill, sediment dredging and exchange	Starvation downstream and headcutting upstream Disposal problems
In-stream habitats	Implementation of slack water zones, addition of boulders and woody debris	Long-term monitoring Extreme slowdown of the flow velocity
	Creation of fish passages	Cost-intensive approach
Plants	Planting and seeding of indigenous species	Long-term monitoring Spread of exotic species Extensive maintenance
Water quality	Removal of discharge pipes	Difficult localisation Alternative effluent treatment
Riparian catchment	Agricultural extensification	Compensation for farmers

channel reconfiguration meander and riffle creation occurred. The technically more complex treatments, which are occasionally combined with stormwater management operations, showed an increasing tendency in the last two decades. Riparian management aimed at the involvement of the entire river catchment took place in only 2% of the projects and consisted of fencing to exclude cattle and of the establishment of buffer strips with shrubs and trees. The land acquisition method was rarely applied (Alexander and Allan 2006).

For the long-term success of stream restoration monitoring might be of importance. However, in this context many projects seem to fail due to the costs which are incurred over a long period of time. Hence, monitoring in U.S. projects was reported in approximately 10% of all projects (Bernhardt et al. 2005). In the evaluation mentioned above in three Midwest states in the USA monitoring was found in only 11%. The monitoring programmes were usually carried out for cost-intensive restoration measures.

Irrespective of the measure applied, in most of the restoration approaches a relatively short reach of the stream is treated to improve the environmental conditions, which then mostly have an effect over a distance that significantly exceeds the originally treated reach. Accordingly, the small-scale measures serve as stepping stones. In restoration applications in the USA, for instance, the projects were normally on a small scale – less than 1.5 km in length (Bernhardt et al. 2005). Detailed analysis of the database of river restoration projects in the Midwest states revealed project lengths of 0.4 km on average (Alexander and Allan 2006). In particular, in urban areas, where the natural streams have been widened, deepened and straightened to a great extent, restoration measures are often reduced to relatively short reaches, as many examples show. For instance, in Helsinki, Finland, different short reaches of 130–150 m length which are endangered by bank instability and are poor in fish population were treated in different ways such as addition of gravel and stones to stabilise the bank and to create new habitats. The small-scale measures enabled a significant increase of e.g. salmonid reproduction in the entire urbanised area (Aulaskari 2008).

The choice of a few reaches depends on the possibility to affect particularly the biological stream properties over greater distances and on the state of the respective stream reaches. The main factors influencing the choice are the bank stability and the vegetation conditions. Moore and Outhet 2008 suggested a general prioritisation of areas with high bank fragility. The priority decreases with decreasing fragility and with decreasing vegetation conditions. Poor vegetation conditions do not require preferential restoration, since these reaches obviously appear to be stable without vegetation. Therefore, they will not deteriorate if they are not treated.

In some European countries considerably greater restoration projects have been or are being promoted as a result of the EU Water Framework Directive. One of these projects is taking place in the highly industrialised Ruhr area of Germany. The river Emscher, which is located in the northern part of the Ruhr area, has been turned into open wastewater channels since 1906. Its bed was lowered because of mining subsidence. Alongside, dikes, which separate the stream from the industrial complexes located next to the stream, were constructed up to a height of 10 m. The river and its tributaries have lost their ecological function completely. In the early 1990s wastewater transportation in open canals was not accepted any more due to horrible odour and optical inconvenience. It was not possible to carry out a restoration that would bring back the situation in former times with meanders, since the areas adjacent to the channel showed extremely changed topography and densely built-up zones. Nevertheless, it was decided to rehabilitate the 340 km wastewater channel ecologically. Those parts which are finished show riverbanks which permit flooding of the meadow if practically feasible. Diverse flora and fauna started to grow, in the first instance pioneer species (tall forbs) which developed on the created soil. After 8–10 years the first species of hardwood and softwood were able to grow. Simultaneously, the sandy riverbed which had been created formed habitats for a number of animal species such as invertebrates and fish. After approximately 15 years 360 species of invertebrates and fish were identified in the rehabilitated Emscher river. A monitoring programme was developed, including investigation of water quality, biological indicators and alteration of the morphological stream structure. The monitoring must take place over a long period, since the development of stable biocenoses requires some time. Only then can the question be answered whether the restoration has been successful or not (Semrau and Hurck 2008).

References

Alexander, G. G., & Allan, J. D. (2006). Stream restoration in the Upper Midwest, USA. *Restoration Ecology, 14*, 595–604.

Anonymous. (2001). *Issue of a clean-up plan for the prevention of hazards based on §13/6 BBodSchG for the former "VEB Transformatorenwerk Karl Liebknecht" in Berlin-Oberschöneweide*. Diplom thesis, University of Applied Sciences Osnabrück (A. J. Knollmann).

Anonymous. (2006). *Efficiency of the hydrogen peroxide amendment for biological in situ remediation*. Diplom thesis, University of Applied Sciences Osnabrück (M. Göckede).

Anonymous. (2008). *Microcosm studies for the biological degradation of trimethylene cyclohexanone and trimethylene hexahydrophenol*. Diplom thesis, University of Applied Sciences Osnabrück (L. K. Püschel).

Anonymous. (2010). *Redox potential based biological remediation using the example of a VCHC contamination in Rheine*. Bachelor thesis, University of Applied Sciences Osnabrück (A. Markmann).

Aulaskari, K. (2008, June). *Restoration of small urban streams in Finland*. 4th ECRR conference on river restoration, Venice, Italy.

Baldwin, B. R., Nakatsu, C. H., Nebe, J., Wickham, G. S., Parks, C., & Nies, L. (2009). Enumeration of aromatic oxygenase genes to evaluate biodegradation during multi-phase extraction at a gasoline-contaminated site. *Journal of Hazardous Materials, 163*, 524–530.

Bernhardt, E. S., Palmer, M. A., Allan, J. D., Alexander, G., Barnas, K., & Brooks, S. (2005). Restoration of U.S. rivers – A national synthesis. *Science, 308*, 636–637.

Bhandari, A., Surampalli, R. Y., Champagne, P., Ong, S. K., Tyagi, R. D., & Lo, I. M. C. (2007). *Remediation technologies for soils and groundwater*. Reston: American Society of Civil Engineers (ASCE).

Birke, V., Burmeier, H., & Rosenau, D. (2003). Design, construction and operation of tailored permeable reactive barrier. *Practice Periodical of Hazardous, Toxic and Radioactive Waste Management, 7*, 264–280.

Black, H. (2001). Ozone as a clean-up tool. *Environmental Science and Technology, 35*, 283–284.

Bradl, H. B. (2005). *Heavy metals in the environment*. Amsterdam: Elsevier.

Cooke, G. D., Welch, E. B., & Newroth, P. R. (1993). *Restoration and management of lakes and reservoirs*. Boca Raton: Lewis Publishers.

Delleur, J. W. (2007). *Handbook of groundwater engineering*. Boca Raton: CRC Press.

Doyle, M. W., Harbor, J. M., Rich, C. F., & Spacie, A. (2000). Examining the effects of urbanization on streams using indicators of geomorphic stability. *Physical Geography, 21*, 155–181.

DRL. (2008). *Compensation of structural deficiencies in streams by radiation effect*. Deutscher Rat für Landespflege, Bonn, No.81. (in German)

EPA – Environmental Protection Agency. (1994). *How to evaluate alternative cleanup technologies for underground storage tanks sites: A guide for corrective action plan reviewers*. http://www.epa.gov/swerust1/pubs/tum_ch2.pdf Accessed 10 Nov 2011.

EPA – Environmental Protection Agency. (2004). *How to evaluate alternative cleanup technologies for underground storage tanks sites: A guide for corrective action plan reviewers, Chapter XII and XIII*. http://www.epa.gov/oust/pubs/tums.htm Accessed 27 Jan 2012.

EPA – Environmental Protection Agency. (2007). *Technology focus: Bioventing and biosparging*. http://clu-in.org/techfocus/default.focus/sec/Bioventing_and_biosparging/cat/Application/. Accessed 30 Nov 2011.

Fields, K., Gibbs, J., Condit, W., Leeson, A., & Wickramanayake, G. (2002). *Air sparging: A project manager's guide*. Columbus: Battelle. ISBN 1-57477-130-2.

Förstner, U. (1998). *Integrated pollution control*. Berlin: Springer.

Gächter, R., & Müller, B. (2003). Why the phosphorus retention of lakes does not necessarily depend on the oxygen supply to their sediment surface. *Limnology and Oceanography, 48*, 929–933.

Gavaskar, A., Gupta, N., Sass, B., Janosy, R., & Hicks, J. (2000). *Design guidance for application of permeable reactive barriers for groundwater remediation*. Columbus: Battelle.

Genske, D. D. (2003). *Urban land – Degradation, investigation, remediation*. Berlin: Springer.

Güde, H., Schünemann, B., & Trautmann, A. (1995). The reaction of Lake Schlein to the reduction of diffuse nutrient loads in the catchment area. In D. Jäeger & R. Koschel (Eds.), *Verfahren zur Sanierung und Restaurieurng stehender Gewässer* (pp. 39–52). Stuttgart: Gustav Fischer (in German).

Haggerty, G. M., & Bowman, R. S. (1994). Sorption of inorganic anions by organo-zeolites. *Environmental Science and Technology, 28*, 452–458.

Hahn, H. H., & Klute, R. (1990). *Chemical water and wastewater treatment*. Berlin: Springer.

Hiester, U., & Schrenk, V. (2005). *In situ thermal remediation: Ecological and economic advantages of the TUBA and THERIS methods*. International conference on soil-water systems ConSoil, Bordeaux, France.

ITRC – Interstate Technology and Regulatory Council. (2005a). *Permeable reactive barriers*. http://www.itrcweb.org/miningwaste-guidance/to_permeable.htm. Accessed 28 Jan 2012.

ITRC – Interstate Technology and Regulatory Council. (2005b). *Technical and regulatory guidance for in situ chemical oxidation of contaminated soil and groundwater*. http://www.itrcweb.org/Documents/ISCO-2.pdf. Accessed 8 Nov 2011.

Jäeger, D. (1994). Effects of hypolimnetic water aeration and iron-phosphate precipitation on the trophic level of Lake Krupunder. *Hydrobiologia, 275*(276), 433–444.

Khan, F. I., Husain, T., & Hejazi, R. (2004). An overview and analysis of site remediation technologies. *Journal of Environmental Management, 71*, 95–122.

Klapper, H. (2003). Technologies for lake restoration. *Journal of Limnology, 62*, 73–90.

Klein, G., & Chorus, I. (1991). Nutrient balances and phytoplankton dynamics in Schlachtensee during oligotrophication. *International Society of Limnology, 24*, 873–879.

Koschel, R. (1990). Pelagic calcite precipitation and trophic state of hardwater lakes. *Archives of Hydrobiology, 33*, 713–722.

Kronvang, B., Jeppesen, E., Conley, D. J., Sondergaard, M., Larsen, S. E., Ovesen, N. B., & Carstensen, J. (2005). Nutrient pressures and ecological responses to nutrient loading reductions in Danish streams, lakes and coastal waters. *Journal of Hydrology, 304*, 274–288.

Leeson, A., & Hinchee, R. E. (1996). *Principles and practices of bioventing* (Vols. I and II). http://www.afcee.af.mil/shared/media/document/AFD-070926-074.pdf. Accessed 2 Dec 2011.

Lehmann, M., Zouboulis, A. I., & Matis, K. A. (1999). Removal of metal ions from dilute aqueous solutions: A comparative study of inorganic sorbent materials. *Chemosphere, 39*, 881–892.

Ludwig, R. D., McGregor, R. G., Blowes, D. W., & Benner, S. G. (2002). A permeable reactive barrier for treatment of heavy metals. *Ground Water, 40*, 59–66.

Mace, C., Desrocher, S., Gheorgiu, F., Kane, A., Pupeza, M., Cernik, M., Kvapil, P., Venkatakrishnar, R., & Zhang, W. (2006). Nanotechnology and groundwater remediation: A step forward in technology understanding. *Remediation, 16*, 23–33.

Meuser, H. (2010). *Contaminated urban soils*. Dordrecht: Springer.

Mirsal, I. A. (2004). *Soil pollution – Origin, monitoring & remediation*. Berlin: Springer.

Moore, J., & Outhet, D. (2008, June). *River restoration in Sydney, Australia*. 4th ECRR conference on river restoration, Venice, Italy.

Muftikian, R., Fernando, Q., & Korte, N. (1995). A method for rapid dechlorination of low molecular weight chlorinated hydrocarbons in water. *Water Research, 2*, 24–34.

Murdock, J., Roelke, D., & Gelwick, F. (2004). Interactions between flow, periphyton, and nutrients in a heavily impacted urban stream: Implications for stream restoration effectiveness. *Ecological Engineering, 22*, 197–207.

Nathanail, C. P., & Bardos, R. P. (2004). *Reclamation of contaminated land*. Chichester: Wiley.

Neilson, A. H., & Allard, A. S. (2008). *Environmental degradation and transformation of organic chemicals*. Boca Raton: Taylor and Francis Group.

NJDEP. (1998). *Revised guidance document for the remediation of contaminated soils*. Department of Environmental Protection State of New Jersey, USA.

NRC – National Research Council. (1994). Water science and technology board. *Alternatives for groundwater cleanup*. Washington, DC: National Academy Press.

Palmer, M. A., et al. (16 authors) (2005). Standards for ecologically successful river restoration. *Journal of Applied Ecology, 42*, 208–217.

Phenrat, T., Saleh, N., Sirk, K., Tilton, R. D., & Lowry, G. V. (2007). Aggregation and sedimentation of aqueous nanoscale zero-valent iron dispersions. *Environmental Science and Technology, 41*, 284–290.

Roni, P., Beechie, T. J., Bilby, R. F., Leonetti, F. E., Pollock, M. M., & Pess, G. R. (2002). A review of stream restoration techniques and a hierarchical strategy for prioritizing restoration in Pacific Northwest Watersheds. *North American Journal of Fisheries Management, 22*, 1–20.

Rönicke, H., Beyer, M., & Tittel, J. (1995). Possibilities to control blue-greens in eutrophic lakes by restoration techniques. In D. Jäeger & R. Koschel (Eds.), *Verfahren zur Sanierung und Restaurieurng stehender Gewässer* (pp. 133–156). Stuttgart: Gustav Fischer (in German).

Schaumburg, J. (1995). Experiences with restoration measures and limnological long-term investigations at four Bavarian lakes. In D. Jäeger & R. Koschel (Eds.), *Verfahren zur Sanierung und Restaurierung stehender Gewässer* (pp. 309–325). Stuttgart: Gustav Fischer (in German).

Semrau, M., & Hurck, R. (2008, June). *The reconstruction of the Emscher system – An integrated planning process according to the WFD*. 4th ECRR conference on river restoration, Venice, Italy.

Skinner, Q. D., Crane, K. K., Hiller, J. G., & Rodgers, J. D. (2000). *Wyoming watersheds and riparian zones*. Laramie: University Wyoming Cooperative Extension Service.

Sondergaard, M., Jeppesen, E., Lauridsen, T. L., Skov, C., Van Nes, E. H., Roijackers, R., Lammens, E., & Portielje, R. (2007). Lake restoration: Successes, failures and long-term effects. *Journal of Applied Ecology, 44*, 1095–1105.

USACE – US Army Corps of Engineers. (2002). *Soil vapour extraction and bioventing*. Engineering and design. http://www.usace.army.mil/publications/eng-manuals/em1110-1-4001/entire.pdf. Accessed 8 Dec 2011.

USACE – US Army Corps of Engineers. (2006). *In situ treatment technologies for contaminated soils* (US-EPA report 542/F-06/013). www.epa.gov/tio/tsp/download/542f06013.pdf. Accessed 30 Sep 2011.

van Agteren, M. H., Keuning, S., & Janssen, D. B. (1998). *Environment and chemistry: Handbook of biodegradation and biological treatment of hazardous organic compounds*. Dordrecht: Kluwer Academic Publisher.

Wiedemeier, T. H., Swanson, M. A., Mouton, D. E., Gordon, E. K., Wilson, J. T., Wilson, B. H., Kampbell, D. H., Haas, P. E., Miller, R. N., Hansen, J. E., & Chapelle, F. H. (1998). *Technical protocol for evaluating natural attenuation of chlorinated solvents in groundwater* (EPA/600/R-98/128). Washington, DC.

Zhang, W. X. (2003). Nanoscale iron particles for environmental remediation: An overview. *Journal of Nanoparticle Research, 5*, 323–332.

Chapter 8
Approaches Without Complex Technical Applications

Abstract In order to remediate contaminated soils and groundwater, not only must complex technical applications be used. In particular, in the case of limited contamination levels, protective and restrictive measures ordered by the public authority can also mean a reduction of the danger to humans and the environment. They are summarised in this chapter. Moreover, due to limited budget or because of the enormous difficulties to apply successfully one of the remediation strategies introduced the idea of natural attenuation can be an adequate solution, particularly related to groundwater hazard. The approach contaminant attenuation in a natural way, expanded by the possibilities of enhanced natural attenuation (ENA) and monitored natural attenuation (MNA), is discussed in this chapter in detail.

Keywords Monitoring • Natural attenuation • Protective measures • Public authority • Restrictive measures

8.1 Protective and Restrictive Measures

Long-term procedures from the first planning processes to the final implementation of the containment approach or decontamination measures must usually be differentiated from rapid protective and restrictive measures arranged by the public authority for immediate response action purposes (see Sect. 4.1.2). These reactions can be temporary, and accordingly they will be altered or finished after realisation of the definite solution for the contaminated land. In a number of examples, however, they have the same status as the definite solution, in particular in the case of a limited budget.

H. Meuser, *Soil Remediation and Rehabilitation: Treatment of Contaminated and Disturbed Land*, Environmental Pollution 23, DOI 10.1007/978-94-007-5751-6_8,

The public authority can arrange different actions. First of all, fencing-in is an appropriate method which is frequently applied. Fences are constructed or, at the very least, prickly shrubs such as *Rubus* sp., *Rosa canina* and *Pyracantha coccinea* are planted, preventing humans from entering the contaminated land easily. Information boards are put up simultaneously. It should be noted that fencing-in cannot be considered as a long-term solution, since damages to the fences and information boards must be expected (vandalism).

Furthermore, the public authority can publish and send public policies relating to soil handling and exposure to the contaminated soil. It recommends careful cleaning and peeling of fruit, vegetables and potatoes. The public authority informs the residents who cultivate vegetable plants about careful consumption of plants containing toxic substances. Moreover, the inhabitants can be informed about reducing a heavy metal transfer from soil to the edible portion of the plants. This can be in the form of a recommendation to lime the soil and to amend well-degraded organic manure like compost (see Sect. 4.1.4). The authority gives indications that people should refrain from using ashes and some problematical mineral manure such as cadmium-containing phosphate fertilizers.

If little success is expected from this advice, the authority can provide information about the required change of the land-use type in writing. Vegetable and kitchen gardens must be transformed into ornamental gardens, thus avoiding consumption of edible plants by the residents.

Temporarily, the dispersion of contaminated dust originating from the topsoil may play an important role. To protect adjacent goods from soiling the use of plastic sheets or the planting of dense vegetation can prevent deflation and sedimentation of contaminated dust. In warm and dry seasons irrigation is recommended, achieving the same objective.

While the measures mentioned are not usually associated with comprehensive technical input, the public authority is bound to order and control simple technical applications. For instance, planter construction in relation to vegetable plots is an adequate solution in order to enable cultivation of vegetables and potatoes in small-scale garden areas and schoolyards, where vegetable cultivation takes place periodically, for example in association with educational projects. The soil surface is artificially raised, the planters are filled with uncontaminated soil and subsequently the rooted soil allows vegetable cultivation. In this way, it is possible to exclude contaminant uptake by plant roots and the plant and fruit consumption remains free of risk.

In contrast, the construction of geogrids is more complex. This approach appears to be necessary in gardens exhibiting contaminated soil in deeper horizons, while the upper portion of the soil does not show enhanced values. The uncontaminated soil above is excavated and stockpiled. Afterwards, the geogrid is put in place and the material is backfilled (see Sect. 5.1.1). In future, gardeners will be aware of contaminated soil below the digging depth, if, for instance, they want to plant shrubs and trees with deep roots. The geogrids are usually coloured, indicating immediately the problem in the subsurface.

8.2 Natural Attenuation

8.2.1 Definitions

Natural attenuation focuses on the reduction of contaminants in soil and groundwater using naturally occurring processes *in situ*. The main field of action relates to the groundwater contamination. The processes used are not technological applications in the narrow sense but the processes can be stimulated or activated by humans. If providing agents are applied, the term Enhanced Natural Attenuation (ENA) is favoured. After application particular attention should be paid to the mobilising agents, since they could mobilise previously bound or adsorbed contaminants which have not been taken into consideration at the preliminary stage. In general, to control and estimate the efficiency of natural attenuation monitoring must be included. For this reason, most natural attenuation applications are closely linked to monitoring in a sustainable manner defined as Monitored Natural Attenuation (MNA).

With reference to the discussion about natural attenuation this approach is often characterised as an alternative term for taking no action, because no technical application is involved, apart from the monitoring equipment. In fact, it is a proactive approach focused on the verification and monitoring of natural remediation rather than relying on engineered processes. It is important that natural attenuation should not result in an opportunity for the polluter to shirk their duty to implement an adequate decontamination or containment strategy. The public authority is requested to address the problem as the case arises.

Natural attenuation is generally aimed at the avoidance of waste usually generated by *ex situ* measures and at the reduction of hazards to human beings and the environment. An advantage of the approach is that it can be assessed as a relatively environmentally friendly method, because neither the soil is disturbed nor the vegetation is removed. Moreover, even aboveground structures are not affected, because natural attenuation can be carried out even under buildings.

The process is distinguished into two approaches. The first one focuses on the degradation or destruction of the contaminants, while the second should decrease their mobility:

- Destructive approaches meaning the biological or abiotic degradation or transformation of pollutants including radioactive decay and
- Non-destructive approaches including sorption, immobilisation, dilution, dispersion and volatilisation of contaminants.

In principle, natural attenuation is only successful, if the long-term contaminant input is smaller than the output. The input is based on the emission from the source zone, e.g. leachates and contaminant desorption. The output is defined by diffusion and dispersion lowering the concentration, sorption onto the matrix, volatilisation, biological transformation and degradation as well as plant uptake. With reference to the soil, volatilisation and biodegradation are the most important factors, while plant uptake is only effective by cutting and removal of the plants, but this approach is part of the concept of phytoremediation (see Sect. 6.4). Referring to the groundwater diffusion, dispersion, transformation and degradation are predominant, but volatilisation as well as sorption may also play an important role.

With reference to the natural attenuation in the groundwater the process must reduce the critical concentration of the contaminants of concern for the down-gradient groundwater users to a level which excludes a risk to human health. The objective will be only achieved, if acceptable values are measured and the plume spreading is limited in controlled conditions. The down-gradient groundwater quality must be carefully monitored to fulfill this aim. The best way is the implementation of several down-gradient observation wells at different distances to the source. After achievement of the objectives the operation time of the wells can be stopped.

Research projects in different countries (Canada, Denmark, Germany, The Netherlands, USA) dealing with plumes of contaminated sandy to gravelly aquifers associated with landfill leachates have presented an overview of the dimensions of the plumes. They showed plume lengths between 100 and 3,000 m and plume widths ranging from 100 to 600 m. Subsequently, the observation wells should cover the total area of the plumes, usually requiring a greater number of wells (Christensen et al. 2000). To rely on hydro-geological data without consideration of analysed observation wells might be dangerous. The plume can be much wider than evidence furnished by hydro-geological data, because the development of different flow directions occasionally remains undiscovered from a geological perspective. A dense observation well network must be set up to get sufficient information about the three-dimensional leachate plume, in particular in the presence of a heterogeneous aquifer.

To estimate the potential for natural attenuation groundwater flow velocity, porosity of the aquifer and the site specific recharge rates must be estimated. Due to the distinct contaminant behaviour in relation to degradation and migration a high number of monitoring data must be available. For this reason, a lot of observation wells located at different distances to the source and a continuous sampling procedure are necessary in the course of time. It is of importance to conduct long-term and continuous monitoring, since the factors are transient, i.e. they alter over time. The source emission can change, hydraulic conditions may show changing flow directions and velocities, electron acceptors can be consumed and sorption potentials for the organic pollutants are influenced by the constant adsorption/desorption kinetics. It is surely not feasible to investigate all relevant factors continuously – a certain amount of simplification must take place. Therefore, the dominant parameters must be extracted and analysed in order to predict the plume development.

8.2.2 *Contaminants to Be Treated*

Natural attenuation is usually applied to organic pollutants that are generally biodegradable. Therefore, contaminants such as volatile chlorinated hydrocarbons (VCHC) and petroleum hydrocarbons (TPH) are widely treated in the context of this approach.

Table 8.1 Expected likelihood of success using natural attenuation (Data from MacDonald 2000, modified)

Contaminant	Attenuation process	Likelihood of success (%)
Petroleum hydrocarbons (gasoline, fuel oil)	Biotransformation	>50
BTEX aromates	Biotransformation	>75
PAH	Biotransformation, immobilisation	>25
Halogenated aliphatic compounds (e.g. tetrachloroethylene, vinyl chloride)	Biotransformation	>25
PCB, PCDD/F	Biotransformation, immobilisation	>25
Nitroaromates (e.g. TNT)	Biotransformation, abiotic transformation, immobilisation	>25
Heavy metals Cd, Cu, Ni, Pb, Zn	Immobilisation	>50 (Cd>25 %)
Heavy metals Cr and Hg	Immobilisation, biotransformation	>25
Nitrate	Biotransformation	>50
Radionuclides ^{137}Cs, ^{90}Sr	Immobilisation, decay	>50
Radionuclides ^{99}Tc, $^{238,239,240}Pu$, $^{235,238}U$	Immobilisation, decay	>25

Some less degradable pollutants such as PAH, PCB and nitrified compounds (explosives) may also play a role associated with the degradation of the compounds (Pennington et al. 1999). Since natural attenuation is applied to soils contaminated with organic pollutants, it is difficult to draw a clear line between natural attenuation and bioremediation (see Sect. 6.3). Heavy metals and metalloids are also contaminants involved in natural attenuation. The main target, however, is the immobilisation of the parameters. For instance, it has been observed that mobile and toxic CrVI reduces to immobile and less problematical CrIII in anaerobic conditions. The likelihood of success varies from parameter to parameter (Table 8.1).

The wells are analysed with regard to either the mass of the contaminants or to indirect parameters, namely the electron acceptors such as oxygen and the products of the degradation. For instance, chloride is used as evidence of degradation regarding chlorinated hydrocarbons unless chloride is a contaminant of the plume itself (Christensen et al. 2001).

In particular, the results from lowering of the redox potential appear to be a good indicator for investigations of the down-gradient plume contaminated with VCHC, since factors like reduction of sulphate, iron and manganese as well as denitrifying conditions give information about success on the basis of reductive processes like dehalogenation. The zones of anaerobic conditions are well-known and consequently helpful for the interpretation of the degradation sequences (Fig. 8.1) (see Sect. 7.1). The down-gradient wells will show the zones of an anaerobic environment, e.g. the sulfate reducing zone, the manganese reducing zone, etc. but the differentiation does not always appear distinctly, since the different zones of the leachate plume may overlap each other.

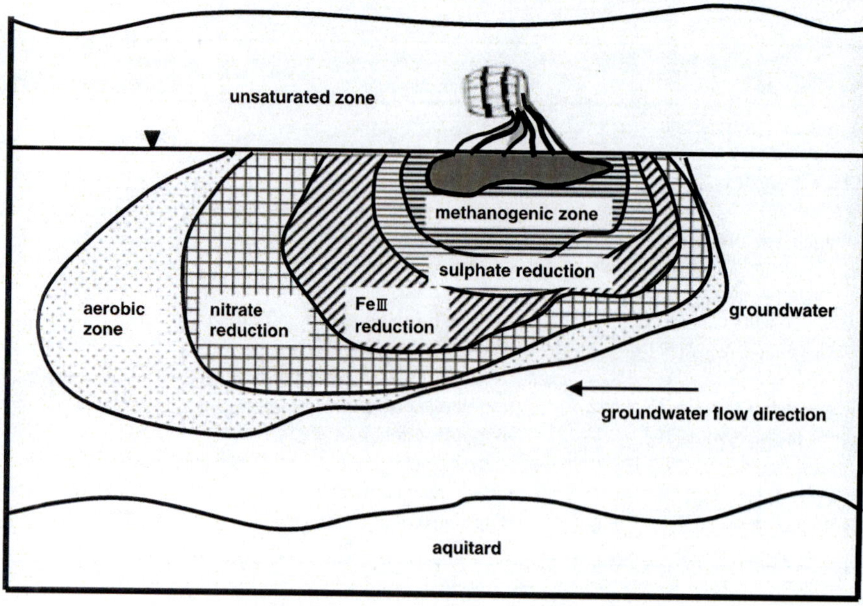

Fig. 8.1 Zoning for the down-gradient anaerobic dehalogenation of VCHC

By way of example, natural attenuation with respect to petroleum hydrocarbons (TPH) contaminating the aquifer is introduced. Some items are strongly associated with this type of contamination. The contaminant group consists of aliphatic compounds which may preferentially constitute the soil vapour contamination as well as aromatic compounds including BTEX aromates being the main contaminants in the groundwater. Furthermore, the petroleum hydrocarbons contain LNAPL that are relatively difficult to handle, because interrelations occur between the phase contaminants and the soluble portion. Some relatively soluble contaminants tend towards a fast degradation and migration, thus extending the plume, while other less degradable contaminants reveal a slow movement off the source.

A second example relates to the volatile chlorinated hydrocarbons (VCHC). Besides the contaminants such as trichloroethylene and tetrachloroethylene as well as their metabolites such as dichloroethylene and vinyl chloride which must be monitored continuously, the geochemical environment where the anaerobic degradation happens must be integrated. The bacteria consume the electron acceptors in well-known sequences, starting with oxygen, followed by nitrate, dissolved iron, dissolved manganese, sulphate and ultimately methane. With regard to the plume development, the redox sequence from methanogenic properties close to the source until the aerobic zone at the fringe area of the plume is expected, and accordingly it is possible to allocate the zones of VCHC degradation (Fig. 8.1). However, this picture is of a theoretical nature, because disturbing factors may impact the zoning. For instance, uncontaminated groundwater may flow to the plume, thus improving the oxidative conditions. Hence, the oxygen-related aerobic degradation would exhibit an increasing tendency and the reductive de-chlorination would be hampered. Normally, the electron acceptors are rapidly depleted, leading to a predominantly reductive degradation.

The estimation of contaminant reduction in line with natural attenuation includes some uncertainties. The source strength is a function of the time-dependent amount of pollutant emitted and the solubility in water. The migration of the pollutants from the source to the observation wells does not always occur in steady conditions. Chemicals appear batch-wise, if, for instance, closed containers or barrels in landfills start leaking after years because the corrosion is a long-term process.

The time span will play an important role during the entire natural attenuation process. There are several periods revealing different contaminant behaviour. In the first period after the hazard has occurred the released contaminant amount might be more intensive than the degraded amount, since the microorganism population is insufficiently adapted to the site conditions. Therefore, a natural attenuation effect will not be recognised. In the following second period usually lasting 2–5 years between contaminant release from the source and degradation more or less stable conditions prevail in which an alteration of the plume does not take place. The third period is characterised by a slow and constant shrinkage of the plume associated with the attenuation process. Theoretically, a fourth period continues with stable conditions at a low pollutant level caused by the remaining persistent and non-biodegradable parameters.

Furthermore, ageing effects which reduce the bioavailability of some organic pollutants may prevent the continued degradation later on. For instance, it has been found that phenanthrene becomes increasingly less available to bacteria in the long term, leading to a strong reduction or complete interruption of the degradation process in the context of long-term natural attenuation (Hatzinger and Alexander 1995)

Problems are also observed in relation to the extraction of the correct contaminant parameters which should be involved in the observation of natural attenuation. It is easy to extract proper parameters, if the source is associated with contaminants which are simple to identify such as TPH in the context of leachates of petroleum stations. With reference to landfill leachates, however, a cocktail of chemicals can be expected and the extraction of some critical parameters to act as indicators is much more difficult. Moreover, during the process of natural attenuation some harmful metabolites will arise which are not of concern at the beginning. For example, the degradation of volatile chlorinated hydrocarbons (VCHC) may accumulate vinyl chloride that has shown very low concentrations at an earlier time. Consequently, attention should be paid to that parameter with an increasing tendency in the course of the observation (Peter et al. 2011).

Problems will arise in relation to possible toxicity of the transformation products. This can be higher than the toxicity of the originally contaminated substance. Another problem may arise with reference to the dissolved organic carbon (DOC) that acts as electron donor. The potential, however, depends on the type of DOC (natural, technogenic). Thus, it is hard to predict whether the DOC suffices to ensure the degradation rates of the microorganisms required. Otherwise, co-substrates must be artificially added (ENA).

On the other hand, NAPL such as oils and tars and, in particular, technogenic carbon such as ash, soot and coal residues increase the sorption capacity and decrease the migration of the pollutants into and in the groundwater. By comparison, natural organic matter reveals a fast and linear absorption behaviour which has a tendency to be reversible but technogenic carbon (combustion residues) exhibits a slow, non-linear and irreversible absorption of organic pollutants (and metals) (Luthy et al. 1997). Some contaminants such as chlorobenzene and PAH even appear to prefer sorption onto technogenic soot-like material. In turn, the desorption of organic pollutants is minimised, if micropores of precipitated mineral and organic material are blocked.

The formation of bubbles associated with the chemical reactions due to the degradation of organic pollutants, which produces reductive gases such as methane and carbon dioxide, may influence the hydraulic conductivity in the aquifer. Furthermore, a vertical gradient of the contamination can take place because of the different specific gravity of the contaminants. For instance, contaminants with higher densities tend to infiltrate into deeper layers of the aquifer where the redox conditions are different from the conditions in the upper portion of the aquifer. Consequently, it is possible to underestimate or overestimate the effects of natural attenuation.

With reference to the metals natural attenuation is predominantly focused on adsorption, precipitation and occlusion processes. The sediment in the aquifer has a higher buffer capacity than the groundwater. Accordingly, the chemical characterisation has to be involved in sediment analyses. The sorbing or precipitating constituents such as the organic matter, Fe hydroxides or calcium carbonate should be investigated prior to the implementation of natural attenuation in order to quantify the expected mobility of the contaminants. The mobility of heavy metals is influenced by the sorption to the organic matter and fine-textured soil particles as well as the formation of insoluble solids. For instance, a redox potential to a level of sulfate reduction to S^{2-} leads to the formation of completely insoluble sulfide precipitants which exclude metal migration. One opportunity to evaluate the different sorption potentials of the matrix is the use of the sequential extraction method. In general, sorption and precipitation can decrease the metal mobility by a factor of 100 and more (Förstner and Gerth 2001).

In principle, a permanent reducing status might prevent the metal mobility in soil and aquifer, while the bioavailability, for example, which determines the decline of organic pollutants, depends mainly on the organic matter content. The sorption potential, however, might be disturbed by the presence of mobilising compounds such as chelating agents (e.g. DOC). Thus, apart form typical contaminants, further indicator parameters should be integrated such as the dissolved organic carbon. For instance, in the case of the mass reduction of metals this parameter might be of interest, since it is possible for DOC to enhance the metal mobility caused by the complexation. It should be noted that the hydro-geological conditions can change over time, so that the previously stabilised metals may tend to be mobilised once again. In contrast, drastic changes in the composition of groundwater entering the contaminated zone occur only exceptionally.

In conclusion, the interpretation of the sorption processes for metals and organic pollutants is difficult to determine, since a lot of soil properties affect the sorption kinetics and are in part competitive.

8.2.3 Field of Application

The acceptance of natural attenuation varies from country to country. If comprehensive investigations are challenged, which possibly exceed the measures of technical remediation, the realisation of this approach appears to become ineffective. The background to this problem derives from the fear of the public authority that uncontrolled dilution and dispersion of the contaminants occur as negative side effects. Nowadays, natural attenuation is only accepted as a stand-alone approach, if alternative methods are too expensive or not proportional. In the USA the idea of natural attenuation often fails due to the long timeframe expected, in Germany regulations exacerbate the practical application and in the Netherlands the approach will not be adopted, if an unacceptable risk to human health cannot be excluded (Peter et al. 2011).

Another problem of natural attenuation is the time span needed. This is difficult to predict and consequently it may take decades. The prediction depends on the long-lasting presence of the reaction partners, which is difficult to calculate. To plan the timeframe of natural attenuation a number of soil and groundwater characteristics must be well-known. Redox processes have a strong significance, influencing other chemical properties such as pH value, sorption and the existing microorganisms to a great extent. Nevertheless, in comparison with operating clean-up methods the time period of natural attenuation is longer but the costs are mostly decisively lower. The time span required depends on the lifetime of the source. If a removal of the source does not take place, the lifetime of the source might be longer than a human lifetime, spreading over hundreds of years (Eberhardt and Grathwohl 2002). Hence, natural attenuation of the plume might be only reasonable in association with source elimination.

It is possible to evaluate the success of natural attenuation using microcosm studies, laboratory batch tests or field investigations. In particular, microcosm studies under more controlled conditions and dealing with native bacteria may demonstrate the expected degradation that depends on the pollutant of interest and the intended aquifer conditions (aerobic, anaerobic). By way of restriction, some organic pollutants such as PAH reveal a very slow degradation process in anaerobic conditions and consequently successful microcosm investigations are mostly limited.

The idea of natural attenuation is selected, if the destruction or immobilising of the contaminants is feasible in the course of time. Subsequently, it requires a site-specific individual decision at each time. In summary, prior to the application of natural attenuation in soil and aquifer (sediment) different chemical information is essential:

- Characterisation of the organic matter (natural, technogenic)
- Sorption and desorption behaviour, particularly of the hydrophobic organic pollutants
- Assessment of the bioavailability of aged hydrophobic organic compounds
- Characterisation of the long-term reactivity of metals
- Estimation of irreversible sorption processes to modify the expected sorption quantity.

In principle, natural attenuation in relation to groundwater contamination is not recommended, if the exact contamination source and spreading is unknown, persistent pollutants are present which can impair highly sensitive uses such as catchment areas of waterworks for drinking water purposes or if the aquifer consists of karst and fissured bedrock with seriously determinable hydraulic conductivity.

To perform the natural attenuation approach a number of characteristics should be checked. Firstly, the prerequisites must be assessed, e.g. in the case of a groundwater-related application the hydro-geological conditions and the chemical properties with reference to the degradation processes. Next, the effectiveness should be checked using, for example, laboratory tests. As a consequence, it is possible to plan and to implement the design of the monitoring system. Based upon the chemically determined effectiveness it should be feasible to predict theoretically the time-related success of the measure. Simulation models are helpful tools. Afterwards, the process of MNA, accompanied by the monitoring programme, commences. Regarding the groundwater different observation wells must be installed down-gradient of the source.

References

Christensen, T. H., Kjeldsen, P., Bjerg, P. L., Jensen, D. L., Christensen, J. B., Baun, A., Albrechtsen, H. J., & Heron, G. (2000). Biogeochemistry of landfill lecheate plumes. *Applied Geochemistry*. http://cues.rutgers.edu/bioreactorlandfill/pdfs/15-Kjeldsenetall2002CritRevEnvSciLandfillLeachat.pdf. Accessed 10 Jan 2012.

Christensen, T. H., Bjerg, P. L., & Kjeldsen, P. (2001). Natural attenuation as an approach to remediation of groundwater pollution at landfills. In R. Stegmann, G. Brunner, W. Calmano, & G. Matz (Eds.), *Treatment of contaminated soil*. Berlin: Springer.

Eberhardt, C., & Grathwohl, P. (2002). Time scales of organic contaminant dissolution from complex source zones: Coal tar pools vs. blobs. *Journal of Contaminated Hydrology, 59*, 45–66.

Förstner, U., & Gerth, J. (2001). Natural attenuation – Non-destructive processes. In R. Stegmann, G. Brunner, W. Calmano, & G. Matz (Eds.), *Treatment of contaminated soil*. Berlin: Springer.

Hatzinger, P. B., & Alexander, M. (1995). Effect of aging of chemicals in soil in their biodegradability and extractability. *Environmental Science and Technology, 29*, 537–545.

Luthy, R. G., Aiken, G. R., Brusseau, M. L., Cunningham, S. D., Gschwend, P. M., Pignatello, J. J., Reinhard, M., Traina, S. J., Weber, W. J., & Westall, J. C. (1997). Sequestration of hydrophobic organic contaminants by geosorbents. *Environmental Science and Technology, 31*, 3341–3347.

MacDonald, J. A. (2000). Natural attenuation for groundwater cleanup. *Environmental Science and Technology, 34*, 346–353.

Pennington, J. C., Gunnison, D., Harrelson, D. B., Brannon, J. M., Zakikhani, M., Jenkins, T. F., Clarke, J. U., Hayes, C. A., Myers, T., Perkins, E., Ringelberg, D., Townsend, D., Fredrickson, H., & May, J. H. (1999). *Natural attenuation of explosives in soil and water systems at Department of Defence sites*. Technical Report, U.S. Army Engineer Waterways Experiment Station, Vicksburg, MS, USA.

Peter, A., Held, T., Hüsers, N., & Swartjes, F. A. (2011). Natural attenuation. In F. A. Swartjes (Ed.), *Dealing with contaminated soils*. Dordrecht: Springer.

Chapter 9
Remediation Planning

Abstract In relation to the choice of one or more site-specific measures for an adequate soil, soil vapour and groundwater treatment, different characteristics must be taken into consideration. In order to recognise suitable approaches, the influence of soil properties such as texture, organic matter content and pH value and the composition of contaminants which are present (metals, organic pollutants, cyanides) should be well known. In this chapter, the relevant parameters which are decisive in the context of evaluating the feasibility for remediation techniques are summarised and compared. Moreover, other aspects influencing the decision on the treatment version, namely, the soil protection factor, which is aimed at very minimal disturbance of natural soil conditions such as the horizonation and the edaphon, the possible time schedule as well as the cost factor, are also dealt with. These topics must be considered with regard to the entire soil remediation planning process, which can be considerably facilitated and improved in the case of a complete integration of all important aspects mentioned above. Irrespective of the parameters which determine the choice of adequate remediation methods, information about general aspects relating to the planning process of contaminated land is additionally given in this chapter.

Keywords Remediation cost • Remediation planning • Soil protection aspect • Time schedule • Technical feasibility

9.1 General Aspects of the Planning Processes

The arrangement of the clean-up plan is able to be combined with the land use planning. Before implementation of a legally binding land use plan the remediation processes should be taken into consideration (see Sect. 2.1.2). Sites intended for highly sensitive utilisation such as playgrounds and kindergarten areas require intensive and complete soil remediation. For this reason, it is not recommended to

H. Meuser, *Soil Remediation and Rehabilitation: Treatment of Contaminated and Disturbed Land*, Environmental Pollution 23, DOI 10.1007/978-94-007-5751-6_9,

construct these land uses at sites that are influenced by high soil pollution. Sites used for moderately sensitive residential areas, however, should not be located at areas with high soil contamination, while industrial plots are possible to locate at areas indicating a high level of contamination (see Fig. 4.13 in Sect. 4.2.4).

Special attention must be paid to the necessary remediation time. In a multitude of areas that are intended for brownfield redevelopment the period of time is strongly limited, because the investment must be carried out as soon as possible. Investors and landlords want to redevelop the contaminated land in order to start the business and to generate financial yield at once. Thus, the private investors are interested in a fast and smooth running of the clean-up process. Consequently, apart from the costs, the remediation time becomes a decisive factor in the decision-making, so that in a number of cases fast treatment is preferred to a long-term solution, though the former is more expensive than the latter.

The real soil clean-up plan should include a number of important features to enable a smooth running process (BBODSCHV 1999). The initial situation must take into account the soil properties and, in particular, the contamination level. This is usually already well-known, since the former investigation phases, namely the oriented and the detailed investigations, have been finished. According to the future land use the clean-up criteria must be defined and decisions which have been ordered by the public authority should be integrated into the clean-up operations planned.

The exact treatment must be described in detail in relation to the area of concern and the operation chart. In general, the necessary demolition of buildings (see Sect. 4.3) and the soil excavation processes can be differentiated. In particular, a quantitative calculation of the excavated material should be included (see Sect. 4.2). With reference to the actual measures the technical application of protection measures (see Chap. 5) as well as the applicability of decontamination measures (see Chaps. 6 and 7) must be examined. Working safety and immission control are also very important factors to be considered (see Sect. 4.4).

Because many measures need long-term monitoring, the clean-up strategy should also answer questions about the necessity for soil, groundwater and vapour sampling procedures and analyses. Besides, the monitoring is carried out by long-term operating equipment and the maintenance of the equipment must also be considered. Ultimately, a favourable clean-up plan is characterised by exact schedules about remediation time and costs.

The clean-up process, which can last months or years, demands patience of the inhabitants confronted with the plan. Thus, it makes sense to involve the inhabitants at the beginning of the procedure. An advisory council should be established and should implement at least those framework recommendations, if political decisions cannot be taken. The advisory council should hold meetings at regular intervals and include, apart from the affected inhabitants, members of the public authority and the politicians.

In general, the council should get the knowledge listed below:

- Detailed and understandable information about the remediation plan
- Information about the governmental or municipal course of action

- Opportunity to be involved in the scientific investigation methods, the background of the chosen assessment system and the drawing-up of the remediation plan
- Participation in the decision-making of the public authority
- Call for open councils and hearings.

9.2 Evaluation of the Feasibility for Soil Remediation Techniques

9.2.1 Influence of Soil Properties

In Table 9.1 the soil properties texture, organic matter content and pH value, which represent the most important features of the soil quality, are assessed with reference to the potential opportunities to apply the different remediation technologies. The texture as main component of the soil physics and mechanics includes aspects of the soil, air and water household as well as technical and constructive characteristics, whereas the organic matter and the pH value refer to the nutrient capacity and biological activity of the soil. Both texture and organic matter content are related to the adsorption potential of pollutants. In order to select the best remediation approach these properties must always be investigated and related to the technology preferred.

The reasons for the allocation to texture, organic matter and pH value classes are described in Chaps. 5 and 6, Sects. 7.2 and 8.2 in detail. Some general conclusions are introduced as follows:

Gravelly and sandy soils are more appropriate to the different remediation strategies compared to more cohesive soils. Most of the techniques listed in Table 9.1 allow a treatability of coarse grained soils. Limitations are visible in association with soils only consisting of gravel and coarse sand and the remediation approaches vitrification, phytoremediation and natural attenuation. The texture classes loamy sand, sandy loam and sandy clay loam are normally treatable if the percentage of silt plus clay does not exceed approximately 40%. These texture classes are less beneficial in relation to *in situ* bioremediation, since the deficient oxygen concentration already leads to reduced biodegradability. Bioventing can degrade less permeable soils because a reduced air volume is required (Khan et al. 2004). Apart from the excavation procedures, surface cover, thermal treatment, electroremediation, stabilisation using organic matter and steam-enhanced extraction (SEE), cohesive soils cause a number of problems for various remediation techniques such as swelling-shrinking processes (encapsulation), bad miscibility (solidification, asphalt batching), a too low percentage of sand (soil washing, soil extraction), lack of oxygen (every type of bioremediation, natural attenuation), detrimental growth conditions for the vegetation (phytoremediation) and strongly reduced air permeability (soil vapour extraction).

Table 9.1 Feasibility of remediation techniques in relation to soil properties

	Texture				Organic matter				pH value		
	Gravel, coarse sand	Medium to fine sand	Loamy sand, sandy loam, sandy clay loam	Cohesive soils (silt plus clay >40%)	<1%	1–5%	>5–30%	Peaty soils	<5.5	5.5–7.5	>7.5
Excavation, transportation and landfilling	++	++	++	+	++	+	o	–	o	++	++
Surface cover (geotextile-based and bentonite-based)	++	++	++	+	++	+	–	–	+	++	++
Sealing	++	++	++	–	++	+	–	–	–	++	++
Barriers installation at the sides	+	++	++	o	++	+	o	–	–	++	++
Encapsulation	+	++	+	–	++	+	o	–	–	++	++
Solidification (cement-based, *in situ*)	++	++	+	–	++	+	–	–	–	+	++
Asphalt batching	++	+	+	–	++	o	–	–	–	+	++
Vitrification	o	++	+	o	++	+	–	–	+	++	++
Stabilisation (liming)	+	+	++	o	++	+	–	–	o	++	++
Stabilisation (organic matter addition)	+	++	++	+	++	++	o	o	o	++	++
Soil washing	++	++	+	–	++	+	–	–	++	o	o
Soil extraction (acids, detergents)	++	++	+	–	++	++	o	–	++	+	+
Bioremediation (*in situ*)	+	++	o	–	–	++	o	–	–	++	o
Bioremediation (landfarming)	+	++	+	–	o	++	+	o	o	++	+
Bioremediation (regeneration pile)	+	++	+	–	o	++	o	–	o	++	o
Bioremediation (bioreactor)	++	++	++	–	+	++	+	–	+	++	+

Phytoremediation (extraction)	o	+	++	–	–	++	o	o	++	o	o
Phytoremediation (degradation/ volatilisation)	o	++	+	–	–	++	o	–	–	++	o
Thermal treatment	+	++	++	+	++	++	++	++	++	++	o
Electroremediation (*in situ*)	++	++	+	++	++	++	o	–	++	+	o
Natural attenuation	o	+	+	–	–	+	o	–	–	++	o
Soil vapour extraction (SVE)	++	++	+	–	++	+	o	–	++	++	++
Bioventing	+	++	+	o	o	++	o	–	–	++	o
Steam enhanced extraction (SEE)	++	++	++	+	++	++	+	–	++	++	++

Assessment of the feasibility:
++ very high efficiency (generally recommended)
+ high efficiency (recommended)
O low efficiency/problematical consequences (rather not recommended)
– no efficiency (not recommended)

Similar proportions can be detected on examination of the organic matter content. Soils exhibiting low humus concentration up to maximum 5% are relatively well treatable. Strong limitations are present for soils revealing only less than 1% organic matter regarding *in situ* bioremediation, landfarming, biopile technique, phytoremediation, bioventing and natural attenuation due to the detrimental living conditions for the microorganisms. Soils with accelerated humus content have a tendency to offer less favourable opportunities in the context of clean-up procedures. For instance, surface cover and sealing appear to be inappropriate, since the organic matter is slowly degraded under the cover system, as a result of which the stability is disturbed. The biodegradability of the substrate means restrictions regarding solidification and asphalt batching as well. Stabilisation based on liming might also be unsuitable due to the enormous lime application rate required. The effectiveness of soil washing and soil extraction is reduced because of the high adsorption potential of the organic matter for pollutants. The same reason might be important on examination of phytodegradation. Peaty soils are the least treatable substrates of all. The negative aspects mentioned in relation to soils with an organic matter content ranging from 5 to 30% are transferable to soils exhibiting more than 30% organic matter. Nearly all remediation strategies fail and the best way to remediate peaty soils is thermal treatment.

The influence of the pH value is lower than the other soil properties discussed. While soils indicating a pH value higher than 5.5 are generally well treatable with most of the techniques, acid soils reveal clear limitations. The barrier installation at the sides, encapsulation, solidification and asphalt batching need a neutral to alkaline environment because acids may cause chemical weathering of the structures installed. *In situ* bioremediation is less successful, since the living conditions for the microbes are not favourable and they can only be compensated for in the uppermost horizon to a moderate extent. For the same reason, natural attenuation, phytodegradation and bioventing are less recommended in acidified soils. It should be noted that very alkaline properties also mean drastic limitations for soil washing and phytoextraction (low solubility of metals), nearly all bioremediation strategies, natural attenuation, phytodegradation and bioventing (detrimental environment for microorganisms), electroremediation (obstruction of the acid front) and thermal treatment (extreme alkalinity after treatment and subsequent bad re-use opportunities).

9.2.2 *Influence of the Pollutant Composition*

The composition of the pollutants is the most decisive factor for the remediation choice. As summarised in Table 9.2, there are enormous differences between the most important pollutants regarding the applicability of remediation measures. In detail, the treatability of the pollutants in relation to the clean-up procedures is explained in Chaps. 5 and 6, Sects. 7.2 and 8.2.

Table 9.2 Feasibility of remediation techniques in relation to soil and soil vapour contaminants

	Metals					Organic pollutants						CN
	Cd, Ni, Zn	Cu, Pb	Cr	As	Hg	TPH (mineral oils)	BTEX	Phenols	VCHC	PAH	PCB	Cyanides (free)
Excavation, transportation and landfilling	++	++	o	++	+	+	+	+	+	++	++	++
Surface cover (geotextile-based and bentonite-based)	++	++	++	++	o	o	+	+	o	++	++	++
Sealing	++	++	++	++	o	o	o	o	–	+	++	++
Barriers installation at the sides	++	++	++	++	o	o	+	+	o	++	++	++
Encapsulation	++	++	o	o	o	o	+	o	o	++	++	o
Solidification (cement-based, *in situ*)	++	++	o	o	–	–	o	–	–	o	o	o
Asphalt batching	++	++	++	++	+	+	+	+	+	++	++	++
Vitrification	+	+	++	++	+	+	+	+	+	++	++	++
Stabilisation (liming)	++	++	++	–	o	+	+	o	o	–	–	–
Stabilisation (organic matter addition)	++	++	+	–	o	+	+	o	o	+	+	–
Soil washing	+	+	++	++	o	o	o	+	o	–	–	+
Soil extraction (acids, detergents)	++	++	++	++	+	+	+	+	+	o	o	+
Bioremediation (*in situ*)	–	–	–	–	–	+	+	+	+	–	–	o
Bioremediation (landfarming)	–	–	–	–	o	++	++	++	+	+	–	+
Bioremediation (regeneration pile)	–	–	–	–	o	++	+	++	+	o	–	+
Bioremediation (bioreactor)	–	–	o	–	–	++	++	++	++	+	o	++
Phytoremediation (extraction)	+	o	–	o	–	–	–	–	–	–	–	–

(continued)

Table 9.2 (continued)

	Metals					Organic pollutants						CN
	Cd, Ni, Zn	Cu, Pb	Cr	As	Hg	TPH (mineral oils)	BTEX	Phenols	VCHC	PAH	PCB	Cyanides (free)
Phytoremediation (degradation/ volatilisation)	–	–	o	–	+	+	+	+	+	o	o	+
Thermal treatment	–	–	–	o	++	++	++	++	++	++	++	++
Electroremediation (*in situ*)	+	+	+	+	–	–	+	+	o	–	–	+
Natural attenuation	–	–	–	–	o	+	+	+	o	–	–	o
Soil vapour extraction (SVE)	–	–	–	–	+	+	o	o	++	–	–	–
Bioventing	–	–	–	–	o	++	++	++	+	–	–	o
Steam enhanced extraction (SEE)	–	–	–	–	+	+	++	++	++	o	–	–

Assessment of the feasibility:
++ very high efficiency (generally recommended)
+ high efficiency (recommended)
O low efficiency/problematical consequences (rather not recommended)
– no efficiency (not recommended)

Some heavy metals (Cd, Cu, Ni, Pb, Zn) are treatable in a number of technical devices apart from methods of bioremediation including natural attenuation, treatments dealing with the gaseous components in the soil (phytovolatilisation, soil vapour extraction, steam-enhanced extraction, bioventing) and thermal treatment. Phytoextraction focuses preferentially on the elements Cd, Ni and Zn. The opportunities to treat chromium are more restricted, since in aerobic conditions chromium tends to change into toxic CrVI. With reference to the detoxification of chromium, which would mean a reduction to CrIII, measures which include temporary loosening of the soil structure such as excavation, encapsulation and solidification exacerbate the treatability. On the other hand, approaches using water and subsequent causing anaerobic properties (soil washing, bioreactor technique) increase the opportunities to detoxify chromium containing soils. Because of the anionic character of arsenic some clean-up methods which are suitable for cationic heavy metals are more problematical, since anions are not adsorbed, in particular in alkaline materials. For instance, encapsulation, solidification and stabilisation (liming and addition of organic matter) cannot completely prevent arsenic leaching and are consequently not recommended. The non-biodegradability of metals also excludes a biological treatment of chromium and arsenic. Mercury (Hg) behaves in a distinct way, so that the technical solutions to remediate Hg-contaminated soils are not comparable with the metals already mentioned. The volatilisation impedes technical constructions such as surface cover, sealing, side barrier installation and encapsulation. The expected gaseous migration might also be the reason for the limited applicability of solidification. Open-air operations such as stabilisation, *in situ* bioremediation, phytoremediation and electroremediation are generally difficult to realise, particularly in the presence of buildings adjacent to the polluted site of concern. In contrast, soil vapour extraction, steam-enhanced extraction and thermal treatment are adequate strategies for removing mercury from the contaminated soils effectively.

While metals are not treatable using bioremediation strategies, the decontamination of most organic pollutants and free cyanides is based preferentially on biological disintegration. TPH, BTEX, phenols, VCHC and cyanides are predestined for biological treatment, whereas more complex organic pollutants (PAH, PCB) exhibit clear limitations. Natural attenuation is also generally possible but often rejected because of the long remediation time. Moreover, organic pollutants including cyanides are treated in different ways. Excavation, asphalt batching, vitrification and thermal treatment can nearly always be carried out.

Surface cover, sealing, side barriers and encapsulation have disadvantages with regard to volatile components, in particular VCHC. Solidification also runs the risk of long-term gaseous release. Moreover, the anionic character of cyanides means leaching problems in the case of solidification and stabilisation (liming). Soil washing and soil extraction are not useful for complex organic pollutants, namely PAH and PCB. In contrast to phytoextraction, which is not of interest due to the very low solubility of the organics pollutants, phytodegradation might play a role for the well-degradable contaminants. The approaches dealing with the gaseous components (soil vapour extraction, steam-enhanced extraction, bioventing) are partly taken into consideration, especially for TPH, BTEX and VCHC.

In conclusion, for every pollutant various solutions can be found in an analogous way to the treatability for different soil properties presented in Sect. 9.2.2. In order to extract the best method both soil properties and pollutant composition must be combined. There is no doubt that the opportunities will decrease to a few remediation procedures, in particular in consideration of other important aspects which must be involved in the clean-up planning, e.g. requirements of soil protection, time schedule and cost.

9.2.3 Consideration of the Soil Protection Factor

From the soil protection point of view the remediation process should be as soil-protective as feasible. In Table 9.3 important soil characteristics are listed in association with the different clean-up procedures.

The most important soil-protective physical property influenced by soil treatment is the soil structure. The valuable chemical properties are defined as pH value, macronutrients (nitrogen, phosphorus, potassium) and the organic matter content, while the biological value of the soil is identified by the fauna community (edaphon). Furthermore, the natural horizonation should be taken into consideration, because this parameter is predominantly connected to the naturalness of the soil profiles.

Horizonation, soil structure and edaphon are in danger of being completely destroyed, disturbed or marginally influenced. The influence, however, can be both negative and beneficial. In some respects the soil conditions are improved, an amelioration is even detectable, e.g. in the case of bioremediation approaches the edaphon abundance and distribution show an increasing trend. The nutrient capacity, the organic matter content and the pH value can be strongly or marginally impaired but it is also possible to obtain improved results.

In general, *in situ* measures reveal less soil disturbance and destruction than *ex situ* techniques. Regarding *ex situ* measures the horizonation is completely destroyed and subsequently the soil structure is more or less different from the initial situation.

In detail, the soil treatments show the following impacts on the soil characteristics of the contaminated layers:

- Excavation that is combined with a multitude of remediation methods leads to destruction of soil horizonation and structure as well as having a strong negative influence on the living organisms in the soil.

Table 9.3 Feasibility of remediation techniques in relation to the requirements for soil protection

	Horizonation	Soil structure	Edaphon	pH value	Nutrient capacity	Organic matter content
Excavation, transportation and landfilling	––	––	–	x	x	x
Surface cover (geotextile-based and bentonite-based)	–	–	o	x	x	–
Sealing	–	–	o	x	x	–
Barriers installation at the sides	o	o	x	x	x	x
Encapsulation	–	o	x	x	x	x
Solidification (cement-based, *in situ*)	––	––	––	–	o	–
Asphalt batching	––	––	––	o	o	–
Vitrification	o	–	––	o	o	–
Stabilisation (liming)	x	+	+	–	o	o
Stabilisation (organic matter addition)	o	+	+	o	+	+
Soil washing	––	––	–	x	–	–
Soil extraction (acids, detergents)	––	––	–	–	–	–
Bioremediation (*in situ*)	x	o	+	o	+	o
Bioremediation (landfarming)	––	–	+	o	+	+
Bioremediation (regeneration pile)	––	–	+	o	+	+
Bioremediation (bioreactor)	––	––	o	–	+	o

(continued)

Table 9.3 (continued)

	Horizonation	Soil structure	Edaphon	pH value	Nutrient capacity	Organic matter content
Phytoremediation (extraction)	o	x	x	o	+	+
Phytoremediation (degradation/ volatilisation)	o	x	+	o	+	+
Thermal treatment	––	––	––	o	–	–
Electroremediation (*in situ*)	x	o	––	–	–	x
Natural attenuation	x	x	x	x	x	x
Soil vapour extraction (SVE)	o	o	o	x	x	o
Bioventing	x	x	x	x	x	o
Steam enhanced extraction (SEE)	o	–	–	o	o	o

Assessment of the contaminated soil/material:
Horizonation, soil structure, edaphon
–– destruction
– disturbance
O low influence
X missing influence
+ improvement (amelioration)
pH value, nutrient capacity, organic matter content
– strong influence
O low influence
X missing influence
+ improvement (amelioration)

- Surface cover, sealing and encapsulation are also linked to the disturbance of horizonation and structure (compaction) and they are responsible for the deterioration of the living conditions for the edaphon. The impact on horizonation and soil structure can be recognised on the periphery of the contaminated land to be treated when side barrier installation takes place.
- Solidification, asphalt batching and vitrification lead to extreme consequences with reference to nearly all soil characteristics. The original soil is transformed into a new solid material that should not be named "soil". Consequently, these techniques occur in a way that is not soil-protective at all.
- The *ex situ* operations soil washing, soil extraction and thermal treatment also result in the generation of new material that is completely different from the original one. This is caused by separation and agent addition (washing, extraction) and by heating (thermal treatment). The thermal treatment causes a completely destructed edaphon and organic matter, clay minerals and the nutrient status reveal considerable disturbance in any case. Again, the techniques are assessed to demonstrate a solution that does not take soil protection into consideration.
- *In situ* methods, however, can also decrease the soil quality. For instance, electroremediation strongly disturbs the living conditions of the edaphon and some anionic nutrients such as nitrate and sulphate are mobilised and leached to a great extent.
- The biological treatment can result in some beneficial effects, for instance higher microbial activity combined with increased nutrient and humus content. The inoculation of microbes might lead to increased activity of the edaphon, fertilizing can be responsible for beneficial effects regarding soil structure, nutrient capacity and perhaps the pH value. The stimulated degradation, however, could cause a decrease in the organic matter content. Furthermore, the pH value can be altered, since the living conditions for the microbes have been adjusted during the operation. In particular, *in situ* bioremediation might be a soil-protective approach.
- Phytoremediation has no detrimental impact on the soil or only a negligible one. In contrast, the amendment of fertilizers, in particular organic manure, possibly leads to improved soil conditions (stabilised structure, enhanced biological activity, accelerated nutrient and humus content). The impact on the horizonation is reduced to the development of a ploughed soil horizon.
- The treatments oriented towards the gaseous pollutants show some detrimental characteristics. For example, the long-term vapour extraction alters the soil physics and, in particular, the pore system, resulting in a modified soil structure. Moreover, the edaphon can be damaged in the long term. The steam-enhanced extraction technology results in more clearly negative influences on soil structure and the edaphon.
- It is beyond controversy that natural attenuation is the only approach without any adverse effect.

9.3 Evaluation of the Feasibility for Groundwater Remediation Techniques

In Sect. 7.1 remediation strategies regarding contaminated groundwater were described intensively and in detail. As shown in Table 9.4, the groundwater remediation techniques focus on homogeneous and permeable aquifers consisting of the texture classes gravel, coarse sand, medium sand and fine sand. The only approach accepting less permeable aquifers (sandy silt, sandy loam) is the drain-and-gate procedure.

Most of the techniques are related to the plume treatment but there are some agents which are directly applied to the contaminant source, because they are capable of attacking the pollutants due to their chemically aggressive reactivity.

Mineral oils (TPH) are treated by a simple pump-and-treat (P&T) technique, which can be effectively supported with agents such as bubble-free oxygen, H_2O_2 and MgO_2, or based on the reactive wall technologies (PRB, F&G, D&G). The BTEX decontamination resorts to more technical solutions compared to TPH. Nearly all agents can improve the P&T approach and additionally the reactive wall technology is an appropriate method. Moreover, for TPH and BTEX biosparging might be an alternative for the decontamination processes. In contrast, the opportunities to remediate PAH-contaminated groundwater are less and, apart from P&T using oxygen, H_2O_2 or MgO_2, the effectiveness clearly seems to be more limited. In the case of PAH only naphthalene shows a higher decontamination potential with reference to the different agents used.

The VCHC elimination occurs with the help of P&T, which can also be accelerated in the presence of some agents such as oxygen, sulphate, H_2O_2 and Fenton's agent. Furthermore, the reactive wall approaches are an appropriate means to reduce the VCHC groundwater contamination. Special attention should be paid to alcohol and tenside flushing successfully applied to aquifers contaminated with VCHC. The air sparging technique can be used for VCHC removal, too.

Except for some very reactive substances (Fenton's agent, ozone, permanganate, persulphate) the real cyanide treatment in groundwater is rather limited. Similarly, heavy metals and, in particular, soluble CrVI in groundwater can be decontaminated using only the reactive wall methods.

The non-aqueous phase liquids (NAPL) swimming onto aquifers require specific equipment for the removal. P&T, PRB, F&G and D&G are adequate solutions but in many cases a direct source treatment is necessary. This can be done with alcohol or tenside flushing.

9.4 Time and Cost Factor

In association with the decontamination techniques in Fig. 9.1 the relationship between remediation time and achievement of the remediation objective is presented. For instance, excavation of the contaminated material is assessed to be the

Table 9.4 Feasibility of groundwater remediation techniques regarding aquifer permeability, application to source or plume contamination and pollutant composition

	Aquifer permeability		Location		Pollutants						
	Gravel, coarse to fine sand	Sandy silt, sandy loam	Source contamination	Plume contamination	Heavy metals	TPH (mineral oils)	BTEX	PAH	VCHC	Cyanides	NAPL
Pump & Treat (P&T)	X			X		X	X		X		X
Pump & Treat with agent infiltration (oxygen)	X			X			X	X	X		
Pump & Treat with agent infiltration (O_2 bubble-free)	X			X		X	X	X			
Pump & Treat with agent infiltration (nitrate)	X			X			X	(X)			
Pump & Treat with agent infiltration (Fe^{3+})	X			X			(X)		(X)		
Pump & Treat with agent infiltration (Mn^{4+})	X			X			(X)				

(continued)

Table 9.4 (continued)

	Aquifer permeability		Location		Pollutants						
	Gravel, coarse to fine sand	Sandy silt, sandy loam	Source contamination	Plume contamination	Heavy metals	TPH (mineral oils)	BTEX	PAH	VCHC	Cyanides	NAPL
Pump & Treat with agent infiltration (sulphate)	X			X			X	(X)	X		
Pump & Treat with agent infiltration (H_2O_2)	X			X		X	X	X	X		
Pump & Treat with agent infiltration (MgO_2)	X			X		X	X	X			
Alcohol flushing	X		X						X		X
Tenside flushing	X		X						X		X
Application Fenton's agent	X		X			(X)	X	(X)	X	X	
Application ozone	X		X				X	(X)		X	
Application permanganate	X		X				X			X	
Application persulphate	X		X			(X)	X	(X)		X	
Permeable Reactive Wall (PRB)	X			X	X	X	X	(X)	X		X

Funnel & Gate (F&G)	**X**			**X**	**X**	**X**	**X**	**(X)**	**X**		**X**
Drain & Gate (D&G)	**X**	**X**		**X**	**X**	**X**	**X**	**(X)**	**X**		**X**
Air sparging	**X**			**X**					**X**		
Biosparging	**X**			**X**		**X**	**X**	**(X)**			

X possible
(X) possible with restrictions

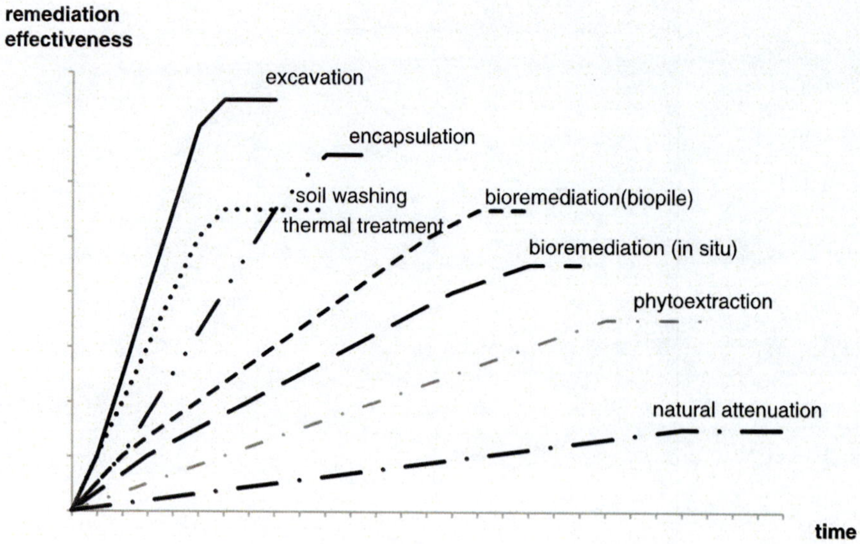

Fig. 9.1 Schematic relationship between remediation time and gained remediation objective

fastest procedure of all, since the site becomes uncontaminated immediately after the excavation process. In contrast, *ex situ* soil washing and incineration need more time because of combined excavation, transportation, treatment and backfill. Moreover, the remediation level is reduced, since a portion of the contaminated material cannot be treated (e.g. the fine fraction during soil washing) or some contaminants are generally not eliminated (e.g. heavy metals during thermal treatment). In the absence of the backfill process, however, the contaminated site is treated in a similarly fast way as in the case of the exclusive excavation procedure. The setting up of technically ambitious containment operations like encapsulation might take more time than excavation or soil washing because of the complexity of the building activities. Nevertheless, a longer remediation time is associated with the *in situ* operations bioremediation and particularly phytoremediation (phytoextraction). Compared to the *in situ* bioremediation, the biopile solution occurs more quickly. The effectiveness of biological measures is considered to be lower, since metals, for example, cannot be treated in this way. Ultimately, natural attenuation is characterised by very long duration and limited efficiency.

In Table 9.5 the time period required for the different remediation technologies are summarised. Based on an approx. 3 ha large contaminated site, fast solutions necessitating weeks to a few months are excavation, less complex containment procedures such as sealing, furthermore asphalt batching, vitrification, stabilisation, thermal treatment, electroremediation and steam- enhanced extraction (SEE). Surface cover can also be a rapid and effective solution for the contaminated land irrespective of the costs which must be calculated. In general, containment measures are often preferred to decontamination procedures due to the quicker response to the soil contamination. They are typical construction activities that are more

Table 9.5 Calculated remediation time for soil and soil vapour based on a contaminated area of approximately 3 ha

	Time				
	Weeks to a few months	More than 4 months to 2 years	3–10 years	Decades	Long-term monitoring
Excavation, transportation and landfilling	**X**	**(X)**			
Surface cover (geotextile-based and bentonite-based)	**(X)**	**X**			**(X)**
Sealing	**X**	**(X)**			**(X)**
Barriers installation at the sides		**X**			**X**
Encapsulation		**X**			**X**
Solidification (cement-based, *in situ*)		**X**			**X**
Asphalt batching	**X**				**(X)**
Vitrification	**X**				**X**
Stabilisation (liming)	**X**				**X**
Stabilisation (organic matter addition)	**X**				**X**
Soil washing	**X**	**(X)**			
Soil extraction (acids, detergents)	**X**	**(X)**			
Bioremediation (*in situ*)		**(X)**	**X**		**X**
Bioremediation (landfarming)		**X**			
Bioremediation (regeneration pile)		**X**	**(X)**		
Bioremediation (bioreactor)	**X**	**(X)**			
Phytoremediation (extraction)				**X**	**X**
Phytoremediation (degradation/volatilisation)				**X**	**X**
Thermal treatment	**X**	**(X)**			
Electroremediation (*in situ*)	**X**				
Natural attenuation				**X**	**X**
Soil vapour extraction (SVE)		**(X)**	**X**	**(X)**	**X**
Bioventing		**X**			**X**
Steam enhanced extraction (SEE)	**X**				**(X)**

X usual treatment time
(X) treatment time exceptionally necessary

time-consuming than the excavation but more time-saving in comparison to most decontamination strategies applied on site and *in situ*. The more complex building activities side barrier installation and encapsulation require more time. This varies between about 4 months and 2 years as in the case of solidification and several bioremediation strategies, namely landfarming, use of biopiles, bioreactor (predominantly due to its low capacity, otherwise a faster process is feasible) and bioventing. In most cases between 3 and 10 years must be calculated regarding *in situ* bioremediation and soil vapour extraction (SVE). The operation time of soil vapour extraction is difficult to predict, leading to a wide variation of the time in practical applications (Barnes 2003). Even decades are forecasted for phytoremediation and natural attenuation.

With reference to the groundwater treatment approaches long time periods must generally be calculated because of the slowness of diffusion processes in the liquid phase. One decade or more than one decade is the typical duration for an application which has been conducted in a great number of case studies dealing, for instance, with pump-and-treat. The remediation process needs similar time periods for the permeable reactive wall technology. In contrast, measures that are reduced to bio-slurping and biosparging might mostly require a shorter duration of 6 months to 2 years (Khan et al. 2004).

With regard to the evaluation of the remediation methods not only the period of time to remediate the polluted area is a decisive factor. Attention should be paid to the usually long-term and cost-intensive monitoring programme which is necessary in a lot of remediation approaches. For instance, unique stabilisation using lime is a very fast process but it must be continuously monitored and repeated many times. Consequently, solutions which seem fast at first sight could require unexpectedly long time periods due to the demand for monitoring. Treatment options requiring long-term monitoring are listed in Table 9.5.

In principle, it is difficult to estimate the costs of the different treatment opportunities, because the remediation case studies are mostly not comparable with each other. Hence, only ranges can be defined. The costs are influenced by the calculated mass to be treated, the transport capacity, the high-tech standard existent and available in the particular country dealing with the contaminated site and by economic features such as currency control. The data, which show average values, give an overview of the costs for remediation activities in the USA and Germany. They derive from Khan et al. (2004) (USA) and LANUV (2005) (Germany). Because of the constantly changing currency rates they are separately expressed in \$ and €. They refer to cubic metres and for conversions from cubic metres to tons the factor 1.5 was used (average specific gravity of soils: 1.5 g cm^{-3}). The volume to be treated is assumed to vary between 1,000 and 10,000 t.

In the USA excavating of 50 m^3 of highly contaminated soil and subsequent landfilling incur costs between \$ 976,000 and 4,148,000 per ha on average. In Germany the landfilling solution is usually not legally acceptable because of the waste legislation regulations. The exclusive excavation process ranges between € 5 and 8 per m^3 for soil and € 9–13 per m^3 for skeleton-enriched material depending on the depth of excavation.

Geotextile-based and bentonite-based surface cover amounts to € 84 per m^2 in conjunction with a planned sensitive land use. Simpler surface cover systems which avoid the implementation of a bentonite layer and are reduced to the construction of geotextiles may cost € 52 per m^2 on average. The costs for a complete sealing of the contaminated land are € 38 per m^2 (asphalt) and € 50 per m^2 (concrete). In relation to a depth of 10 m the side barrier system shows a wide range of costs depending on the technique applied and the distinct soil properties present. For example, the sheet-pile wall of steel amounts to € 100–700 per m^2 of wall surface, bored pile walls € 130–600 per m^2, slurry walls as one-phase procedure € 70–160 per m^2 and slurry walls as twin phase procedure € 150–550 per m^2, respectively. The data prefer to the wall surface instead of soil surface.

In the USA the maximum costs for *ex situ* solidification including excavation (cement-based solidification, asphalt batching) are $73 per m^3, whereas in the case of an *in situ* process (vitrification) values which range from $80–330 per m^3 are realistic. The *in situ* vitrification might be cost-intensive due to the energy costs reaching a maximum of $355 per m^3, as noticed in U.S. case studies. In Germany for cement-based solidification € 55–62 per m^3 must be spent.

Stabilising approaches are generally one of the cheapest methods to treat contaminated soils. For unique liming only € 5–6 per m^2 and for addition of organic matter € 6–7 per m^2 are charged but the procedure must be repeated continuously.

Soil washing is predominantly cost-effective in relation to the quantity of material that remains contaminated and that consequently requires further treatment procedures, because only the gravelly, sandy and in part silty fraction is cleaned. In the USA the average costs are approximately $113 per m^3. Some case studies, however, stated even lower costs of approximately $50 per m^3. In Germany soil washing on the basis of the attrition technique costs € 48 per m^3 (fine fraction <10%) to € 69 per m^3 (fine fraction 10–40%), whereas the costs for high pressure soil washing and soil extraction using tensides might be as high as € 52 per m^3 (fine fraction <10%) and € 79 per m^3 (fine fraction 10–40%).

It should be noted that the soil conditioning processes which must occur prior to soil washing, *ex situ* bioremediation and thermal treatment are not included. € 9 per m^3 is estimated for crushing, € 17 per m^3 for screening and magnetism and € 17 per m^3 for light fraction removal.

Because of the low installation and operation costs the price for landfarming in the USA usually ranges from $20–40 per m^3. In case studies dealing with PAH this was significantly more expensive ($92–107 per m^3) than in examples where soils polluted by TPH and BTEX have been decontaminated (only approx. $7 per m^3). In Germany, where landfarming is rarely applied, the costs for soils containing only <10% fine fraction amount to € 16 per m^3 and for soils with 10–25% fine fraction € 21 per m^3. In a similar way to landfarming *in situ* bioremediation does not require cost-intensive installation devices. The remediation may cost approximately € 35 per m^3 on average.

In contrast, the costs for the technically more ambitious biopile alternative tend to be higher in the USA, ranging from $171 to 342 per m^3. The budgets for some case studies, however, were reduced. For instance, the decontamination of explosives caused operational costs of only \$ 61 per m^3. In Germany the expenses vary significantly depending on fine earth fraction, water content and level of contamination but they are principally lower than in the USA. Soils with a low percentage (<10%) of clay and silt and low moisture content (<15%) cost € 16 per m^3, while soils exhibiting 10–25% fine fraction and 15–40% water content are more expensive (€ 26 per m^3). The costs for soils indicating very high concentration of organic pollutants increase to € 74 per m^3.

The bioreactor solution incurs costs between \$130 and 200 per m^3 on average but there are case studies which cite \$60 per m^3 (sludge polluted with chlorinated aliphatic compounds and PAH). In Germany this technique varies between € 82 per m^3 (fine fraction <10%) and € 124 per m^3 (fine fraction 10–25%).

With regard to a contaminated depth of 50 cm the phytoextraction costs vary from \$15 to 24 per m^2. The calculation in Germany results in € 17 per m^2 in relation to the use of herbs and up to € 51 per m^2 in relation to wooden plants. The prices calculated for a 10 year long treatment include soil tillage, fertilizing, plantation and maintenance.

The costs for thermal treatment in the USA range from \$34 to 228 per m^3. Case studies in the USA indicated costs of \$154–228 per m^3 (chlorinated aliphatic compounds and PAH), \$127 per m^3 (PCB) and \$ 134 per m^3 (pesticides) and have a tendency to be lower than the costs usually paid in European countries. It is likely that the price differences in energy costs are responsible for the differences ascertained. In Germany and the Netherlands the costs for thermal treatment (pyrolysis) are € 129 per m^3 on average. The rehabilitation of deposited material previously decontaminated by thermal treatment increases the cost by € 15 per m^2. For *in situ* steam enhanced extraction (SEE) € 52 per m^3 must be paid.

In Germany the costs for electroremediation range from € 40 to 200 per m^3 and show relatively high levels due to the energy costs involved.

Natural attenuation does not cause high operating costs. These vary between € 6 and 10 per m^2 but additional monitoring costs which are necessary for a long period of time should also be taken into consideration. In Table 9.6 the remediation costs are divided into classes.

The soil remediation costs depend on a multitude of factors but, in particular, the increasing costs for energy consumption might be a problem for most of the treatment opportunities in future. As shown in Table 9.7, the energy consumption of the clean-up technologies is very different. In particular, thermal treatment and electroremediation consume a lot of energy and might become the most expensive remediation techniques. With reference to vapour and groundwater decontamination, attention should be paid to the energy use, since these applications normally operate over a long period of time. For this reason, the real energy expenditure should always cover the calculated time period needed.

According to the Environmental Protection Agency of the USA costs for soil vapour extraction of \$13–34 per m^3 on average must be expected but the differences between various case studies were enormous. For instance, VCHC treatment incurred costs of approximately \$3–4 per m^3 in relation to a relatively short remediation time of 6 months to 1 year. In the presence of non-chlorinated compounds higher costs and longer treatment periods are generally expected. Based on the U.S.

Table 9.6 Classified average remediation costs (€)

	<25 € m^{-3}	25–50 € m^{-3}	>50–75 € m^{-3}	>75–100 € m^{-3}	>100–150 € m^{-3}	>150 € m^{-3}
Excavation	**X**					
Solidification (cement-based, *in situ*)			**X**			
Asphalt batching			**X**			
Vitrification				**(X)**	**(X)**	**X**
Soil preparation	**(X)**	**X**				
Soil washing		**(X)**	**X**			
Soil extraction (acids, detergents)			**X**			
Bioremediation (*in situ*)		**X**				
Bioremediation (landfarming)	**X**	**(X)**	**(X)**			
Bioremediation (regeneration pile)	**(X)**	**X**	**X**	**(X)**	**(X)**	
Bioremediation (bioreactor)			**(X)**	**X**	**X**	
Thermal treatment				**(X)**	**X**	
Electroremediation (*in situ*)			**(X)**	**(X)**	**(X)**	**X**
	<10 € m^{-2}	10–25 € m^{-2}	>25–50 € m^{-2}	>50–100 € m^{-2}	>100–200 € m^{-2}	>200 € m^{-2}
Surface cover (geotextile-based and bentonite-based)				**X**		
Sealing			**X**			
Barriers installation at the sides					**(X)**	**X**
Encapsulation						**X**
Stabilisation (liming)	**X**					
Stabilisation (organic matter addition)	**X**					
Phytoremediation (phytoextraction)		**X**	**(X)**			
Natural attenuation	**X**					

X usual treatment cost
(X) exceptional treatment cost
Exchange rate: 1 € = 1.44 $

Table 9.7 Ranges for energy consumption of different clean-up technologies (Data from DVWK 1996)

Treatment	kWh m^{-3} soil
Soil washing	5–35
Soil washing (high pressure technique)	10–25
Biological treatment (regeneration pile)	1–20 (TPH, BTEX), 20–50 (PAH)
Biological treatment (bioreactor)	0.3
Biological treatment (*in situ*)	1–20
Thermal treatment (low temperature)	250–400
Electroremediation	40–400 (68–720)
Soil vapour extraction and treatment (activated carbon adsorption)	1–6
Groundwater extraction and treatment (activated carbon adsorption)	5–8
Groundwater extraction and treatment (air stripping)	13–24

case studies the treatment of the more complex and less volatile pollutants TPH and BTEX meant operating costs ranging between \$158 and 585 per m^3 within 2 years. Mixed contamination consisting of chlorinated and non-chlorinated compounds cost \$79 per m^3 within 4 years. Based on a theoretical case dealing with three SVE wells, a depth of contamination of 5 m and a remediation time of 1 year, the investment cost might be € 16,700, the operating costs € 13,800 and the costs for vapour purification € 49,800 (sum of annual costs: € 80,300).

For bioventing, which is also related to the vadose zone, costs between \$20 and 60 per m^3 are common but some case studies relating to BTEX and TPH elimination showed higher operational costs, amounting to \$158 per m^3. The annual investment plus operation costs varied from \$24,000 to 177,000.

In general, it is difficult to predict the cost for groundwater remediation. Cost data vary greatly from case study to case study. Apart from the investment for the installation, the operating costs for the aboveground treatment, which are predominantly dependent on the remediation time required, must be included. Case studies in the USA showed annual operating costs, including the installation, as high as \$72,500 (TPH) and \$130,000–240,000 (BTEX, PAH) respectively. A P&T application, including one extraction well to a depth of 10 m and a remediation time of 1 year, would incur investment and operating costs of € 38,400 as well as € 78,000 which must be paid for the purification of the extracted and contaminated water. If a bioslurping module is included, which is responsible for the removal of LNAPL, additional costs of € 21,500 must be reckon with.

In consideration of energy and maintenance, expenses for the treatment of 1,000 L of contaminated water ranged from \$0.26 to 26. Case studies in the USA stated costs of \$0.03–0.20 (chlorinated aliphatic compounds) and \$ 3.65 (mixed contamination of chlorinated compounds, BTEX, TPH and PAH).

The relation between cost and period of time must be taken into consideration in the long run. As shown in Fig. 9.2, the relation will alter with increasing remediation time. With regard to a site contaminated with mineral oils (TPH) mainly in the groundwater the elimination of the source consisting of excavation and final disposal

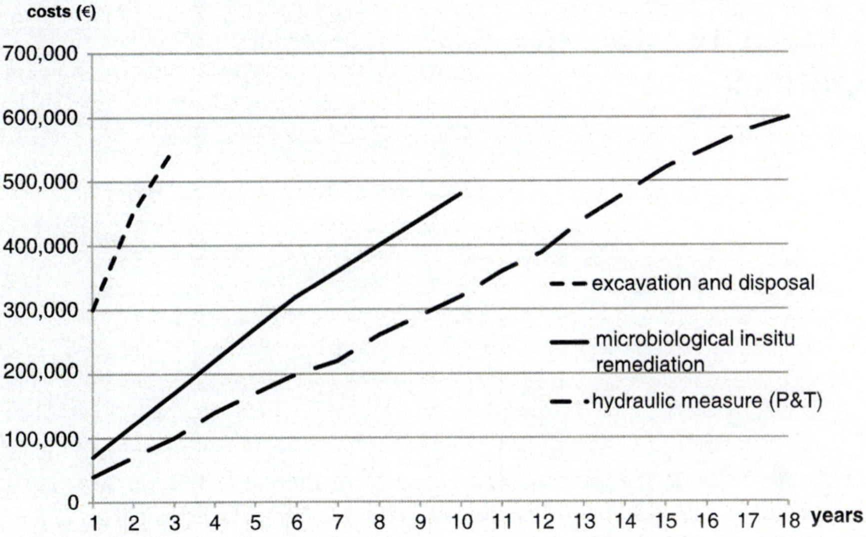

Fig. 9.2 Theoretical example for a time-cost relation for different clean-up strategies in an area exhibiting soil and groundwater pollution with TPH

appears to be very expensive in relation to the first 3 years. In contrast, the costs for microbiological *in situ* remediation of soil and groundwater are less cost-intensive in the same period of time but after about 10 years, which are necessary to achieve a successful remediation result, they increase considerably, reaching approximately the level of the excavation and disposal measures. If the contaminated source is not treated, hydraulic measures, namely pump-and-treat (P&T), can be applied. This solution will be relatively cheap in the first decade but when the next decade is taken into consideration the costs will add up and ultimately exceed the two alternative clean-up strategies presented. Hence, in the context of the clean-up planning, long-term consideration must be given to the actions.

References

Barnes, D. L. (2003). Estimation of operation time for soil vapour extraction systems. *Journal of Environmental Engineering, 129*, 873–878.

BBODSCHV. (1999). *Federal Soil Protection and Contaminated Sites Ordinance*. http://www.umweltbundesamt.de/altlast/web1/berichte/pdf/bbodschv-engl.pdf. Accessed 10 May 2012.

DVWK. (1996). *Remediation of contaminated soils*. DVWK publications, 116, Bonn. (in German)

Khan, F. I., Husain, T., & Hejazi, R. (2004). An overview and analysis of site remediation technologies. *Journal of Environmental Management, 71*, 95–122.

LANUV. (2005). *Leistungsbuch Altlasten und Flächenentwicklung. Materialien zur Altlastensanierung und zum Bodenschutz* (Vol. 20). http://www.lanuv.nrw.de/veroeffentlichungen/malbo/malbo20/malbo20start.ht (in German). Accessed 20 May 2012.

Chapter 10
Outlook

Since the 1980s in the industrialised countries in Europe and North America a systematic characterisation at contaminated sites has been carried out. This is predominantly associated with risk assessment in relation to human health and environmental conditions. Investigation and assessment are based on different phases, the preliminary, the oriented and the detailed investigation. In the case of proven site contamination, which can cover soil, groundwater and soil vapour, a remediation plan is developed and ultimately the remediation measures are executed. The subsequent containment and decontamination procedures should achieve the main target to exclude hazards to human health and to assure environmentally acceptable conditions. Thus, in the course of time, while the systematic investigation programmes took place technically ambitious containment and decontamination approaches were developed and continuously improved. Accordingly, most of the remediation techniques were invented and engineered in the well-developed industrialised countries where the contamination became more obvious and important, e.g. Canada, France, Germany, The Netherlands, Scandinavian countries, United Kingdom and the USA. In relation to soil but also to the treatment of soil vapour and, in particular, groundwater, a remediation market came into existence which treated an increasing quantity of contaminated soil and groundwater.

For instance, in Germany the capacity for *ex situ* soil washing, bioremediation and thermal treatment was about 2.4 million tons in 1995 and in 2008 the capacity had already reached 7.1 million tons. In 2008 71 biological treatment plants existed, decontaminating 4.1 million tons of contaminated soil, whereas 20 soil washing and soil extraction plants treated 2.2 million tons and 9 thermal treatment facilities approximately 800,000 t. In summary, in Germany exactly 100 technology-based providers treated contaminated soils *ex situ* (Frauenstein and Mahrle 2009). In the United Kingdom (U.K.) there were 75 companies providing technology-based solutions in 2007 but the remediation market continuously shows a tendency to expand, since, for example, between 2005 and 2010 it grew by 21%, reaching approximately € 500 million. In 2015 it is assumed that it will increase to € 640 million. In contrast

H. Meuser, *Soil Remediation and Rehabilitation: Treatment of Contaminated and Disturbed Land*, Environmental Pollution 23, DOI 10.1007/978-94-007-5751-6_10,

to Germany, in most of the countries mentioned above, such as the U.K., excavation and removal (dig-and-dump) is the preferred solution, amounting to 76% of the total remediation costs in 2005. Nevertheless, in the period between 2001 and 2005 the market for alternative remediation approaches rose as well in the U.K. Examples are the containment market, which grew by 7% (in 2005: € 30 million), and soil washing, which grew by 7% (2005: € 32 million). Moreover, other technologies which had only a low market share in the past (e.g. bioremediation 5%, solidification 7%, thermal treatment <1%) showed a similarly considerable increase between 2005 and 2010 (2011). This was caused on the one hand by a general increase in remediation project numbers in the U.K. and, on the other hand, by site construction for the Olympic Games in London, which have been held on former industrialised brownfields (Randall 2007).

The market for remediation services, which includes treatment of contaminated soils, groundwater and buildings, grew worldwide between the beginning of the 1990s and 2000/2001. In the USA, the biggest market worldwide, the turnover was € 6.6 billion in 1994 and in 2001 the figure was € 8.9 billion. In Canada the value was € 300 million in 1994 and 7 years later this amounted to € 800 million. In the European countries, including Western, Central and Eastern Europe, € 3.0 billion was recorded in 1994 and in 2000 the turnover was € 6.6 billion. In contrast, the market of the developing countries in Latin America, Asia and the Middle East reached values of € 600 million in 1994 and € 2.6 billion in 2000 (USITC 2004).

After the beginning of the twenty-first century the remediation sector continuously showed an increasing tendency. In Europe (initially in the developed countries in the West and North and since 1990 in the former socialist countries in Eastern Europe such as Poland and the Czech Republic) there is a continuing strong development due to the following reasons:

- The proposed EU Landfill Directive will exacerbate the dig-and-dump solution in the near future; in addition, other important directives such as the EU Water Framework Directive and the EU REACH Programme for chemical hazards will contribute to the increasing necessity to treat contaminated sites.
- Furthermore, some countries (e.g. U.K.) enforce quotas for new housing and commercial buildings on brownfields in order to protect uncontaminated greenfields on the city periphery.
- Land remediation standards across Europe are being developed and harmonised.
- The growing environmental awareness of the public means that contaminated land is considered to be a scare factor in combination with the political decisions in some countries to prioritise environmental issues; moreover, there is greater availability of environmental information to the public in relation to evidence of the impact of contaminants on human health.
- From the economic point of view, public sector funding for remediation expanded, whereby particularly the countries in Eastern Europe benefited from this. Apart from payment of subsidies, tax incentives were established. In general, the property value might rise after remediating the contaminated land.

- The effectiveness of remediation techniques improved due to a high number of research projects which were often approved, particularly in relation to new environmental technologies. They include integrated remediation approaches, which use different methods simultaneously or sequentially, and monitoring technologies. In the context of the increasing number of sites and soil and groundwater quantities to be remediated it was possible to reduce the operational cost of clean-up.

A lot of remediation techniques have originated from other technological processes. For instance, the containment approaches such as side barrier installation stem from the construction sector and thermal treatment is also associated with techniques which are normally used for treating hazardous waste. Some treatment approaches such as stabilisation and phytoremediation are strongly linked to the agricultural sector. However, there are also several remediation techniques which might be preferentially developed in association with the contaminant problem, e.g. bioremediation and most of the groundwater clean-up methods.

Irrespective of the technique available, an expanding remediation market in the developed countries might be expected if one takes the land consumption in urban agglomerations into account. The lack of land for construction and the disproportion between already verified contaminated sites and successfully remediated sites (e.g. in Germany 313,600 registered contaminated sites in 2010/2011 of which only 25,900 sites (= 8.2%) have been completely remediated) require the redevelopment of areas which have been anthropogenically used beyond an agricultural use.

In a similar way the shortage of drinking-water in urbanised areas might require a remediation of contaminated plumes in line with the reduction of discharges. The increasing urbanisation which can be identified not only in the developed world does not allow the presence of highly contaminated source areas from which plumes are dispersed, endangering the well system of the public supply.

The factors shortage of land for construction and contamination of drinking-water derived from groundwater are particularly important in the developing countries such as China and India. Until the beginning of this century there was no awareness of the environmental factor soil at all and only little knowledge about the groundwater contamination problem in the developing countries. In China, for example, in the last decade the importance of the topic handling contaminated soils increased enormously, leading to a new expanding remediation market which can be seen in some urban agglomerations such as densely populated cities like Hangzhou and Suzhou. For this reason, the expected and desired growth of the contaminated soil and groundwater treatment field is probably not restricted to the countries with a relatively long remediation experience. There might be forecasts for great growth in the Eastern European countries and the Asian countries, where a high number of former industry sites are located but remediation measures have only occurred to a very small extent.

A large-scale impact on originally undisturbed land is also associated with mining activities. Open-cast coal mining, quarries, open pit mines for unconsolidated rock and peatland harvesting consume huge terrains where the vegetation is cut and different upper layers of soil and rock are removed. Moreover, ore mining areas leave behind considerable areas, indicating soil contamination. Results from shaft mining are large-scale heaps with differing chemical composition which frequently cause leaching of contaminants associated with groundwater problems. Because of the extent of disturbed land, which is often located close to urban agglomerations, rehabilitation in mining and raw material extraction areas is principally required. This necessity results predominantly from the land consumption and the negative impacts on groundwater and surface waters. It should be noted that the areas used for mining are relatively small at country level (e.g. in Australia only 1,366 km^2 (= 0.02%) are used for mining) but at regional level the extent of disturbed land must be considered as a serious problem. Thus, the most important motivation for rehabilitation strategies is agricultural re-cultivation to assure sufficient land for crop production and re-establishment of woodland to compensate for losses due to climate change. These strategies are complemented by soil decontamination measures as well as the creation of leisure landscapes, particularly in the proximity of urban areas.

In consideration of the fact that the world's population will increase and that subsequently the raw material exploitation will reach a higher level in the future, the demand for site rehabilitation of exploited terrains might continue to remain important. These circumstances apply mainly to the developing countries such as China, India and Russia where the population growth is high and simultaneously a high number of raw material extraction areas are present. Hence, the rehabilitation processes must occur to prevent devastated land without the possibility to grow crops or to plant trees and to avoid moonscapes on our planet.

Whether the predicted development happens quickly or slowly is difficult to forecast. Taking the carbon dioxide and global warming discussion into consideration and the awareness in the public and in politics, the first political decisions were initiated relatively fast (Kyoto Protocol in 1997 and subsequent UN Climate Change Conferences). However, effective decisions have not been taken up to now because some important countries tend to block the development due to their preference for economic interests rather than the required ecological approach, although everybody knows the measures must be taken as fast as possible. Unfortunately, in relation to the soil and groundwater factors a rapid increase in awareness and subsequent political decisions have not occurred to a sufficient extent up to now. However, the aggravation of the ecological situation in association with the shortage of usable land in and outside urban agglomerations will probably necessitate the taking of action more quickly than assumed.

References

Frauenstein, J., & Mahrle, S. (2009). News from being on the prowl for soil. *Altlasten-Spectrum, 2*, 85–98 (in German).

Randall, J. (2007). *The land remediation industry*. EIC Guide to the UK Environmental Industry. www.eic-guide.co.uk/docs/contam.pdf. Accessed 18 May 2012.

USITC – U.S. International Trade Commission. (2004). *Remediation and nature and landscape protection services: an examination of U.S. and foreign markets*. USITC Publication 3727, Washington, DC.

Index

H. Meuser, *Soil Remediation and Rehabilitation: Treatment of Contaminated and Disturbed Land*, Environmental Pollution 23, DOI 10.1007/978-94-007-5751-6,

H

I

M

N

Q

S

U

V